Peter J. Brown

Meteorological Disasters in Medieval Britain (AD 1000–1500)

Historical Catastrophe Studies/ Historische Katastrophenforschung

Edited by/Herausgegeben von
Dominik Collet, Christopher Gerrard, Christian Rohr

Peter J. Brown

Meteorological Disasters in Medieval Britain (AD 1000–1500)

Archaeological, Historical and Climatological Perspectives within a Wider European Context

DE GRUYTER

ISBN 978-3-11-071957-4
e-ISBN (PDF) 978-3-11-071962-8
e-ISBN (EPUB) 978-3-11-071965-9
ISSN 2699-7223
DOI https://doi.org/10.1515/9783110719628

Library of Congress Control Number: 2023939775

Bibliographic information published by the Deutsche Nationalbibliothek
The Deutsche Nationalbibliothek lists this publication in the Deutsche Nationalbibliografie; detailed bibliographic data are available on the internet at http://dnb.dnb.de.

Cover image: Detail from the 'Holkham Bible Picture Book', c. 1327–1335 © The British Library Board, Add MS 47682, f. 8r.
Printing and binding: CPI books GmbH, Leck

www.degruyter.com

Contents

List of Figures — IX

1 Introduction — 1
1.1 What constitutes a disaster? — **4**
1.2 Natural hazards in a British context — **6**
1.3 Approaching the study of disasters in the past — **9**
1.4 The contribution of archaeology to the study of disasters — **17**

2 Disaster and society in medieval Europe — 21
2.1 Living with risk — **22**
2.2 Religious beliefs, superstitions and disaster — **32**
2.3 Summary — **44**

3 A litany of disasters — 45
3.1 The storms and floods of 1287/88 — **48**
3.1.1 Historical background — **48**
3.1.2 The archaeological evidence — **55**
3.1.3 New Romney, Kent — **57**
3.2 Flooding in Marshland, Norfolk, 1287–1349 — **67**
3.2.1 The medieval landscape — **69**
3.2.2 Edwardian floods (1287–1349) — **71**
3.3 The St Maur's Day windstorm of 1362 — **79**
3.3.1 Historical background — **81**
3.3.2 The material evidence — **88**
3.4 Landslides in medieval northern Europe — **92**
3.5 Summary — **96**

4 In the aftermath: practical responses against hazards — 97
4.1 The immediate aftermath — **99**
4.2 The victims of disaster — **101**
4.3 Agricultural management — **104**
4.4 Structural responses — **107**
4.4.1 Bridges — **107**
4.4.2 Ecclesiastical structures — **112**
4.5 Flood defences — **116**
4.6 Financial relief — **124**
4.7 Migration, Abandonment and Decline — **129**
4.8 Summary — **136**

5 **Systems of protection: religious and superstitious defences against hazards — 137**
5.1 Saintly relics — **141**
5.2 Images of the Saints — **145**
5.2.1 St Christopher — **146**
5.2.2 St Barbara — **147**
5.2.3 The Magi (The Three Kings of Cologne) — **148**
5.2.4 Summary — **149**
5.3 Pilgrim badges — **152**
5.4 *Ampullae* — **155**
5.5 Charms and textual amulets — **158**
5.6 Papal *bullae* — **160**
5.7 Church bells — **163**
5.8 Foundation and special deposits — **165**
5.9 Folded coins — **169**
5.10 Graffiti — **175**
5.11 Summary — **180**

6 **Memories of disaster — 182**
6.1 Memory in the past — **183**
6.2 Popular verse — **190**
6.3 Inscriptions — **194**
6.4 Commemorative activities, material culture and other manifestations of memories — **198**
6.4.1 Commemorative events — **198**
6.4.2 Individual commemoration — **199**
6.4.3 Commemorative material culture — **201**
6.4.4 Ex votos — **202**
6.4.5 Memory and structural repairs — **205**
6.5 Summary — **206**

7 **Discussion — 209**
7.1 The causes of disaster — **209**
7.1.1 Climate — **210**
7.1.2 The late 1280s — **213**
7.1.3 1361/1362 — **214**
7.1.4 Summary — **217**
7.1.5 Other factors — **218**
7.2 The impacts of disasters and the responses of society — **220**
7.2.1 Economic impacts — **221**
7.2.2 The management of disasters — **224**
7.2.3 The intangible impacts of disasters — **229**

7.2.4 Summary — **233**
7.3 Physical signatures and material culture — **236**
7.3.1 Disasters as 'watersheds' — **237**
7.3.2 Signatures of disaster — **239**
7.4 Summary — **243**

8 An endless chain of catastrophes? — 245

Bibliography — 251

Index — 282

List of Figures

Fig. 1.1 The disaster cycle —— 3
Fig. 1.2 Conceptual diagram of the relationship between hazard, vulnerability and risk —— 5
Fig. 1.3 Risk from natural hazards in northern Europe today —— 8

Fig. 2.1 Archaeological evidence for medieval grain storage —— 28
Fig. 2.2 Medieval graffito inscriptions depicting demons —— 36
Fig. 2.3 Scenes from the Pricke of Conscience window, All Saints', North Street, York —— 40
Fig. 2.4 The celestial conjunction and disasters widely predicted to occur in 1524 —— 42

Fig. 3.1 The extreme weather recorded occurring in 1287 and 1288 —— 54
Fig. 3.2 New Romney and Romney Marsh at the time of the storms of 1287/88 —— 59
Fig. 3.3 The impact of the storms of 1287/88 at New Romney —— 62
Fig. 3.4 Section 5 from the Church Road site, New Romney —— 65
Fig. 3.5 Building 2, Southlands School site, New Romney —— 66
Fig. 3.6 Map of Marshland, Norfolk, with parish boundaries —— 70
Fig. 3.7 Flooding in medieval Marshland, Norfolk, 1287-1349 —— 73
Fig. 3.8 The impact of flooding in Marshland by decade between 1287-1349 —— 74
Fig. 3.9 The west front of the parish church of Walpole St Peter, Norfolk —— 76
Fig. 3.10 The impact of the floods of 1338 in Marshland against taxable wealth in 1334 —— 78
Fig. 3.11 Oxhow Clough in the present day —— 80
Fig. 3.12 Reconstruction of the timing of the storm of 15th January 1362 —— 83
Fig. 3.13 View of Norwich Cathedral's clerestory. —— 89
Fig. 3.14 Exterior view of Ashwell St Mary, Ashwell, Hertfordshire —— 90
Fig. 3.15 Four maps exploring the impact of the 1362 windstorm in the British Isles —— 92
Fig. 3.16 Map of Chapel Cleeve, Somerset —— 94
Fig. 3.17 The cliff collapse at Cleeve in 1466 —— 95

Fig. 4.1 Pit containing the remains of cattle excavated at Shapwick, Somerset —— 104
Fig. 4.2 Map of the location of Heigham Bridge, Norfolk, with a schematic diagram —— 110
Fig. 4.3 The succession of medieval bridges at Hemington, Leicestershire —— 111
Fig. 4.4 Plan of St Mary's Church, Hitchin, Hertfordshire —— 113
Fig. 4.5 The gatehouse of St Albans Abbey —— 115
Fig. 4.6 Structural evidence thought to pre-date the 1362 storm at St Albans —— 115
Fig. 4.7 A ditch interpreted as a flood defence from Flempton, Suffolk —— 117
Fig. 4.8 A 13th century culvert excavated at Newton, Cambridgeshire —— 119
Fig. 4.9 The communal nature of dike maintenance —— 121
Fig. 4.10 Schematic diagram showing different types of breach repair —— 123
Fig. 4.11 Disruption and continuity in New Romney after 1287/88 —— 134

Fig. 5.1 Maps of medieval finds in the PAS database and medieval population —— 141
Fig. 5.2 Four reliquary crosses from the PAS database —— 143
Fig. 5.3 The geographical distribution of reliquaries recorded in the PAS database. —— 145
Fig. 5.4 The chronological distribution of reliquaries recorded in the PAS database. —— 146
Fig. 5.5 Images of Saint Christopher and Saint Barbara —— 148
Fig. 5.6 Alabaster relief depicting the Adoration of the Magi —— 150

https://doi.org/10.1515/9783110719628-201

Fig. 5.7 The chronological distribution of evidence relating to the cult of St Christopher — **152**
Fig. 5.8 The chronological distribution of artefacts in the PAS database related to the cult of Saint Barbara — **153**
Fig. 5.9 The geographic distribution of artefacts in the PAS database related to saintly cults — **154**
Fig. 5.10 Material culture relating to saintly cults from the PAS database — **154**
Fig. 5.11 The number of pilgrim badges reported by the PAS database — **156**
Fig. 5.12 The geographical distribution of pilgrim badges from the PAS database — **156**
Fig. 5.13 The number of *ampullae* reported by the PAS database — **157**
Fig. 5.14 The geographical distribution of *ampullae* from the PAS database — **159**
Fig. 5.15 Three *bullae* from the PAS database — **162**
Fig. 5.16 The chronological distribution of Papal *bullae* in the PAS — **163**
Fig. 5.17 The geographical distribution of Papal *bullae* in the PAS — **164**
Fig. 5.18 Distribution map of the folded coins in the PAS database — **170**
Fig. 5.19 Three folded coins from the PAS database — **171**
Fig. 5.20 Types of folds found on folded coins in the PAS database — **173**
Fig. 5.21 Chronological distribution of folded coins in the PAS database — **174**
Fig. 5.22 Chronological distribution of folded coins in the PAS database (*continued*) — **174**
Fig. 5.23 The structural phasing of St Nicholas Church, New Romney — **177**
Fig. 5.24 The graffiti on pillar D2, St Nicholas Church, New Romney — **178**
Fig. 5.25 Three examples of medieval ship graffiti — **178**

Fig. 6.1 The 'lifecycle' of a memory — **184**
Fig. 6.2 The transmission and reception of memories — **189**
Fig. 6.3 The graffito at St Mary's, Ashwell, Hertfordshire — **193**
Fig. 6.4 The left and right panels of the St Elizabeth's Day flood altarpiece — **204**

Fig. 7.1 Summer temperature variance across Europe in the late 1280s and early 1360s — **216**
Fig. 7.2 An index of environmental instability in Britain 1300–1500 — **217**
Fig. 7.3 The disaster cycle applied to the disasters considered in Chapter 3 — **236**
Fig. 7.4 Disasters at the intersection of climate and society — **241**
Fig. 7.5 Artefact chronologies and population — **242**

1 Introduction

The impacts of natural hazards on human societies can be acute, far reaching and at times surprising. As much as the products of hazards arising entirely by natural means, however, the disastrous circumstances created by such events are socially-created phenomena. Not only are natural hazards themselves the product of a complex web of Earth-system processes, their effect on human society is the result of a constellation of inter-related and unpredictable variables. The impact of any given hazard on a particular human society, therefore, can be estimated but never predicted with absolute certainty. We are frequently reminded of this when natural hazards strike human populations today – though it is usually only the short-term 'moment of disaster' that captures the attention of the media rather than the multitude of processes that caused the event to occur and influence its longer-term consequences. By turning to the 'laboratory' of the past, however, it is possible to investigate the impact of extreme natural events, both during these tumultuous moments and across longer timespans. Additionally, the different ways in which human societies have responded to these types of event – and whether measures that were adopted ameliorated or exacerbated the situation – may be explored. The subject of this book, therefore, is the relationship between later medieval society (AD 1000–1550) and natural hazards. What follows explores meteorological disasters – a subset of natural disaster – focussing on their impact in Britain, though with a comparative eye to the medieval world more generally. The aims of this study, therefore, are:

- To reconstruct, in detail, a number of case studies to explore the impact of extreme natural events on medieval society
- To characterise how medieval society responded to these types of event (both physically and through religiously motivated and superstitious practices)
- To assess to what extent exposure to recurrent hazards affected the resilience and/or vulnerability of medieval society (either positively or negatively)
- To place the role of disasters as drivers of cultural change within a wider historical and archaeological context.

In order to accomplish these objectives a variety of historical and archaeological sources of evidence are investigated in order to explore different aspects of these interactions between nature and culture. Disasters as a subject of academic enquiry are highly interdisciplinary[1] combining elements from the physical sciences, such as physical geography and climatology, with the humanities and social sciences. The investigation of disasters in the past, therefore, requires the integration of many disparate fields including studies of historical documents, archaeological sites and standing buildings, which preserve evidence of the occurrence of disasters, as well as material culture

1 Juneja, Mauelshagen 2007: 4.

https://doi.org/10.1515/9783110719628-001

related to how contemporaries reacted to these events. The range of different types of hazard, their varying impacts in different locales, and their occurrence within varying historical contexts, necessarily means that there can be no 'one size fits all' approach to analyse the occurrence and impact of historical natural hazards. Rather, the approach adopted must be tailored to the available sources of evidence. These must be analysed creatively in order to tease out as much information as possible relating to the occurrence of the hazard, and its impact on contemporary society. The later medieval period is especially conducive to this type of research due to the significant quantity of extant documentary source material, which in most areas of Europe is orders of magnitude richer than from any preceding period. From an archaeological perspective, some types of natural hazard can be ephemeral, meaning that historical evidence is invaluable as a record of the occurrence and impact of these events. In addition, the documentary record assists in interpreting evidence for the responses these events provoked as well as their less tangible impacts on society.

Geographically, this volume focuses on Britain – in particular exploring a number of disasters that struck the east coast of England. While Britain is not widely considered to be a zone which is especially affected by natural hazards – with no significant exposure to seismic or volcanic hazards – meteorological, or weather-related, hazards occur with some regularity. Furthermore, the documentary evidence which survives from the medieval period in Britain is amongst the most complete in Europe. However, where appropriate, archaeological and historical evidence and research is brought in from many other regions, including Scandinavia, Germany, the Low Countries, Italy and France, to contextualise the experience of disasters in medieval Britain.

The format of this book is arranged around the 'disaster cycle',[2] a conceptual framework which describes the 'lifecycle' of a disaster and is applicable to any event. This framework encompasses the physical damage caused in the moment(s) of disaster, phases of repair and reconstruction in the immediate aftermath and the periods of quiescence in the, sometimes lengthy, gaps between the occurrence of hazards. Following the Braudelian model of historical time,[3] the effects of natural disasters can be envisaged over multiple overlapping timescales. In his first layer of historical time, Braudel generally viewed environmental systems as gradual and cyclical and judged them to be 'almost imperceptible in nature'.[4] With respect to rapid-onset natural hazards, however, although it is possible to conceive of many as occurring cyclically over the *longue durée*, the short-term effects unleashed by these events on the individuals and communities they affected were not just extremely noticeable but could be decisive in determining the fates of these communities during and immediately after

2 Alexander 2002: 6; see Fig. 1.1.

3 In which historical time is envisaged over three layers encompassing relatively gradual changes in environmental systems, more rapid changes in political and social systems and the more immediate history of the individuals which populated these overarching systems.

4 Braudel 1972: 20–21.

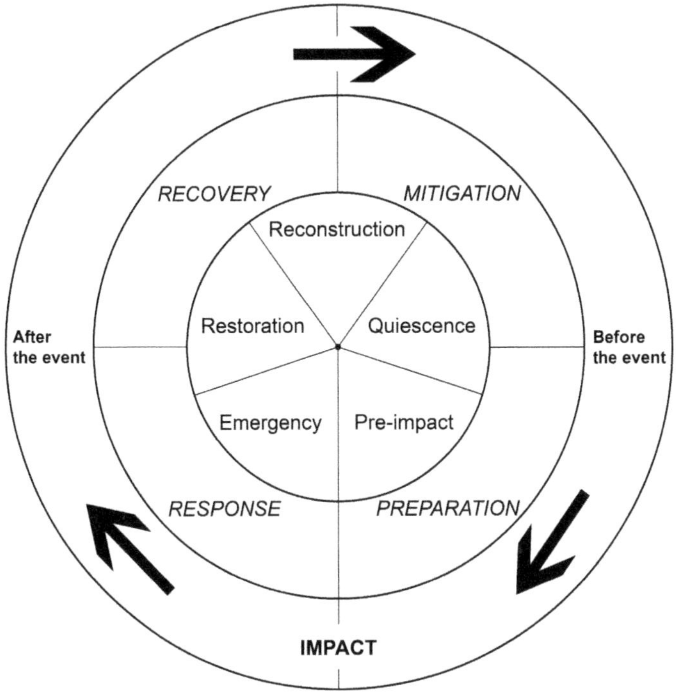

Fig. 1.1: The disaster cycle. Although each event will play out differently to its predecessors, the effects of a given disaster on a given society can be classified into the same fundamental stages. Redrawn by the author after ALEXANDER 2002: 6.

their occurrence. While Braudel's model is not well suited to the immediate impact of environmental hazards, therefore, as the remainder of this volume demonstrates, conceiving of the impacts of these events on the different spheres of life encompassed by Braudel's layers – individuals, society more generally and the wider environment in which medieval people lived – remains a valuable approach which offers a holistic understanding of the impacts generated by these events over different timescales.

Thus, following an introduction of key concepts and an overview of existing research in the remainder of this chapter, Chapter 2 approaches how medieval populations perceived anomalous natural events during periods of relative quiescence, particularly how their beliefs about disasters were informed by, on the one hand, religious instruction, and on the other, practical experience. Chapter 3 presents a number of detailed case studies to explore what actually happened, both physically and socially, during the occurrence of rapid-onset natural hazards during the medieval period. This includes a variety of types of hazard which affected society in different ways and at different scales. Turning to the aftermath of these events, Chapter 4 analyses the physical responses adopted by society in order to understand what 'real world' steps were taken to mitigate and protect against disastrous events during the period under consideration. Chapter 5 augments this evidence by examining how people sought protection from these events through spiritual means, guided by both the Christian world-view and traditional, sometimes superstitious, beliefs. Next, Chapter 6 considers to what extent disastrous events lived on in the memories of individuals and communities long after their occurrence, as well as whether these memories fed back into reducing vulnerability from recurrent hazards. Chapter 7 draws the evidence presented in the preceding chapters together in order to consider the broader nature of the relationship between human society and disaster during this period. The is followed by Chapter 8 which sums up the discussions presented in the preceding chapters and draws some conclusions about the nature of disasters in the medieval period, their impacts on society and how contemporaries lived through and understood these events.

1.1 What constitutes a disaster?

Natural hazards – the instigators of any 'natural' disaster – encompass a wide gamut of different types of event which are caused by diverse processes. Such phenomena include wildfires, earthquakes, avalanches, tsunamis, volcanic eruptions, hurricanes and meteor impacts as well as many other potential hazards. Of course, many of these types of event are rare occurrences or restricted to particular geographic zones. As Britain is the focus of this volume, the emphasis is on the types of hazard which most commonly afflict this part of the world: meterological hazards. Compared to 'natural hazards', 'meteorological hazards' is a more restricted category; including all those hazards of which the underlying cause relates to weather systems. The main categories of event are, therefore: floods, storm surges, droughts, wind and thunderstorms but not geophysical hazards such as earthquakes or tsunamis. Non-meteorological hazards are discussed throughout what follows only to provide parallels or where a causal relationship exists with a meteorological hazard – such as a landslide caused by high precipitation (rather than seismic activity).

The occurrence of a natural hazard, however, does not automatically trigger a 'natural' disaster. We may imagine that, on another planet, similar to Earth but devoid of life, completely natural fluctuations in sea level, precipitation and weather patterns might lead to the inundation of areas that are usually dry, long droughts, severe rainfall or the occurrence of violent atmospheric storms. To an astronomer watching from afar, these events would, most likely, not register as disasters; only as inevitable natural processes. It is only, therefore, by introducing the presence of human society, and its interests, that such events, and their impacts, come to be considered as disastrous. This simple distinction between presence and absence is what causes the natural action of a hazard to precipitate a 'natural' disaster. A simplistic definition of a 'natural' disaster, therefore, is an unwelcome change from an accepted norm experienced by a human community as a result of the action of a natural hazard. More clinically, the most

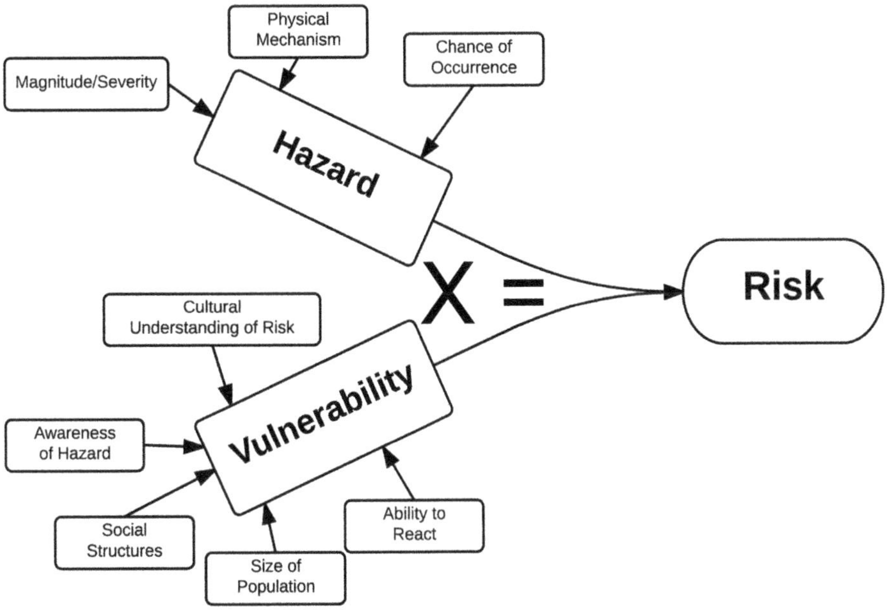

Fig. 1.2: A conceptual diagram of the equal roles played by hazard and vulnerability in generating risk. A number of the factors which play into the creation of vulnerability and the attributes of a given hazard are shown. Created by the author.

comprehensive database of modern disasters defines a disaster as an event which either results in over 10 casualties or negatively impacts at least 100 people.[5] Importantly then, for a natural hazard to result in a 'natural' disaster, an impact on human life, assets or interests is required.

The term 'natural disaster' can be problematic however. Although, as above, a human component is a prerequisite for any 'natural' disaster, in certain cases human decisions, rather than the inevitable and natural occurrence of a natural hazard, have been held almost entirely responsible. A good example is the Fukushima Daiichi incident in Japan in March 2011 in which a high magnitude earthquake caused a tsunami wave which flooded the nuclear power station at Fukushima leading to reactor meltdown, explosions and the leakage of radioactive material. As a result, over 100,000 people were evacuated and a 20km exclusion zone was enforced around the nuclear power station.[6] Another commonly cited example is the devastating flooding and loss of life which occurred when Hurricane Katrina struck the US Gulf Coast in August 2005. The ensuing disaster was a product of poorly designed flood defences and widespread

5 Smith, Petley 2009: 24.
6 Hasegawa 2012.

poverty which increased vulnerability, with many inhabitants unable to escape to safety.[7] In such examples, while the conditions themselves arose naturally: the earthquake and tsunami which hit Fukushima and the hurricane conditions of Katrina, the main factors which made these events 'disasters' were economic, social or political issues both internal to the societies they affected and the result of human decisions. While these two examples sit at the extreme end of the scale, and have received copious amounts of both scholarly and popular analysis, all disasters incorporate a human element meaning that, at a semantic level, none can truly be considered 'natural'. Clearly then, the risk created by a 'natural' disaster is composed of multiple elements. The two most fundamental of these are: the natural hazard – encompassing the physical mechanism behind its occurrence, its magnitude and area of effect – and the vulnerability of the human society – a product of factors such as their awareness and understanding of the hazard, their location in relation to the hazard's area of effect and their resilience when the hazard strikes. This is often conceptualised as an equation, as in Fig. 1.2, in which these two factors combined produce the risk from any given disaster.[8] Accepting that 'natural' disasters are the results of the interplay between a natural hazard and human decisions, the term 'natural disaster' is used throughout what follows in reference to these events.

1.2 Natural hazards in a British context

While meteorological hazards are the primary focus of this book, it is useful to provide a brief overview of the main natural hazards which pose a risk in and around Britain. The types of natural hazards which typically impact northern Europe more generally are relatively uniform. The combination of long, exposed, coastlines and the interactions between the Atlantic Ocean and climate systems strongly influence both the weather and many of the hazards experienced across this zone. Extra-tropical cyclones frequently track across northern Europe – indeed, the storm tracks which occupy this zone are among the most active in the northern hemisphere.[9] These events cause high magnitude windstorms and/or precipitation and most commonly occur during the autumn and winter months. At a continental scale, westward coastal areas between latitudes of 50° and 70°, which are exposed to the full force of cyclonic systems from the Atlantic, Ocean are worst affected.[10] Strong storm winds can also produce other hazards such as aeolian sand,[11] when sand deposits become mobilised, and storm surges, during which high wind speeds drive bodies of water against coastlines causing the water

7 Comfort 2006: 503.
8 Wisner et al. 1994: 49.
9 Lamb, Frydendahl 1991: 3.
10 Bartholy et al. 2006; see Fig. 1.3, B.
11 Clarke, Rendell 2009.

to 'pile-up' bringing about a localised rise in sea-level. The low pressures associated with storm systems also precipitate a rise in sea-level amplifying the flood risk posed by storm surge events. The North Sea basin, and adjoining coasts, are particularly vulnerable to this effect (see Fig. 1.3, A) due to the basin's relatively shallow bathymetry, funnel-like shape and the 'bottleneck' of the English Channel which slows the escape of water driven from the north. The risk to surrounding coastlines is epitomised by devastating modern events such as the storm surge and resulting flood which struck the UK and the Low Countries on 31st January/1st February 1953.[12]

Floods caused by other factors routinely occur throughout northern Europe. While there are many different categories of flood, including those caused by ice-jams, mass movements or the failure of levees and dams, the most common, in addition to storm surge flooding, are river floods and flash floods.[13] River floods occur when sustained or intense precipitation, over a timespan of days to weeks, swells rivers beyond their normal bounds. These events become particularly severe when flooded soils become waterlogged or frozen as this prevents floodwaters from draining into the soil. River floods are common throughout northern Europe (see Fig. 1.3, C), especially during the winter months when atmospheric depressions cause warm fronts bearing moisture to pass over the zone from the west. Flash floods are localised extreme precipitation events which occur over a short time frame, usually under six hours. These events most commonly affect mountainous areas, although lowlands can also fall victim, and can also be caused, or exacerbated, by rapid snowmelt. Precipitation supplied by moisture from the Atlantic Ocean means that an excess of precipitation is more common than a deficit. Drought, however, does occur across the region although not to the same extent as in southern and eastern Europe.[14]

Turning to geo-tectonic hazards, active volcanoes within northern Europe are limited entirely to Iceland[15] which is somewhat anomalous for the zone in geological terms. Risk from earthquakes across the majority of northern Europe is low, with a slightly elevated probability of seismicity in parts of Belgium, northern France, western Norway and Wales, as documented by the 2013 European Seismic Hazard Map produced by the SHARE project.[16] Again the exception is Iceland which sits atop the Mid-Atlantic Ridge, a major fault line between two tectonic plates. In spite of the relatively low seismic risk across northern Europe, damaging earthquakes do occur; modern examples include the magnitude 4.4 17th February 2018 Cwmllynfell earthquake which affected Wales and England[17] while medieval cases include the tremors of 21st May 1381, the epicentre of

12 Baxter 2005; Gerritsen 2005.
13 Barredo 2007: 130.
14 Lloyd-Hughes, Saunders 2002.
15 The only exception is Beerenberg on Jan Mayen Island, Norway, located in the North Atlantic.
16 Giardini et al. 2013; see Fig. 1.3, D.
17 BGS 2018.

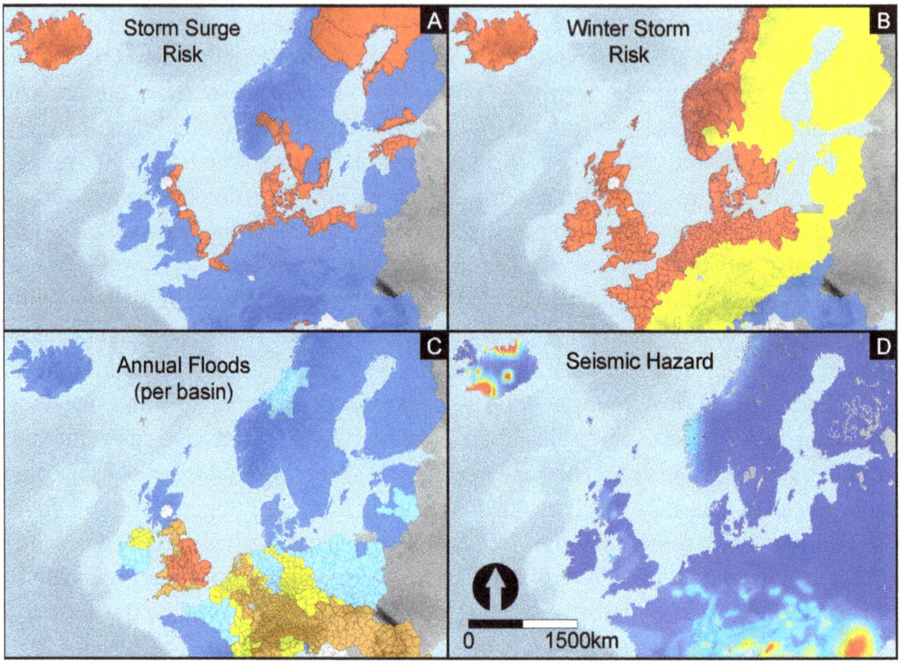

Fig. 1.3: Risk from different natural hazards in northern Europe today (Red=high, Blue=low). Created by the author using data from the European Spatial Planning Network available at: http://rimap.espon.eu. Seismic hazard map from GIARDINI et al. 2003.

which is estimated to have been located in the Straits of Dover, which caused structural damage in south-east England and the Low Countries.[18]

Another hazard to which the entire zone is exposed are mass movements such as landslides. These can be caused by seismic activity, although, as above, excepting Iceland, there is only a low risk of such an occurrence across northern Europe. High rainfall on the other hand, can also precipitate mass movements and is a common occurrence across the zone. In the majority of areas however, such occurrences only have the potential to be low magnitude events due to the relatively flat topography of much of the zone – including most of England, much of Ireland, northern France, the Low Countries, northern Germany and eastern Scandinavia. The exception are mountainous regions including parts of Wales, the English Lake District, the Scottish Highlands and parts of Scandinavia. JAEDICKE et al. identify the Scottish Highlands, two distinct areas of eastern Norway and two separate areas on the north and south coasts of Iceland as hotspots for landslides caused by precipitation within northern

18 HOFFMAN 2014: 308.

Europe.[19] Perhaps the zone worst affected by rainfall-induced mass movements is the western coast of Norway where steep mountains and high precipitation as a result of the arrival of humid air transported by Atlantic low pressure systems coalesce.[20]

One of the most severe, although rare, hazards to affect the region are tsunami. These can be caused by either submarine earthquakes, landslides or by meteor impact. The former appears to have affected the British Isles in 1755 when tsunami waves were generated by the Lisbon Earthquake.[21] The impact of undersea landslides in the region, on the other hand, are revealed by investigations into the 8200 BP Storegga Slide event, in which the land bridge between Britain and the European continent was inundated, undoubtedly with disastrous consequences for those dwelling in between.[22] Outside of such events which are either relatively recent, and thus historically documented, or of extremely high magnitude, and thus identifiable in the geological record, there are difficulties in distinguishing between tsunami and severe storm events. The occurrence of tsunami affecting the British Isles during the medieval period has been suggested but never proved beyond doubt.[23] BAILEY, for example, discusses the possibility, without concluding it to be the case, that increased submarine earthquake activity was the cause of the heightened occurrence of coastal flooding during the 14th century in East Anglia.[24]

1.3 Approaching the study of disasters in the past

Specific disasters, both mythical and historical, have always figured prominently in the popular imagination of the past. One need only think of the Atlantis myth, the Biblical flood of Noah or, more recently, archaeological discoveries such as Roman Pompeii which seem to preserve the cataclysmic realities of a moment of disaster from long ago. Throughout the early modern period, specific disasters spurred research into their underlying causes and, in some cases, how they might be mitigated. The 'Great' storm of 1703, for example, inspired the writer Daniel DEFOE to collate contemporary eyewitness testimony as well as trying to understand, through the literature available at the time, why such an event had occurred.[25] Most famously, the 1755 Lisbon earthquake gave rise to early research into the causation of the disaster and what might be done to mitigate damage on the same scale occurring in the future.[26] The Verdalen landslide

19 JAEDICKE et al. 2014: 333.
20 JAEDICKE et al. 2009.
21 BANERJEE et al. 2001.
22 WENINGER et al. 2008: 16–17.
23 HASLETT, BRYANT 2007a; HASLETT, BRYANT 2007b.
24 BAILEY 1991: 189; BAILEY 2007a: Fig. 16.
25 DEFOE 1704.
26 ARAÚJO 2006.

in Norway in 1893 drove research into the mechanics behind landslides – including historical analysis of past events – though "as memory of the disaster faded, so did the funding for further … investigations".[27] As demonstrated in the latter case, these early studies were affected by an element of 'amnesia'[28] – although, for a time, the occurrence of a disaster inspired research into why they happened, as well as their impacts, up until the mid 20th century, this was generally a temporary phenomenon which lapsed as memory faded of the event which had originally sparked interest.

Academic research into disasters as a discrete category of event worthy of consideration is a relatively recent phenomenon.[29] In the United States, social scientists first began investigating disasters in detail during the post-war years, using natural disasters as proxies for plausible social responses to military emergencies such as a nuclear attack by a foreign power.[30] During this period, natural disasters were, somewhat fatalistically, treated as unpredictable events which could not be averted. Only from about 1980 has a view emerged which, as previously described, treats disasters as events created by an interplay of natural processes and cultural decisions.[31] In this context, from the 1990s, but especially since the early 21st century, the sub-discipline of historical disaster studies – which seeks to approach the study of disasters, as a combination of naturual and cultural concerns from the historical perspective – has emerged. Prominent scholars within this sub discipline include Gerrit SCHENK, who, as well as investigating a number of detailed case studies,[32] has also concentrated on refining concepts and methodology[33] while promoting the value of studying disasters from a historical perspective[34] and Christian ROHR, whose research especially focusses on hydrological and seismic disasters in central Europe[35] as well as methodological questions.[36] Others include Christian PFISTER, who has particularly focussed on reconstructing climate related hazards[37] as well as looking at risk culture in historical societies more generally,[38] Franz MAUELSHAGEN, who has investigated cultural responses to disaster during the post-medieval period,[39] and Greg BANKOFF who has studied the cultural component of disaster and risk in both the present-day Philip-

27 ROKOENGEN et al. 2001: 58–59.
28 SCHENK 2007a: 10.
29 JUNEJA, MAUELSHAGEN 2007, SCHENK 2007a.
30 QUARANTELLI 1987.
31 OLIVER-SMITH 1996: 304.
32 SCHENK 2007b; 2017.
33 SCHENK 2007a.
34 SCHENK 2015.
35 ROHR 2003; 2013.
36 ROHR 2007.
37 E.g. PFISTER 2017.
38 PFISTER 2011.
39 MAUELSHAGEN 2012.

pines [40] as well as in medieval and post-medieval England.[41] For the Low Countries, the research groups of Tim SOENS and Bas VAN BAVEL have worked on medieval and post-medieval flooding, as well as other disasters – particularly exploring how social and economic conditions contributed to creating the risk to which these populations were exposed.[42]

Beyond this exclusively historical sub-discipline, however, most other disciplines have done little to engage with the study of historical disasters. The exception is historical climatology, which finds much common ground in tracing the occurrence of specific events in order to provide data for reconstructions of historical climate[43] and in some cases to investigate the impact of climatic changes on society.[44] Medieval economic historians, on the other hand, although well placed to contribute toward the study of disasters in the past have generally ignored disasters as a research question. There are a number of reasons for this. CURTIS ascribes the general drift of the historical discipline away from the social sciences as one of the main reasons.[45] A prevailing view is that medieval disasters bear no relevance to modern issues due to the fact that medieval culture, particularly religious thinking, was so far removed from modern beliefs that no useful comparisons are possible. Perhaps most pervasively, natural disasters are still regarded by many as entirely natural occurrences. This is problematic as the historical discipline, as Bruce CAMPBELL has argued, has largely rejected nature and the environment as a 'protagonist' emphasising instead the primacy of human actions and decisions in precipitating cultural change.[46]

Archaeology as a discipline is well placed to investigate the human-environment relations between natural hazards and vulnerable societies. Indeed, this was among the 'grand challenges' identified as priorities that archaeologists should aim to tackle according to an international gathering of prominent archaeologists in 2012.[47] However, as with economic history, archaeology as a discipline has done little to engage with wider debates in disaster studies, a fact highlighted by some national research frameworks.[48] This is crudely illustrated by a search of the major disaster studies journals (Disasters, Natural Hazards, Journal of Mass Emergencies and Disasters) with the keyword 'archaeology'. Very few of the research papers returned through this method include anything more than a passing reference to archaeological evidence.[49]

---

40 BANKOFF 2003.
41 BANKOFF 2013.
42 E.g. SOENS 2011; 2013; VAN BAVEL et al. 2020.
43 E.g. OGILVIE, FARMER 1997; BRÁZDIL et al. 2005.
44 E.g. BEHRINGER 1999; ANDERSON et al. 2016.
45 CURTIS et al. 2016.
46 CAMPBELL 2010b: 282–284. See also HOFFMAN 2014: 342–351.
47 KINTIGH et al. 2014: 11, 18.
48 HALL, PRICE 2012: 31.
49 BROWN 2017.

The sparsity of interaction between archaeologists and other disciplines working on disasters is certainly not indicative of archaeologists' avoidance of studying sites and landscapes affected by natural disasters. Indeed, a conference of the Society for Medieval Archaeology held in 2016[50] sought to explore the contribution that archaeologists can make to the study of disasters for the medieval period.[51] In most instances where archaeologists have encountered evidence for the occurrence of natural disasters, however, the disaster is not the primary research objective. The relatively rare cases where this has occurred include early work by SHEETS which provided a broad but brief overview of archaeological evidence for the occurrence of natural hazards.[52] Since then, a number of researchers have produced synthetic and comparative work on archaeology and disasters,[53] but these have generally been limited in scope and have, necessarily, focussed on either specific periods, geographic regions and/or types of hazard. A number of researchers have examined specific case studies in detail; FERNÁNDEZ et al., for example, examined archaeological evidence for the impact of a high magnitude flood event on a medieval village in Asturias, NW Spain,[54] while FORLIN and GERRARD conducted investigations into the impact of a devastating landslide in 1522 which buried the town of Vila Franca do Campo on São Miguel in the Azores, Portugal.[55] Looking at how historic communities responded to the risk posed by hazards, GARDINER and HARTWELL reconstructed the phasing and chronology of medieval and post-medieval flood defences located in English wetland environments.[56] Turning to the longer-term implications of living with the risk of natural hazards, Stephen RIPPON has made the flood risk inherent to wetland environments a central part of his research into human occupation of these locales[57] while MENOTTI et al. interpreted the flood risk experienced by Bronze Age lake shore settlements in the central European Alpine region as a motivation for the development of complex ritual practices.[58] RIEDE has examined the impact of prehistoric volcanic eruptions in Europe and called for further integration of archaeology into the study of past disasters.[59] The above examples demonstrate that one of SHEETS' key conclusions over 40 years ago that: "archaeologists are almost completely unaware of the hazard research conducted by social scientists within the

50 Waiting for the End of the World: perceptions of disaster and risk in medieval Europe, held from 2nd–4th of December 2016 at Rewley House, Oxford, United Kingdom.

51 Note that this conference has given rise to an edited volume. GERRARD et al. 2020. The contributions contained within present analogous and complementary subject matter to this book and, as such, many of these contributions are referenced throughout the remainder of the text.

52 SHEETS 1980.

53 E.g. GUTTORMSEN 2008; GERRARD, PETLEY 2013.

54 FERNÁNDEZ et al. 2019.

55 FORLIN, GERRARD 2017; GERRARD et al. 2021.

56 GARDINER, HARTWELL 2006.

57 E.g. RIPPON 2000a; 2000b; 2001; 2004.

58 MENOTTI et al. 2014.

59 RIEDE 2014.

past few decades"[60] can no longer be said to represent the *status quo*. His other primary conclusion, however, that: "whatever piecemeal knowledge has been generated by archeologists with respect to natural disasters largely has been ignored by hazard researchers"[61] certainly still holds some truth.

The area in which archaeology has contributed most significantly to the study of disasters, beyond the limits of the discipline itself, is archaeoseismology. Archaeological evidence can be used to estimate the physical characteristics of past seismic hazards which, in turn, can enhance knowledge about the nature of future seismic activity. Conducted overwhelmingly with a natural science rationale, the data obtained from archaeological sites relating to past earthquakes can inform models of contemporary and future seismic risk. The value of this data for evaluating modern-day risk has, however, had an unfortunate blinkering effect meaning that archaeoseismological research rarely engages with the relationship between past societies and the seismic events they investigate. This is a short-coming recognized by archaeoseismologists themselves.[62] and was the subject of a recent research project at Durham University which sought to redress this situation for the medieval period in Europe through the investigation of a number of key case studies.[63]

Compared to seismic hazards, meteorological hazards have not seen the same intensity of research. Archaeologists frequently encounter evidence for the occurrence of these type of events on archaeological sites; floods are attested through layers of alluvium or marine sediments, discrete layers of wind-blown sand attest to past storm activity[64] and even rainstorms occasionally leave an archaeological signature.[65] Of course, certain hazards, such as windstorms, droughts or lightning strikes are ephemeral and rarely leave lasting impressions in the archaeological record. For those hazards which do leave an identifiable material signature, analogous to the way in which archaeoseismologists have studies seismic hazards, such data has been interrogated to answer practical scientific questions – for example investigating changes in river flood regimes.[66] Synthetic and comparative research on the impact of these events on contemporary society, however, has been lacking. In the rare cases where archaeologists have considered such events as a primary research topic, the focus is often on archaeological methods and research is usually published in discipline specific journals which may not be read by readers beyond the boundaries of the archaeological discipline.

---

60 Sheets 1980: 25–26.

61 Sheets 1980: 26.

62 Sintubin 2011: 8.

63 The Leverhulme Trust funded project 'Risk and Resilience: exploring historic responses to earthquakes, 1200–1755'. See also Forlin, Gerrard 2017.

64 Brown 2015.

65 Hinzen et al. 2013.

66 Kiss, Laszlovszky 2013.

Part of the reason that archaeology has not figured more prominently in interdisciplinary approaches to disaster can be explained by some of the difficulties peculiar to archaeological techniques and sources of evidence. By definition, disasters are short-term occurrences which alter conditions from what is considered 'normal' at a given location. In most cases therefore, only a short window of time exists for the deposition and accumulation of an archaeological signature. In the case of some types of hazard, such as volcanic eruptions, landslides, floods and some earthquakes, the deposition of large volumes of ash, sediment or debris in short spaces of time (hours and days) is a common occurrence, providing a recognisable archaeological signature. However, other types of hazard such as windstorms, snowfalls,[67] and droughts rarely leave behind any long-term material evidence. Even in the case of those hazards which are more recognisable archaeologically, later site clearance and the reworking of material can often truncate and erase evidence for the occurrence of earlier hazards.

It is also difficult, and indeed one of the major challenges encountered in the following chapters, to combine archaeological data with data relating to the past from other disciplines.[68] Climatic data obtained from proxies such as tree rings or ice cores usually operate at interannual time-scales while documentary data can, in some instances, be securely dated to a particular calendar date.[69] On the other hand, the chronological resolution applicable to most archaeological data – in the absence of well dated material culture, such as coins, or absolute dating methods, such as tephra- or dendro- chronology – rarely allows individual contexts and artefacts to be precisely dated beyond a date range less than *c.*100 years.[70] This poses a problem when dealing with specific historical events, such as a natural disaster, as it becomes difficult to convincingly demonstrate that archaeological evidence can securely be connected to a specific, documented disaster.[71] This is exemplified by the documentary evidence which records the burning of the Cistercian Abbey of Strata Florida, Ceredigon, Wales, as the result of a lightning strike in 1284.[72] Although the archaeological evidence corroborates the written description – the solidified molten lead unearthed during excavation is a detail specifically mentioned in the historical source[73] – it is impossible to definitively prove that the melted roofing lead recovered was a product of that particular blaze and not another fire, of which there are a number of possible, historically documented,

67 Schofield 2009.

68 Cooper, Peros 2010: 1226.

69 Although, of course, historical documentary data brings with it its own set of chronological issues relating to calendrical systems and source reliability, a topic discussed fully by Brázdil et al. 2005.

70 This is exemplified by the high-energy deposit discussed in Chapter 3.3.1 in relation to the storm of 1362.

71 Galadini et al. 2006: 408.

72 Christie 1887: 115–117.

73 Williams 1889: 153–154.

candidates.[74] These chronological issues may explain to some degree why medieval archaeology has particularly engaged with disasters in Norse Iceland. Here tephra layers from volcanic eruptions provide precisely dated reference points which can be used to anchor archaeological layers in time. This allows analysis of the material changes that took place in the aftermath of a particular eruption, with chronological evidence of their association to the hazard. This research is often conducted by, or in association with, volcanologists with primarily scientific aims but a number of studies have explicitly focussed on the social impacts and consequences of these disasters.[75]

Another challenge in approaching the study of disaster from an archaeological perspective lies in the controversy that surrounds the impact of exogenous natural forces on human societies. As with the traditional historical viewpoint, discussed above, archaeologists are also wary of straying too far into the realms of environmental determinism. When researchers have proposed a link between environmental factors and cultural change, they invariably attract criticism – a fact which has likely impeded research into human environment relations within the discipline.[76] The recent periodisation of geological time based, to some extent, on the impact of disasters on human societies,[77] for example, provoked criticism for misinterpreting the evidence and ascribing cultural changes to climatic factors when other plausible explanations are available.[78] In an example from medieval archaeology, within which the impact of environmental fluctuations on society has not traditionally been a major research theme, the desertions at the medieval villages of Barton Blount, Derbyshire, and the village identified as Goltho, Lincolnshire, were interpreted as largely a result of the impact of the declining climate brought about by the onset of the Little Ice Age. The excavator, Guy Beresford argued that, under the deteriorating climatic conditions, the clay soils would have become unworkable for long periods – making continued occupation at the sites untenable.[79] Swift rebuttals followed, dismissing Beresford's interpretation as simplistic environmental determinism, pointing to the continuity of settlement at neighbouring village sites as evidence that "raindrops [are not] locally selective".[80] In addition, further detractors invoked the post-Black Death economic and demographic situation as a more plausible explanation for abandonment.[81]

---

74 Williams 1889: 154.

75 Dugmore et al. 2007: 7–8.

76 Rippon et al. 2014: 236.

77 Namely the proclamation of the Meghalayan Age based on an extreme drought which occurred *c.*4200 years BP and is interpreted as having significant implications for societies throughout the Near East, the Indian subcontinent and China.

78 Middleton 2018.

79 Beresford 1975: 51-52.

80 Beresford, Hurst 1971: 21.

81 Wright 1976.

Beresford's original counter argument was that differential abandonment between nearby villages might be explained by relatively minimal differences in local soil composition[82] and, indeed, soil chemistry and geology are now being recognized as major determining factors in the development of medieval villages.[83] Furthermore, 40 years after the publication of Beresford's interpretation of the deserted medieval village sites at Barton Blount and Goltho, similar conclusions were drawn regarding the desertion of the medieval settlement at Cedars Park, Stowmarket, Suffolk. At this site it is suggested that increased wetness during the climatic decline of the 14th century provoked the cutting of increasingly large enclosure ditches to drain surface water while cobbled surfaces may also have been a response to the difficulties of traversing wet clay. Such conditions, it is argued, would have made the surrounding clay soils difficult to work. 'Puddling' on clay soils requires time to dry out before the soil can be worked so increased wet weather could have, therefore, reduced the number of days when soils could be worked below a viable threshold making abandonment a favourable option.[84]

Comparably, PLATT has made the case that the sudden appearance of homesteads equipped with moats in medieval England, during the early 14th century, relates to the climatic decline known as the 'Dantean Anomaly' (*c.*1315–*c.*1322).[85] This climatic aberration was marked by unprecedented rains throughout 1315[86] leading to a severe famine which was swiftly followed by a cattle panzootic in the 1320s. PLATT, therefore, argues that these environmental fluctuations created conditions – dearth and hunger – among the populace that drove those with something to protect to dig moats around their homesteads for security.

Of course, in all of these examples, economic and social considerations are just as paramount as climatic and environmental factors. Although modern scientific climatic reconstructions call into question received wisdom surrounding the role of climate in facilitating and constraining human activity[87] it remains controversial to equate developments in human affairs with changes in nature and environment. The above examples not excluded, all other potential avenues should be fully explored before climatic and environmental factors can be considered as plausible drivers of cultural change. Where many in the past have entirely eschewed ascribing any impact to these forces, however, the potential impacts of changes in the natural environment on contemporary populations is becoming increasingly clear.

As we will see, there are many cases in which medieval populations chose not to relocate in the aftermath of severe disasters and continued to occupy locations which

82 BERESFORD 1981: 36.

83 WILLIAMSON et al. 2013: 79–80; RIPPON et al. 2014: 200–201

84 WOOLHOUSE 2016: 122.

85 So called due to the fact that its abatement more or less coincided with the death of the Italian poet Dante Alighieri. See PLATT 2010; 2012.

86 JORDAN 1996: 17–18.

87 See for example CAMPBELL 2016b.

had been drastically altered by natural hazards. Even when extreme natural forces were unleashed, therefore, many choices remained undetermined allowing human populations considerable leeway in which to choose how to respond. Consciously or unconsciously, no doubt, the cost/benefit calculations which medieval populations entered into in these situations factored in not only economic and social considerations – such as the costs of setting up anew compared to resettling the original place of habitation, dependencies to local lords and the proximity of family members and friends – but also intangible factors encompassing ideas of 'place' and familial ties to the land. The various pathways open to medieval populations in the aftermath of fluctuations in the environment and climate, therefore, were constrained by both the realities of the physical changes in the environment as well as economic and social considerations. From two sides, therefore, these two categories of factors affected what decisions people made and how they chose to react. While environmental forces never 'determined' what people did, abrupt changes, to a greater or lesser extent, did influence decisions and what options were viable.

1.4 The contribution of archaeology to the study of disasters

Despite the difficulties in approaching disasters from an archaeological perspective, archaeologists can make a valuable contribution to the study of disasters. Perhaps the most obvious point is that archaeology can extend our knowledge of disasters far back in time. In most parts of the world, reliable scientific data relating to hazards is rarely available before the 19th century. As a result, hazards with a long return period often sit beyond the period covered by instrumental records. In these cases, information on the causes, magnitude and effects of such hazards must be collected through alternative means – including historical sources and investigations of the archaeological and geological records. A good example are high magnitude earthquakes in the Himalaya. By correlating extant historical records with geological evidence and radiocarbon dates derived from trenching,[88] together with standing building evidence – in this case medieval temples[89] – seismic activity in the region can be investigated over longer time periods than those covered by instrumental records. This permits both the identification of undocumented seismic events and the characterization of modern day risk if similar events were to recur in the present day.

Another important area which can benefit from an archaeological contribution is the provision of precise information concerning the impact of a rapid-onset hazard. While historical sources often provide descriptions of the occurrence of hazards, these sources are usually low in detail and cannot always be taken at face value – often

88 Stolle et al. 2017.
89 Rajendran et al. 2013.

dates and details were misreported or elaborated[90] while the coverage of documentary evidence, at least for medieval Europe, is biased towards the literate and land-owning classes. In rare cases, where high magnitude rapid-onset hazards such as floods, landslides, earthquakes or volcanoes cause the destruction of a settlement or structure, this can promote the long-term preservation of in-situ remains. This, in turn, preserves a record of the final abandonment and destruction of the site by the hazard and, perhaps, the last-minute responses of the affected population. While a number of high profile sites of this category, mostly from the Classical world, are known such as the Roman settlements of Pompeii and Herculaneum and the Minoan town of Akrotiri on the Greek island of Santorini, medieval case studies are also known. One of the most fully investigated is the case of the castle of Saranda Kolones, Paphos, Cyprus, where, during the 1222 earthquake, as the structure collapsed, most of the inhabitants appear to have made a hasty escape leaving behind objects of value as they fled.[91] Remains of one unfortunate individual who perished after escaping down a latrine shaft only to find his only exit blocked indicate at least one human casualty,[92] with faunal remains crushed beneath fallen masonry attesting to further losses.[93] In the aftermath of the earthquake the archaeological evidence suggests salvage attempts either to recover the bodies of casualties or to claim and reuse the fallen masonry for the repair of the town of Paphos, which had also suffered severe damage during the earthquake.[94] Evidence from a nearby cave may indicate the presence of refugees made homeless following the earthquake while newly built structures suggest attempts to remedy this situation by constructing new housing in the earthquake's aftermath.[95] Such an example demonstrates the rich level of detail that archaeological data can lend to an event which, although documented by contemporary sources to some degree – the earthquake itself was recorded but little mention was made of the castle – would be otherwise unknown.

The demographic and economic impacts of hazards can also be investigated through archaeological evidence. While medieval chroniclers often record the number of fatalities lost to specific natural hazards – 50,000,[96] for example, were reported to have been lost in a 13th century flood in the Netherlands[97] – these cannot be trusted at face value. Although as above, archaeological evidence can confirm the presence of

90 Rohr 2003: 136–137; Brázdil et al. 2005: 373–374.
91 Rosser 2004: 39–40.
92 Rosser 1986: 47.
93 Megaw 1957: 49.
94 Rosser 2004: 47–48.
95 Rosser 1985: 94.
96 This number is certainly an exaggeration as it seems doubtful the chronicler would have had access to any accurate figures if these were even produced. High figures such as these were merely used by medieval writers to illustrate that a very high number had died. Ziegler 1969: 51–53.
97 Pertz 1861: 215.

fatalities,[98] it is impossible to quantify exact numbers in any given event. Over a longer timespan, however, demographic decline can be inferred through material remains. Systematic test-pitting in eastern England, for example, provides material evidence for the acute decline which followed the Black Death allowing an estimate of the percentage of demographic change in the locales studied.[99] This approach could theoretically be applied to landscapes or settlements affected by wide-area hazards, such as tephra falls, landslides or aeolian sand inundations, in order to gauge what, if any, impact these hazards had on demography and economic activity over the medium-long term.

The occurrence of a hazard itself is also only one aspect of the impact of a disaster which may be investigated archaeologically. In the case of extreme events, sites may be entombed and preserved in-situ by the action of the hazard itself. In other cases, the occurrence of a disaster may lead to later site abandonment or a period of reduced activity. In the case of a protracted abandonment at some point after the occurrence of a hazard, it would be difficult to demonstrate the causation behind the abandonment with certainty. However, abandonment in advance of or in the immediate aftermath of the occurrence of a hazard may leave a distinctive archaeological signature. In the case of 'normal' site formation, an assemblage comprising a limited number of old, low-value, damaged or cumbersome objects would be expected with more valuable or useful items being removed during the abandonment phase. When a site is abandoned for 'catastrophic' reasons, however, fewer items are likely to be removed, with a greater incidence of valuable, personal and functional objects. As a result, the nature of the assemblage should be easily distinguishable from a 'normal' site.[100] An example comes from the medieval farmstead at Eckweek, Somerset, which appears to have been abandoned rapidly in the mid-late 14th century.[101] The factors behind the abandonment of this particular site are unknown although plague (most likely the 1348–52 or the 1361–62 outbreaks), including its indirect effect on the mobility of workers, is considered a possible contender.[102]

Archaeological evidence can also provide valuable insights into internal processes within societies before, during and after disasters. In some cases, for example, archaeological evidence may allow something of the role of authorities in the post-disaster rebuilding stage to be inferred. Where it is possible to compare the layout of structures

98 The absence of fatalities from an archaeological site affected by a disaster may not automatically contradict historical sources stating that high casualties occurred as, where possible, bodies might be removed in the aftermath of an event in order to carry out proper burial rites.

99 Lewis 2016.

100 Young 2020: 275.

101 In support of a rapid abandonment are the high relative abundance of artefacts, including domestic artefacts, structural metalwork, horse gear, craft-related items and personal belongings, including jewellery and coins. At the time of abandonment the site also appears to have achieved a zenith of prosperity compared to earlier periods. Young 2020: 260–267.

102 Young 2020: 298–299.

before and after the occurrence of a disaster, for example, the alignment of structures may attest to whether rebuilding was planned centrally or if individuals were left to make repairs of their own volition. The latter is demonstrated in the aftermath of conflagrations in medieval Bergen by the permanence of property boundaries, suggesting individual property, and thus the responsibility to repair the damage, were unaffected by the repeated fires which razed the town to the ground.[103] Archaeological evidence can also illuminate disasters' positive and negative economic impacts. As an example, archaeological excavations at Vila Franca do Campo, São Miguel, Azores, reveal that, following the landslide of 1522 in the relatively newly settled Portuguese Azores, the destruction caused by the landslide, rather than tipping the population into total poverty, invigorated the local economy by forcing the surviving population to produce their own roof tiles, where previously they had relied entirely on imports from the Portuguese mainland.[104] Clearly, the landslide disrupted the established economic order but it may have, in fact, had some positive economic repercussions.

The archaeological record can also provide evidence for ritual activities which may have been stimulated by the occurrence of natural hazards. Although ritual and belief are notoriously difficult to infer through material remains alone, a number of practices can be connected to beliefs surrounding disasters. For example, one interpretation of burnt marks in churches and vernacular architecture, which seem to have been deliberately created, is that they were believed to bestow protection on the structure from lightning.[105] Similarly, the distribution of *ampullae*, vessels obtained through pilgrimage containing dust, holy water or oil blessed at the shrine of a saint, in agricultural fields across medieval England has been interpreted as evidence for belief in saintly protection against extreme natural events such as hail and drought.[106] This type of evidence can be profitably combined with the historical record which, for medieval Europe, is rife with descriptions of processions, prayers and ritual acts which were believed to provide communal or personal protection against natural hazards.[107]

As this section has outlined, archaeology has much to 'bring to the table' but a holistic understanding of disasters is only possible through the integration of strands of evidence from a wide variety of disciplines. What follows, therefore, while attempting to capitalise on the opportunities provided by the archaeological record, integrates research from other disciplines in order to investigate how disasters affected populations in medieval Britain in as much detail as possible.

103 Hansen 2015.
104 Forlin, Gerrard 2017: 104.
105 Lloyd et al. 2001.
106 Anderson 2010.
107 Hanska 2002.

2 Disaster and society in medieval Europe

Before any disaster befalls a community, much of its potential to cause damage is 'pre-set' by conditions already present within the society itself. Vulnerability to specific hazards, is created by physical factors, such as the proximity and relative level of a home to a source of flooding, but also less-tangible cultural considerations such as socio-economic status, the availability of charitable relief to the poor and religious beliefs. To understand the impact of specific disasters on medieval society, it is first necessary to consider how risk as a concept was conceived by populations during the Middle Ages – what mitigative measures were available and the extent to which the medieval mindset regarded natural disasters as avoidable or inevitable events against which efficacious measures could be taken. A key stage in the 'disaster cycle' (see Fig. 1.1), therefore, is to what extent medieval society was prepared for the occurrence of a disaster prior to its occurrence. While in modern terms, this might constitute 'battening down the hatches' in response to the forecast of a violent storm, such reliable weather predictions were totally unavailable to medieval populations.[1] This meant that short-term preparations against unforeseen events were rarely possible – and where high magnitude natural hazards struck in this way the results were likely to be devastating. This being the case, medieval people practised a wide array of longer-term strategies in order to guard against risk posed by natural hazards and future uncertainty more generally. This chapter, therefore, explores the extent to which medieval populations were prepared for disaster and what factors peculiar to the medieval period made people more or less vulnerable to the occurrence of natural hazards. As we shall see, while archaeology can make an important contribution to this debate, much of the evidence, at present, comes from the documentary record.

Of course, medieval Europe was not an entirely culturally homogeneous zone. Important distinctions existed between populations, in both spatial and temporal terms, encompassing the modes of landholding, the roles of institutions and specific religious beliefs. It is impossible, therefore, to cover how all these considerations varied, both geographically and over time, in a single chapter. Accordingly, this chapter adopts a broad brush approach which is refined in respect to more detailed case studies in chapter 3.

1 Medieval weather predictions were based either on traditional knowledge based on observed weather patterns and temperatures, which offered a certain degree of reliability, or popular folk beliefs which ranged from observing the behaviour of animals to the practice of prognostication, which claimed to be able to predict the weather based on such unlikely events as the day of the week on which Christmas or New Year fell, which month brought the first thunder clap of the year or on which days of the 12 days of Christmas the sun shone brightly. See Jones 2013: 39–40.

https://doi.org/10.1515/9783110719628-002

2.1 Living with risk

In their influential *longue durée* history of human-environment interactions in the Mediterranean, HORDEN and PURCELL emphasised the importance of cultural practices to deal with the risk posed by natural hazards such as the sudden storms, flash floods and droughts to which the Mediterranean climate is particularly prone.[2] Of course, such practices are not, and have never been, exclusive to the Mediterranean world and comparable, equivalent practices have been relied upon by northern European populations throughout history. For the medieval period, one of the most well-known is the practice of fragmented land-holding associated with the open field system, common across much of midland and central southern England, in which peasants living in a nucleated settlement farmed small strips of land scattered throughout the lands of the parish, usually in 2 or 3 large and unenclosed 'open' fields.[3] As an adaptation against risk, scattered holdings have been most consistently championed by MCCLOSKEY[4] who argued that the practice reduced the likelihood that any single land-holder would suffer the total loss of their crops as a result of unfortunate circumstances, such as flooding, hail or pests – which would be unlikely to affect more than a small portion of their diverse holdings.[5] Despite the inherent inefficiencies of scattered holdings, mainly the increased time spent travelling between parcels of land, MCCLOSKEY calculated that the practice was an economical method of insuring against risk to agrarian produce, especially when compared to alternatives such as grain storage which, in tying up potential earnings, could result in a reduction in earnings by as much as one third.[6]

MCCLOSKEY's interpretation of medieval peasants' practices to guard against risk has proved divisive and many scholars have proposed variations or entirely alternate explanations for how risk was managed by medieval society. According to COSGEL, for example, scattered holdings were attractive because they enabled lords and peasants to deliberately distribute risk – rather than simply trying to minimise its consequences.[7] For FENOALTEA, diversified land holdings were important not for their geographical distribution of risk but for the fact that they increased efficiency by facilitating the distribution of labour – allowing each peasant the opportunity to work on the portion

2 HORDEN, PURCELL 2000: 175–230.

3 WILLIAMSON 2003: 1–5.

4 See for example MCCLOSKEY 1989: 34–46; 1991

5 "The land and weather of England are notoriously variable, even over the two square miles or so of the typical village. A place with sandy soil on a rise would shed some year's excessive rain, yet one with clay soil in a valley would hold another's insufficient rain. An exposed place would have wheat likely to become tangled by rain and high winds at harvest but free of mold in a generally wet year. A sheltered place would be relatively immune from windy disasters but less dry and more moldy on account". MCCLOSKEY 1989: 34.

6 MCCLOSKEY 1989: 46–48.

7 COSGEL 1990.

of their holdings most suited to the present conditions at any given time.[8] BEKAR and REED, meanwhile, argue that the insurance provided by scattered holdings, rather than spreading risk across different locations, soils and terrain types, related to the fact that land could serve as an attractive savings instrument – with scattered holdings allowing small portions of land to be bought and sold as and when budgets required.[9] On the other hand, scattered holdings have been interpreted as a pragmatic method of ensuring resources were divided equitably amongst tenants.[10] Such models highlight the variety of priorities that must have coloured medieval calculations of risk, consciously or not, when organising their agricultural affairs. Although they probably arose primarily as a bi-product of the open field system,[11] scattered holdings likely did contribute to reducing vulnerability or were, at least, inherently risk averse – if open field farming had been vulnerable to commonly occurring and predictable natural hazards it would seem strange that, as a practice, it enjoyed such longevity.[12] Open field farming, and the scattering of holdings within these fields, however, saw significant variation in different areas, for example in south west England,[13] and was not the only agriculture model practised during the medieval period. In some areas, for instance, cohesive parcels of land around isolated farmsteads were held by single tenants.[14] The existence of viable alternative land holding practices, other than the open field model, implies that any advantage offered by scattered holdings was either relatively minor or limited to specific situations.

Additionally, the model of scattered holdings within the open field system was almost certainly motivated by other considerations rather than being simply a practice to minimise risk. For example, by dividing up the productive landscape into different parcels of land, which were then allocated to different members of the village community, scattered holdings ensured that the areas worked by each person were relatively equitable – each member of the community had a more or less equal share of the most agriculturally productive areas as well as those areas which might have been undesirable due to their poor fertility, unfavourable situation or distance from the village. This can be seen clearly in the annual allocation of strips of common meadow which, in most villages in medieval England, were shared between every member of the community who held rights to the hay crop.[15] As BRIAN has explored, this was

8 FENOALTEA 1976: 141–144.

9 BEKAR, REED 2003.

10 WILLIAMSON 2003: 181.

11 The factors behind which are numerous and debated. RIPPON et al. 2006: 66–67.

12 Adopted from about the 8th century until its dismantling during the post-medieval enclosure movement. WILLIAMSON 2003: 6; RIPPON et al. 2006: 67–68.

13 RIPPON et al. 2006: 59–63.

14 RIPPON et al. 2006: 58–59.

15 BRIAN 1999.

often accomplished through the casting of lots[16] which ensured, over the long term, a random, and therefore, relatively fair distribution of the resources of the village. By the later Middle Ages, outside areas of common meadow and pasture, land holdings had generally become crystallised and were passed down through family lineages or privately transferred between peasants rather than being re-distributed *en masse* amongst the village community on a regular basis.[17] However, when the system was first established, ensuring that members of the village community received a relatively equitable share of the resources within a particular parish provides an alternate explanation for the development and spread of scattered holdings rather than simply as a practice which reduced exposure to risk.

Beyond scattered holdings, other scholars have advanced entirely different explanations for how medieval peasants coped with risk. CULL et al., for instance, have argued that crop diversification provided significantly greater insurance against uncertainty than any offered by the scattering of holdings.[18] Certainly, the evidence for crop diversification throughout both the documentary and archaeological records leaves no doubt that crop rotation and the tailoring of planting strategies, including the selection of crop varieties and the percentages sown, to suit specific soils and situations, formed an important part of medieval agricultural regimes.[19] In south west England, for example, the regional practice of 'convertible husbandry' – which involved the alternation and rotation of grain and grass crops – can be traced in the pollen record from nearby peat bogs.[20] On a larger scale, the stark variations in medieval crop and animal husbandry between different surface geologies, revealed by analysis of archaeobotanical and zooarchaeological evidence in aggregate from a swathe across southern England, attest to medieval farmers' deliberate selection of appropriate crops and livestock, in part based on their environmental setting.[21] Additionally, medieval communities dealing with crises seem to have been acutely aware of the need to carefully select crops and organise agricultural regimes to suit local conditions.[22] Relatively simple decisions relating to arable and pastoral farming, therefore, such as what crops to grow,

16 The system used for casting lots was based on symbols associated with either different individuals/families/estates or different parcels of land. These symbols were then drawn at random – either at the parcel of land, in which case the symbol indicated who would harvest the crop from that particular parcel, or, alternatively, the symbol stood for a specific parcel of land in which case the crop from a particular parcel of land associated with that symbol was allocated to the individual/family or estate.

17 DYER 1998: 118–127.

18 CULL et al. 1992.

19 See for example CAMPBELL, OVERTON 1993: 62–66; FYFE et al. 2004: 1711–1712; DAVIES 2007: 2057.

20 RIPPON et al. 2006: 55–58.

21 RIPPON et al. 2014. It should be noted, however, that more detailed comparisons at the macro scale are hindered by the small size of the majority of existing archaeobotanical datasets as well as the limited number which contain weed seeds – as the ecology of different weed species are key to investigating different agricultural strategies VAN DER VEEN et al. 2013: 171–172.

22 RIPPON 2001: 27–31. This is discussed in greater detail in Chapter 4.3.

which animals to rear and where best to carry out these tasks, were fundamental to the economic prosperity of the individual and/or community and, as such, were also key to minimising potential risks.

Land was central to McCloskey's view of medieval insurance – it provided the means to support oneself, both for subsistence and through the production of a surplus that could be saved and sold. Sufficient land-holdings to provide for these requirements, therefore, afforded tenants a significant buffer against slipping into poverty. According to Braid, the sale of accumulated land holdings was instrumental in the survival of many during the Great Famine.[23] When land-holdings were insufficient to raise sufficient capital, however, tenants became very vulnerable. The occurrence of such problems rose starkly from *c.*1290, when population in England reached its medieval zenith. Growing morcellation of land-holdings, as more people sought to own the same amount of land, meant that tenants, especially at the lower end of the scale, were forced to eke out a living from ever smaller parcels of land.[24] Between 1279–1332, 50% of the rural peasantry occupied holdings of a size which was insufficient to provide for a family.[25] As a result of such pressure, it is unsurprising that when environmental and social shocks occurred during the early 14th century spikes occurred in both the rate of land transactions, as those unable to support themselves were forced to sell, and crime, as those pushed to a tipping point were left with no where else to turn.[26] Although land did provide security, there were certainly those who were unable to take advantage of, or forced to give up, its benefits.

Although holding land provided security and insurance, even landholders could ill afford to be complacent about their ability to ride out a crisis. There were many other ways in which medieval communities could provision themselves against future risks. An important aspect to any such strategy was the storage and stockpiling of resources during times of plenty to see out times of need. While McCloskey argued that grain storage was a rare and prohibitively expensive practice,[27] a growing body of evidence demonstrates that grain storage was, in fact, common.[28] Although the underground silos common in the Mediterranean world and prehistoric Britain are unknown in the medieval period,[29] evidence from both the historical and archaeological records attests to a variety of types of structure which provided storage space for agricultural produce. Simple stacks, for example, provided a method of drying and storing hay, corn and sometimes cereals – although these were usually used only for short periods rather than

23 Braid 2010: 362.
24 Campbell 2016b: 182–191.
25 Dyer 2012: 43.
26 Campbell 2016b: 192–196; Fig. 3.18.
27 McCloskey, Nash 1984.
28 Claridge, Langdon 2011.
29 Gardiner 2013: 34.

for long-term storage.[30] Barns provided an indoor space where produce could be dried and stored. The form of these structures is relatively well known. Some have survived into the present as standing structures – as is the case with the four monastic barns of Glastonbury Abbey which remain standing in Somerset.[31] Within these structures, which in some cases exceeded 40x9m, produce would typically have been stacked from ground level to the eaves – with the space above providing additional capacity when required.[32] In some cases barns were reserved for the storage of specific crops – as was the case at the preceptory of the Knights Templar at Cressing Temple, Essex, which had separate barns for wheat and barley.[33] There are difficulties in estimating the precise capacity of such barns,[34] however, and, in any case, the size and magnificence of some barns may have related more to the social status of their owners rather than their practical storage requirements.[35] The main function of barns was to provide a space where crops could be dried and threshed after which, produce was typically moved either to a specific area of the barn or to a separate granary.[36] These structures, which were often supported by wooden posts to keep the contents off the ground and free from damp, are known through textual evidence[37] and, although encountered relatively rarely, also through the archaeological record.[38] Gardiner suggests that the relative scarcity of granary structures on medieval sites in England may be explained by the fact that, on many smaller farms, storage chests or large jars, rather than dedicated granary buildings, were the primary method of storing produce.[39]

Distinguishing between barns and granaries and other functional structures archaeologically is not an easy task. In rare cases this has been possible when archaeologists have happened on barns or granaries that had burned down – preserving their contents *in situ*.[40] This, however, hints at one of the risks associated with stockpiling – that large quantities could be lost in one-off incidents.[41] Without the *in situ* preservation of the contents, distinguishing barns or granaries from other types of

30 Gardiner 2013: 24–25.
31 Bond, Weller 1991.
32 Bond, Weller 1991: 83–85.
33 Brady 1997: 80–81.
34 Bond, Weller 1991: 85.
35 Brady 2018: 266–268.
36 Brady 1997: 81.
37 Claridge, Langdon 2011: 1251–1252.
38 Claridge, Langdon 2011: 1244–1245; Gardiner 2013: 32–34.
39 Gardiner 2013: 32.
40 E.g. Ruas et al. 2005.
41 Indeed, not only were barns and their produce destroyed by natural disasters, another example of which comes from the coastal floods that struck the English Fenland in 1338, stockpiled resources also created a draw for would be thieves, as occurred at Shapwick in 1595. Dugdale 1662: 254; Gerrard 2007b: 995.

structure poses a challenge.[42] This point is reinforced by a trawl of developer-funded 'grey literature' for medieval rural settlement in a specific English county, in this case Suffolk,[43] through which very few structures actively interpreted as barns or granaries are definitively revealed – with those that can be identified with certainty all standing buildings.[44] However, many archaeological interventions in the county did encounter evidence which may relate to grain storage. At a site in Great Barton, for example, the excavators suggest that the high numbers of burnt grains suggest grain was dried in preparation for storage[45] while at Hadleigh Quarry analysis of the grains recovered through environmental sampling suggested some may have become spoiled during storage.[46] Additionally, at a number of sites investigated through excavation, the evidence was interpreted as relating to possible barn or granary structures such as the large post-pad, interpreted as a possible barn roof support, at Chevington Hall,[47] the 'barn-like' structures at Cedars Park, Stowmarket,[48] or the posthole alignments, interpreted as evidence for granaries, at Stoke Road, Clare[49] and Sizewell Wents, Leiston.[50] Even where likely candidates for storage structures are identified, their often ephemeral remains shed limited light on exactly how these structures were used[51] or to what extent storage was a ubiquitous part of medieval peasants' daily lives. Gardiner cites the transition from subsistence agriculture to more market-oriented cash cropping as an important factor behind the development of storage infrastructure while, additionally, the rise of powered milling often made it more efficient to process a large quantity of grain in a single episode, rather than milling smaller quantities by hand on a daily basis – as had previously been the norm.[52] While adequate stores of agricultural produce would undoubtedly have provided a buffer against the threat of harvest failure, such considerations suggest that this may not always have been the primary motivation behind storage.

When reserves of agricultural produce proved insufficient to see out a crisis, some relief could be found through alternative foods.[53] Peasant houses would have, invari-

---

42 Gardiner 2013: 29.

43 This search included all sites with medieval evidence (AD 1066–1539), located outside medieval towns, which had seen an archaeological intervention since November 21st 1990 (the introduction of PPG16) and had been entered into the County Council Archaeology Service's Historic Environment Record by October 2017.

44 E.g. Alston 2013.

45 House 2017: 66.

46 Sommers 2016: 32.

47 Gill 2004: 6.

48 Mundin, Woolhouse 2006: 27; Fig. 2.1: A, B.

49 Brooks 2015: 36–37; Fig. 2.1: C.

50 Gill et al. 2013: 25–26; Fig. 2.1: D.

51 Bond, Weller 1991: 83.

52 Gardiner 2013: 34–35.

53 Gerrard, Petley 2013: 1070.

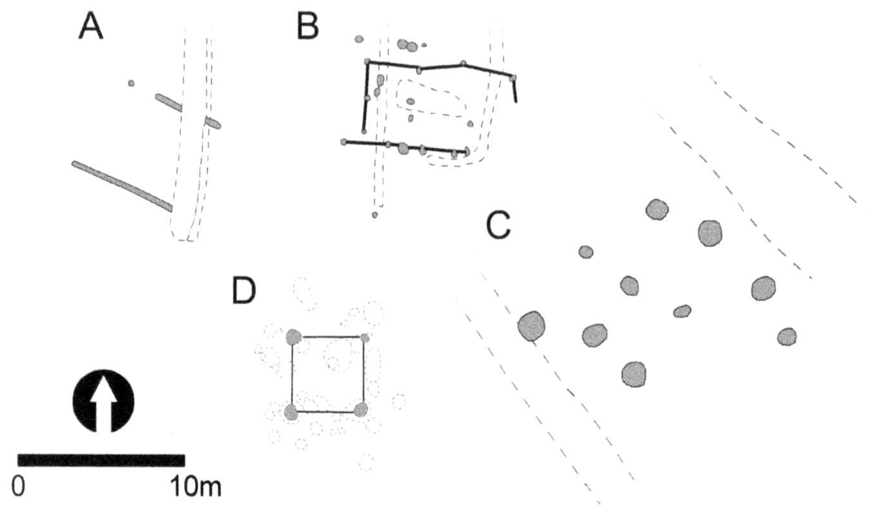

Fig. 2.1: Evidence from excavations in Suffolk relating to grain storage. A and B: 'Barn-like' structures from Cedars Park, Stowmarket. C and D: Posthole groups interpreted as the remains of granary structures respectively at Stoke Road, Clare, and Sizewell Wents, Leiston. Redrawn by the author after MUNDIN, WOOLHOUSE 2006: Fig. 5, BROOKS 2015: Fig. 7 and GILL et al. 2013: Fig. 9.

ably, had at least a small area of ground set aside as garden in which fruits and vegetables could be grown to augment the staple crops and, in times of dearth, this produce would have increased resilience – although it would probably have been insufficient to act as a fall-back for any length of time.[54] Such gardens are widely attested by the archaeological record; returning to rural medieval sites in Suffolk, medieval contexts interpreted as relating to gardens or orchards were encountered at Moat Farmhouse, Dennington,[55] Spexhall Manor, Spexhall,[56] and Church Street, Peasenhall,[57] – the latter of which fell within tenement boundary ditches dating to either the 13th or 14th centuries. According to GERRARD, the evidence from the well-studied medieval village at Shapwick, Somerset, suggests that many of the tofts contained open areas of pasture, vegetable garden and orchard which provided the inhabitants with a valuable and varied source of sustenance.[58] The lack of material culture or structural evidence associated with such remains, however, and the fact that, typically, only small areas are uncovered within the limits of archaeological trenches, means that accurately dating

54 DYER 2006: 36–38.
55 NEWMAN, BOULTER 1997.
56 SOMMERS 2002: 4.
57 GARDNER 2005: 20–21.
58 GERRARD 2007b: 987.

and understanding the full extent of these features is rarely possible. Moreover, the scarcity of waterlogged or carbonised plant remains from medieval rural sites in Britain limits the extent to which archaeobotanical evidence can shed light on the types of crops grown in such settings – particularly in the case of fruits and vegetables.[59]

Where the produce from such open areas proved insufficient, as was presumably the case under severe famine conditions, scarcity bred desperation driving people to search for sustenance from any available source. During a famine in Novgorod, Russia, in 1230 for example, a local chronicle reports that people were forced to eat moss, snails, pine-bark, lime-bark, lime and elm-tree leaves, cats, dogs, horseflesh and even the flesh of other humans.[60] The remains of foods identified as 'famine foods' are sometimes detected on archaeological sites, as in the case of limpets (*Patella vulgata*) at Eldbotle, East Lothian, which, as the excavators note, "are not commonly consumed as part of the human diet".[61] Without supporting contemporary documentary evidence, such interpretations project modern or anthropologically derived views of certain foodstuffs onto archaeological remains which might not accurately reflect the views of medieval people. One possible interpretation of the mussel middens at the Sands of Forvie, Aberdeenshire, for example, holds that the shellfish consumption was a direct response to environmental changes caused by wind-blown sand incursions, which could have severely impacted local agricultural production. On the other hand, the middens find a close parallel in Iron Age and early medieval middens found in coastal Denmark – where shellfish were gathered, processed and possibly preserved, on an 'industrial' scale as part of a specialised strategy.[62] Evidently, it is challenging to differentiate between foods consumed as part of a routine diet,[63] rather than a discrete response to a specific case of environmental stress or disaster.

As well as relying on resources – land, stockpiled agricultural produce or, *in extremis*, forageable famine foods – society itself also provided a safety net in a number of ways. Charity and support were available to those who fell on hard times through accident or illness (the deserving poor) through donations or financial assistance from local benefactors, while poverty deemed to be the result of idleness attracted universal condemnation.[64] Landlords and the exchequer frequently exercised leniency in the collection of rent and taxes in light of genuine difficulties experienced by tenants – when

59 Van der Veen et al. 2013: 164, 173.

60 Michell, Forbes 1914: 76. It should be noted, however, that the more extreme references, such as those to acts of cannibalism, may be cases of Biblical tropes being slipped into the reporting of contemporary events – as was certainly the case in some 14th century sources documenting the Great Famine in England which also made mention of cannibalism. See Marvin 1998.

61 Hindmarch, Oram 2012: 283.

62 Noble et al. 2017.

63 Or, indeed, resources used for another purpose entirely; at Eldbotle the limpets may have been intended for use as fishing bait rather than human consumption. See Hindmarch, Oram 2012: 283. Another possibility is that they were primarily used as a flavouring rather than as a source of sustenance.

64 McIntosh 1988.

sea floods inundated land at Broomhill, Kent, in the early 15th century, for example, the tenant successfully petitioned the landowner, Christ Church Priory, for a reduction in rent.[65] In some cases, however, failure to keep up payments meant forfeiture of land.[66] Further help came from religious fraternities – lay guilds which were usually centred around the worship of a particular saint – which dispensed financial assistance to their members, and sometimes others, in times of need. Local laws enforced by manorial courts also frequently granted concessions to the poor such as the right to take peas growing in the fields or unreaped grain left behind after the harvest.[67] The Church too provided assistance to the needy through almshouses, hospitals and charitable donations. Hospitals established to care for the ill, infirm and poor, known through the historical record, have been widely excavated; for example at Skriðuklaustur in Iceland[68] and throughout medieval England.[69] Documents specify that certain hospital foundations were especially orientated towards caring for the poor,[70] but, while the ill and infirm were often buried in the hospital burial grounds, the transient nature of the poor means that the evidence they left behind was usually ephemeral. Bioarchaeological evidence, however, does attest to high levels of 'stress' among medieval populations. 35% of the individuals considered by Roberts and Cox exhibited evidence for enamel hypoplasia – a condition which relates to episodes of malnutrition, disease, vitamin A deficiency or emotional stress which are likely to correlate with poverty.[71] Evidently, despite the assistance provided to the poor through various channels, many struggled to subsist at a comfortable level. As Dyer has observed, the motivations behind medieval charity differed from modern philanthropic principles because charitable acts and bequests were largely driven by benefactors' desires to accrue spiritual benefits, such as a reduction in the amount of time spent in purgatory, rather than the elimination of poverty through the most efficient means available.[72]

One of the most important mechanisms for coping with risk was undoubtedly credit. Where otherwise profitable enterprises were disrupted by disasters or other unexpected events, credit offered a means by which short-term problems could be solved. Moreover, the knowledge that such lending was available in the event of an unexpected disaster must have affected how medieval people planned for the future and what measures they took to mitigate risk. Although such relationships do not leave a physical signature unless related documentation survives – which is rarely the case – credit was far from

65 Draper 2007: 224.
66 Soens 2011: 349.
67 Richardson 2005: 406–409.
68 Kristjánsdóttir 2010.
69 Roffey 2012; Huggon 2018.
70 Hall 2006: 89–90.
71 This research considered the remains of 3,758 excavated individuals dating to the late-medieval period from across the UK. Roberts, Cox 2003: 264–265.
72 Dyer 2012: 67.

an alien concept during the Middle Ages. Those wishing to undertake a pilgrimage, for instance, could obtain financial assistance from groups such as the Knights Templar who issued 'letters of credit', appropriating the income from the pilgrim's lands for the duration of their journey.[73] Specifically relating to disasters, DYER discusses the importance of raising credit against the potential future profits of agricultural land, specifically in relation to the case studies of the 1483 floods at Henbury-in-Salt-Marsh, Gloucestershire, and the devastating fire of 1478 in Shipston-on-Stour, Warwickshire, in allowing disaster-struck populations to recover from these catastrophes.[74] Even in such comparatively well documented cases, however, there is no firm evidence for the credit relationships those stricken by disasters entered into. Despite the lack of concrete evidence, credit doubtless allowed individuals a good chance of recovery if a wealthy lender willing to bet favourably on an individual's chances of returning to profitability could be found.

While there were a number of ways by which medieval populations could guard against risk over the longer-term (months and years) – either averaging out risk spatially, through land holding and crop diversification, or temporally, through storage and the utilisation of credit – compared to modern society, medieval communities were relatively poorly equipped to deal with disasters over the shorter-term (hours and days). This is borne out in reference to fire disasters. Although these events were probably rarer than is often suggested, and medieval fire precautions were largely effective,[75] when fires did break out their effect was acute and responses were limited. Archaeology widely illustrates this fact through a variety of different forms of evidence. At Einbeck, Lower Saxony, for example, which was razed by fire in July 1540, well-dated evidence for widespread destruction has been excavated at many sites across the town[76] while evidence for multiple damaging episodes of fire between *c.*1120 and 1476 have been excavated in medieval Bergen.[77] Damaging fires in major structures are occasionally evidenced through scorched stonework – allowing not only an understanding of the fire's spread but also the reuse of, at least superficially, damaged fabric in later repairs – as is the case at Norwich Cathedral.[78] Excavation of low-status domestic structures at Fuller's Hill, Great Yarmouth, and Pennard, Swansea, on the other hand, provide archaeological evidence for the effects of fires which appear to have been caused when storm winds, and perhaps aeolian sand, caused structures to collapse in upon open hearths.[79] It is clear from the destruction such fires caused that extinguishing them was no easy task. The historical sources reveal that when

---

73 STOPFORD 1994: 66.
74 DYER 2020.
75 GARRIOCH 2016.
76 HEEGE 2005.
77 HANSEN 2015: 161.
78 GILCHRIST 1998.
79 ROGERSON 1976: 159; MOORHOUSE 1985: 5.

fires threatened medieval Bergen all available manpower was called upon to assist in bringing the fire under control.[80] This is further borne out by an episode from Keyingham, East Yorkshire, in which a lightning bolt struck the parish church, igniting a fire. The fact that the *ad hoc* fire-fighters, who must have volunteered immediately and instinctively, were actually able to extinguish the blaze was interpreted as a miraculous occurrence, made possible only through the intercession of a former rector, Philip Inglebard, whose tomb had been found to exude a sweet oil in the blaze which was later itself held responsible for a number of miracles.[81] Thus, fire mitigation in the 'moment of disaster' was an extremely challenging task and for most categories of hazard an appropriate immediate response was even less obvious or available. Accordingly, the management of flooding,[82] grain shortages[83] and earthquakes in southern Europe,[84] was reactive – taking place after the 'emergency' phase had passed and without the involvement of bodies resembling modern emergency services.

As this section has explored, medieval society was equipped with a relatively wide array of measures to combat the long term challenges that an unpredictable world could throw up. This was particularly the case concerning strategies against perhaps the most universal and familiar hazard – harvest failure. As Marston has observed, the majority of these risk adaptive strategies have left little or no obvious traces in the archaeological record with the important exception of grain storage and, to a lesser extent, reliance on 'famine foods'.[85] The practices and concepts outlined above, however, are generally enmeshed in modern concepts of risk and reward, and, while medieval people demonstrably participated in this system – and must have well understood the risks and potential gains – they are unlikely to have conceptualised the dangers of the world, and the methods to mitigate against them, from the same viewpoint or in the same terms as we do today. Therefore, it is important to consider how medieval people interpreted disaster beyond the cold calculations of economic cost, loss and potential profits. The next section of this chapter, therefore, considers the religious and, at times, superstitious framework through which disasters were interpreted during the Middle Ages.

2.2 Religious beliefs, superstitions and disaster

In terms of religion, although the majority of medieval Europeans were Christian, widespread Jewish minorities existed with Muslim occupation to varying degrees throughout

80 Hansen 2015: 159.
81 Bond 1868: 194–195.
82 Galloway, Potts 2007: 376.
83 Braid 2010: 348–351.
84 Figliuolo 2020.
85 Marston 2011: 195.

southern Europe[86] and other distinct groups such as the Pagans of Livonia.[87] By the 12th century, all of northern Europe had been officially converted to Christianity although Jewish and Pagan minorities persisted throughout the period and Pagan beliefs and rituals continued to be practised by nominal Christians – whether or not this was their intention. In any society, beliefs about natural hazards are built-up over the *longue durée* meaning that, despite the Christianisation of the continent, earlier beliefs represent an important facet to how people perceived these types of event during the Middle Ages.[88]

A wide variety of historical evidence relating to beliefs associated with extreme weather events and natural hazards exists from the Middle Ages.[89] At least from the early medieval period, connections between the weather and the supernatural became common in literary sources. The Icelandic Sagas, for example, contain numerous references to control over the elements through supernatural means[90] while later literature often contains references to earlier beliefs. In the 14th century chivalric romance *Sir Gawain and the Green Knight*, for example, the snowstorm and misty conditions which occur in the build-up to the tale's pivotal scene have been interpreted as references, conscious or not, to pre-Christian Celtic traditions of fairy magic – with comparable scenes found in numerous Irish folk stories, in which snow storms and, particularly, mist accompany magical and otherworldly experiences.[91] That, at least for some early medieval Europeans, magic in relation to the weather transcended folk stories and literature can be seen in certain primary sources. The 8th century Archbishop of Lyon, Agobard, for example, composed a treatise condemning the beliefs of many among the laity who ascribed hail and thunder to the magic of *tempestarios* – those who conjure up storms through magic.[92] Although the practice was not treated credibly by Church officials, such as Agobard, who emphasised that only God could control the weather, evidently many of the lay population did believe in the efficacy of human magic over the weather.

Comparable beliefs persisted throughout the medieval period. In the widely circulated 13th century *Travels of Marco Polo*, for example, various ethnic and non-Christian groups, invariably described as idolaters, are ascribed the power to control or exert

---

86 Catlos 2014.
87 Selart 2015.
88 Analogously, medieval Islamic scholars often drew on a combination of Islamic teachings and, perhaps without fully appreciating their origin, pre-Islamic traditions in their interpretations of earthquakes. Akasoy 2010: 390–391.
89 This is discussed again in chapter 5.
90 Mitchell 2011: 65–66.
91 Puhvel 1978.
92 Lewis 2001.

influence over weather and natural hazards.[93] Whether Polo himself[94] believed these anecdotes or included them merely to embellish the exotic and otherworldly nature of the far eastern lands he describes is unclear. Within medieval Europe itself, magical control of the weather was an ability which was commonly ascribed to witches. Until the 14th century, witches were often invoked as the progenitors of bad weather but from the 15th century this attitude made the transition from a frequently repeated 'urban myth' to secular and ecclesiastical policy giving rise to organised and brutal persecutions of supposed witches[95] – a phenomena which BEHRINGER has linked to the climatic fluctuations of the Little Ice Age.

As Agobard's criticism of the laity who believed in weather magic demonstrates, not everyone understood hazards in the same way. While not a natural hazard *per se*, an analogy can be found in medieval debates surrounding the causation of eclipses. While the lay population invoked a wide variety of demons, magicians and monsters to explain their occurrence, resorting to shouting and panic when an eclipse actually happened,[96] the educated classes were aware that eclipses were a predictable natural occurrence – although many still believed that eclipses were miraculous occurrences and/or portents of future events.[97] While the ultimate predictability of eclipses convinced some medieval learned men that they were natural events, this appears to have been less the case with unpredictable random events such as storms. Although the oft cited 7th century chronicler Isidore of Seville described storms chiefly as natural events: "storms are created out of a confluence of opposing airs at the midpoint and change of these two seasons [spring and autumn]",[98] even before his time storms had been connected to the devil and, in spite of official attempts to stamp out these beliefs,[99] they persisted into the later Middle Ages. The 13th century encyclopaedist Vincent of Beauvais, for example, described storms as the result of turbulence caused by airborne demons who fell to Earth along with the Devil after being expelled from heaven by God.[100] This belief is repeated in other literary sources such as Jacobus de Voraigne's mid 13th century compilation *The Golden Legend*[101] and the writings of later churchmen

93 The "devilish enchantments" of the people of Kashmir, for example, included changing the weather and bringing on thick darkness while Tibetan astrologers are described as having the ability to summon and halt thunderstorms on command. In addition, the Kashmiris and Tibetans employed by the Great Khan (Kublai Khan, who reigned from 1260–1294) were able to clear the skies above the Imperial Palace through "the arts of the Devil". LATHAM 1958: 78, 173–174, 109–110.

94 Or his co-author, Rustichello da Pisa.

95 BEHRINGER 1999: 345–346.

96 BARTLETT 2008: 57–59.

97 BARTLETT 2008: 62–69.

98 BARNEY et al. 2006: 276.

99 FILOTAS 2005: 95.

100 BELLOVACENSI 1624: 306.

101 RYAN 1993: 288.

such as the 15th century preacher Vincent Ferrer's *Sermones Aestivales*.[102] The logic to these beliefs was that the demons had the power to destroy the earth and were only prevented from doing so by God's protection. When a storm occurred, it was not sent by God but God did allow it to happen by choosing not to intervene and prevent the onslaught of the demons. This reluctance on the part of God, which allowed storms to occur, was often interpreted as a punishment for previous sins committed.[103]

Demonic apparitions, referencing their causative role, are widely reported by medieval writers describing extreme weather events. In the chronicle of John Stone, a monk of Christ Church Canterbury, for example, the damage to the belfry in a storm in 1458 is ascribed to the work of evil spirits.[104] *The Chronicle of Lanercost* reports how during a thunderstorm in York an evil eye was seen within the thunderclouds and the yells of demons were heard screeching through the air.[105] At St Albans in 1251 during a storm, a flaming sword was seen waving about while thunder and murmurings were heard. This caused the onlookers to make the sign of the cross, invoke the holy ghost and chant hymns which caused the storm to pass.[106] The *Chronicle of Melrose* describes how in 1165 a storm arose in Yorkshire in which the devil was sighted in the form of a huge black horse. Accompanied by thunder, lightning and hail this horse galloped out into the sea but left behind enormous hoof prints, one of which at Scarborough remained visible for at least a year.[107] A similar account from 1205 is given by Ralph of Coggeshall in which, in the aftermath of a storm, hoof prints were discovered which were attributed to a demonic presence.[108] That some of these traditions may have had material inspirations may be suggested by a legend from Lancashire in which prehistoric archaeological features, rock art, a cairn and a cup-marked stone, may have contributed to a local legend in which these marks were attributed to the footprints of the Devil.[109] Such traditions persisted beyond the Reformation as, during a thunderstorm in 1577, the appearance of a demonic black dog was reported at the churches of Bungay and Blythburgh, Suffolk. The dog was reported to have killed five of the parishioners and left burn marks on the doors of the church at Blythburgh as a reminder of its appearance.[110] That demons were the root cause of storms, as well as other unwelcome occurrences, certainly appears to have been a widespread belief[111] and, as a variety of material and

---

102 Ferrer 1572: 175.
103 This can be seen in chroniclers accounts of disasters; in the case of the Great Famine, some chroniclers' accounts specifically discussed the poor weather that exacerbated the scarcity of food, as well as rapid onset hazards such as the storm of 1362. Jordan 1996: 22; Schmidt 2011: 176–177.
104 Searle 1902: 74.
105 Maxwell 1913: 103.
106 Giles 1953: 466.
107 Stevenson 1856: 130.
108 Stevenson 1875: 155–156.
109 Barrowclough, Hallam 2008.
110 Stubbs 2011: 37–61.
111 Hanska 2002: 128.

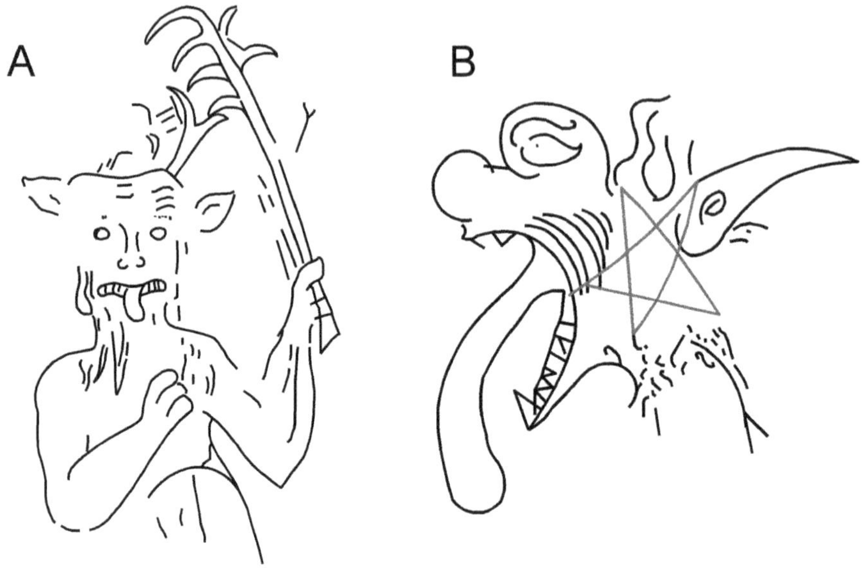

Fig. 2.2: Medieval graffito inscriptions depicting demons. A: Beachamwell St Mary, Norfolk. B: Troston St Mary, Suffolk, 'cancelled out' by the later inscription of a five-pointed star. Redrawn by the author after CHAMPION (2014: 248, Fig. 105; 2015a: 114–115). Scales not given.

artistic evidence demonstrates, figured prominently in the imaginations of contemporaries. Examples come from graffito inscriptions depicting demons found in medieval church settings, such as those at Beachamwell St Mary, Norfolk, and Troston St Mary, Suffolk (see Fig. 2.2), which provide pictorial representations of medieval fears for the harm that demons could cause in the temporal world. Fears of such destruction were so great that numerous superstitious practices, discussed in Chapter 5, were widely relied upon to prevent the harm wrought by these malevolent forces.

Certainly, in the case of some events the belief that disasters were 'acts of God'[112] seems to have been at the forefront of the minds of many. This is widely seen in historical sources describing a high magnitude windstorm which struck England on St Maur's Day in January 1362.[113] The *Eulogium Historiarum*, for example, states that "it was believed by some [the storm] was a scourging from God"[114] – presumably as a punishment for sins. This view is echoed in the contemporary poem *Piers Plowman* in which the St

112 Either specifically orchestrated by God or allowed to occur through his decision not to prevent them.

113 This is discussed in greater detail in Chapter 3.

114 HAYDON 1863: 229.

Maur's Day Storm is described as a punishment for "pride and for no point else".[115] For an anonymous chronicler of Canterbury, the storm was sent by Satan, and presumably purposefully not deflected by God, as a warning against holding a joust which had been scheduled to take place two days later. Jousting was seen as a sinful pursuit – which distracted militarily-able men from crusading – for which the participants were guilty of all seven deadly sins and had been officially banned by the church until 1316.[116] Although by 1362 the ban had been lifted, evidently jousting had failed to shake its sinful image. In this particular joust, which took place despite any protests, the participants seem to have provocatively embraced orthodox disapproval by outfitting themselves in attire symbolising each of the seven deadly sins they were thought to be committing.[117] A somewhat comparable near-contemporary situation can be seen during the flooding of Cologne in 1374 when the inundation of the town led to the cancellation of a planned carnival which was replaced with "public litanies and processions in honour of God and the Saints".[118] In this case, ecclesiastical authorities seem to have been motivated by the presence of the floodwaters to prevent sinful behaviour, such as alcoholism and sexual misconduct which were likely common at carnival celebrations, instead replacing it with acts of communal piety to appease God, who would hopefully speed the return of floodwaters to their normal bounds.

Clearly human behaviour could provoke God's wrath and, accordingly, disasters were frequently blamed on the behaviour, beliefs or practices of specific groups. Those most frequently linked to disasters in this way were Jews and homosexuals. In *c.*1020, for example, the French chronicler Ademar of Chabannes reported the execution of a group of Jews held responsible for bringing on an earthquake in Rome through their irreverent attitude to a crucifix.[119] Comparably, in 1311, James II of Aragon brought charges against Count Pons Hugh IV de Ampurias for alleged acts of sodomy as "such criminal, vile, and horrid [deeds] … [cause] earthquakes, famine and pestilence".[120] These beliefs transcended the medieval period and were exported beyond Europe during the post medieval era.[121]

Despite the view that disasters were punishments from God which is widely gleaned from the historical sources, it must be noted that estimating to what extent such beliefs were representative of the medieval population as a whole is a considerable challenge.

---

115 SCHMIDT 2011: V: 15; 176–177.

116 BARKER 1986: 70–83; 95.

117 TAIT 1914: 151.

118 ROHR 2007: 94.

119 PERTZ 1841: 139.

120 BRUNDAGE 1993: XVII.

121 The Portuguese missionary Gaspar da Cruz remarking on the 1556 earthquake in Shaanxi Province, China, for instance stated that "because this sin is common among them, God was willing to send them a grievous punishment". BOXER 1953: 223. The sins to which da Cruz is referring are 'unnatural vices' a term coined by Aquinas (*Summa Theologica*, II–II, Q.154, Art. 11) comprising bestiality, homosexuality, improper intercourse and masturbation.

Typically, the available sources emanated from institutional, official or high status voices which may not reflect more popular views – such as those about whom Agobard of Lyon bemoaned – which were coloured by superstition and folk-beliefs. Chroniclers' accounts of disasters are often filtered through a layer of religious interpretation which frequently exaggerate or distort the facts. In some chroniclers' descriptions of the Great Famine of 1315–1322, for example, details derived from the Biblical account of the Samarian Famine were intermixed[122] while the English chronicler Matthew Paris,[123] interpreted events in comparison to those he knew through biblical and classical texts. In his *Historia Anglorum*, for example, in two accounts of English floods he references Deucalion, the ancient Greek equivalent of Noah.[124] This comparison was also made by the 15th century humanist Enea Silvio Piccolomini[125] in his description of the destructive floods in Bavaria and Austria in 1445.[126] ROHR interprets the use of this analogy, and similar terminology derived from other classical sources and the Bible, as indicators of the unreliability of these sources which essentially paraphrase earlier texts rather than providing specific details relating to contemporary events.[127] Whilst revealing a potential source of inaccuracy in the documentary record, such references do highlight the interpretive framework through which ecclesiastics interpreted the occurrence of disasters.

The lack of evidence for popular perceptions of natural disasters is paralleled by popular views of other contemporary issues. Discussing the crusades, for example, WARD cautions that the lack of documentation for popular opinions "should not blind us to the causative role played by popular ideas and groups".[128] Nonetheless, Church teachings undoubtedly greatly coloured the way that ordinary people thought about natural hazards. Disasters featured prominently in Biblical stories such as the Flood of Noah, the plagues of Moses as well as widely throughout the lives of the Saints. In addition, catastrophes were also predicted to signal the coming of the Day of Judgement. To medieval Christians, this meant that the occurrence of a flood, storm or earthquake came with the additional fear that such an event merely presaged worse to come. According to one tradition which gained popularity during the later Middle Ages, the Day of Judgement was to be heralded by 15 days each with its own 'sign of doom'

122 MARVIN 1998.

123 b. 1200 d. 1259.

124 In 1256: "Piles of bridges, stacks of hay, the huts of fishermen with their nets and poles, and even children in their cradles, were suddenly carried away, so that the deluge of Deucalion seemed to be renewed", and in 1257: "On the Innocents' day [28th December] in this year such a quantity of rain fell that it covered the surface of the ground, and the times of Deucalion seemed to be renewed". GILES 1954: 175, 212.

125 b. 1404 d. 1464.

126 ROHR 2007: 98.

127 ROHR 2007: 100.

128 WARD 1992: 119.

which included a great flood, earthquakes, and the burning of the world by fire.[129] These apocalyptic beliefs were communicated to the lay populace through a variety of mediums including theatre, stained glass, sermons, verse and texts.[130] The Pricke of Conscience Window in All Saints' Church, North Street, York, for example, which dates to 1410–1420, originally contained 15 scenes accompanied by text explaining the arrival of the Day of Judgement and the various natural hazards that would occur on each day.[131] With such ideas widely circulating, it is understandable that for many lay people, natural hazards were terrifying not just because of the short-term danger they spelled but also because of their fearful eschatological connotations.

In the same way that the Day of Judgement was to be presaged by disasters, advance warning of the occurrence of hazards was also often thought to be presaged by environmental anomalies. Between 1315–1317 for example, the passing of a comet, 'bloody' rain, an eclipse and two earthquakes were all taken as portents of the Great Famine.[132] In the case of the windstorm of January 1362, although no contemporary explicitly made the connection, we can imagine that, in hindsight, the eclipse, 'bloody' rain and appearance of a cross of blood in the air reported in the previous year[133] would have similarly coloured perceptions of the January gale. To contemporaries, such celestial anomalies and miraculous apparitions were far from coincidental. The literary composition of these disturbing occurrences is somewhat formulaic with strikingly similar occurrences – for example, an eclipse, a comet, hurricanes and floods, immediately presaged the First Crusade in the 1090s.[134] Excluding some of the more fantastic visions, many of these events may have actually occurred but their interpretation as signs of greater calamities to come followed a biblical trope connected to the Book of Revelations and the Apocalypse.[135]

That such signs were actively interpreted as harbingers of impending disasters, rather than merely signs which could be interpreted in hindsight, is demonstrated by the non-event of the 'universal flood' of 1524. Following the publication, in 1499, of news that a 'grand conjunction' in the stars would take place in 1524, a popular perception, fuelled by printed pamphlets – such as the astrologer Johannes Carion's *Prognostico*[136] – gradually developed which held that 1524 would be accompanied by calamities, social unrest and a universal flood (See Fig. 2.4). The prognostication gripped contemporary populations with such fervour that some in Italy and Germany began to consider constructing 'arks' to save themselves from the predicted floodwaters.

---

129 Powell 2004: 310–316.
130 Powell 2004: 293.
131 Powell 2004; see Fig. 2.3.
132 Jordan 1996: 22–23.
133 Hearne 1731: 425–426.
134 Ward 1992: 121–125.
135 Ward 1992: 108.
136 Carion 1521.

Fig. 2.3: Two of the scenes in the Pricke of Conscience window at the church of All Saints', North Street, York, which depict natural hazards. Above: damage caused by an earthquake. Below: flood and burning sea. Photographs by the author.

When late 1523 came, some fled to the countryside and the mountains to escape the fated disasters – which of course never came.[137] The flood of 1524 though, attests to both the medieval belief in the interconnectedness of the celestial system with terrestrial natural hazards as well as, from the mid 15th century, the power of widely circulated printed pamphlets, which people seem to have accepted as reliable sources of information, to shape public opinion.

As well as being perceived as indicators of an impending apocalypse, medieval theology also made natural disasters events worthy of additional fear. While anxiety for the safety and survival of oneself and one's family and friends is a natural concern during the moment of disaster, medieval Christians were also lumbered with additional concern for their eternal soul. As DANIELL has explored,[138] death by natural disaster was a particularly problematic end to befall a medieval Christian; certain hazards singled out unlucky individuals as particularly deserving of a heavenly punishment, as in the case of death by a lightning strike or death by drowning, and were reserved, almost exclusively, for non-Christians in medieval literature. In the most severe cases, hazards, such as severe floods, earthquakes or fires, entirely removed the body from circulation preventing proper burial and the administration of last rites. Such unfortunate occurrences were believed to interfere with, if not totally prevent, the soul's ascension to heaven.

Although, as the evidence and considerations discussed above demonstrates, disasters were widely conceptualised as events orchestrated from on high, they were also sometimes seen as entirely natural occurrences. Floods and storms for example, were certainly not always interpreted as anything beyond routine nuisances. The everyday annoyance of such occurrences is reflected by an early 15th century petition – which contains no overbearing theological interpretation – from the parishioners of Adstone, Northamptonshire, to the Pope asking for a baptismal font to be installed and a cemetery to be consecrated near their town, as these important facilities were currently located about a mile away at the Augustinian monastery of Ashby, and were frequently inaccessible due to floods and storms.[139] Similarly, certain bridges, such as those at Godalming and Cobham, Surrey, were especially intended to provide crossings during times of flood, and indeed were only open to the public when the depth of water prevented passage by nearby fords.[140] In these cases, clearly the risk of flood was both frequent and well appreciated, so much so that the provision of contingency crossings was seen as necessary and worthwhile. Such interpretations are also reflected in descriptions of natural hazards found in contemporary literature. In a 15th century Welsh poem by Lewys Glyn Cothi, which describes a flood on the River Tywi, although reference is made to religious figures including St James, Noah, St Peter and God himself, the

137 SCHOENER 2007.
138 DANIELL 1997: 65–71.
139 BLISS, TWEMLOW 1904a: 229.
140 MALDEN 1911: 24–25, 442.

Fig. 2.4: A depiction of the celestial conjunction in the constellation of Pisces (represented by a large fish) widely predicted to bring disasters and a universal flood in 1524. From a pamphlet by the astrologer Leonhard Reynmann printed in Nuremburg in 1523. Sächsische Landesbibliothek – Staats- und Universitätsbibliothek Dresden, Magica.148,56. Public Domain.

flood is not interpreted as divine retribution and the author only invokes God's help by asking him to "place a bridge over the River Tywi",[141] thus clearly accepting the everyday nature of the flooding, which could be expected to be a recurrent problem into the future.

Here we might also consider natural explanations for disasters which were, nonetheless, untrue. While not perceived as a case of divine retribution, the multitude of forest and urban fires which sprang up across central Europe during the 'mega-drought' of 1540 sparked a wave of paranoia around the suspected covert operations of *mordbrenner*, or clandestine arsonists. The conflagration that year in the Protestant town of Einbeck, Lower Saxony, for example, which resulted in over 300 deaths, was immediately seen as a Catholic plot – creating a climate of fear and alarm in neighbouring towns. Correspondingly, mysterious marks on doorways were interpreted as secret arsonist codes, extra guards were appointed and possible arsonist hiding places were scoured.[142] Despite the contemporary conviction that a plot existed, this has largely been dismissed by historians who have argued not only that starting fires against the Protestants would not have matched Catholic interests[143] but also that the widespread occurrence of fires finds a more convincing explanation in the unprecedented drought[144] which caused what may have been the driest summer in many parts of Europe in the last millennium.[145] As PFISTER concludes: "conspiracy theory directed against political opponents offered a means of explaining the seemingly inexplicable".[146] Nevertheless, even the likely misguided attribution of the fires of 1540 to *Mordbrenner* signifies that contemporaries well-recognised that not all disasters were divine in origin.

Offering explanations where no easy one existed was perhaps what united the varying viewpoints on the occurrence of disasters considered above. The exception are those events which were treated as routine natural occurrences but the majority of these were not really disasters – they were low-magnitude, reasonably predictable and, therefore, expected natural hazards. High magnitude unexpected events which caused disasters, on the other hand, were not explicable merely as natural events. Establishing that medieval Europeans believed that disasters were more than simply natural events is important because misidentification of the factor(s) behind the occurrence of a hazard affects how people prepare for, and respond to, such events. Such unpredictable, extreme events provided fertile breeding ground for the development of superstitions and fears. When actions were taken against these poorly understood calamities they were frequently ineffective due to a misunderstanding of the problem. Precautions

141 GRIFFITHS et al. 2017: 98–103.
142 PFISTER 2017: 159–161.
143 STOPP 1970: 205.
144 PFISTER 2017: 164–165.
145 WETTER et al. 2014.
146 PFISTER 2017: 184.

against plague, for example, were influenced by incorrect identifications of the vector by which the disease was spread.[147] Similarly, in common with meteorological hazards, as discussed above, supply and harvest problems were often blamed on scapegoats such as Jewish communities[148] and periods of climatic instability seem to have resulted in an increase in witch hunts.[149] Such misidentifications of the root causes of problems, combined with both the fact that preventative actions rarely seemed to offer long-term safety and the prevalent belief that God allowed or prevented the occurrence of disasters, created an environment in which, for many, a wide array of ritualistic behaviour seemed necessary and logical in order to gain protection against the manifold dangers of the world.

2.3 Summary

As this section has explored, risk was an inevitable part of the lives of medieval populations for which, to some extent, they were prepared. There were a number of well-established strategies medieval people could follow to reduce their vulnerability to the hazards that could be foreseen occurring during an average year. The accumulation of land, maintenance of adequate grain reserves and adherence to best agricultural practice – by rotating crops and, perhaps, scattering holdings – were all important considerations to minimise the chance of harvest failure – and mitigating its consequences. At the same time, however, those truly disastrous events – high-magnitude and unprecedented natural hazards – which nobody saw coming, sat somewhat removed from the natural world. Their occurrence was difficult to explain through natural means while supernatural causes – such as punishments from God, indicators of the coming Judgement or the mysterious magic of malevolent groups – offered explanations that accorded with widely held beliefs. To protect against these types of events, therefore, practical measures alone were insufficient. Even so, in the aftermath of a high-magnitude event, and where such events were regular occurrences,[150] practical measures were often taken to guard against any future recurrence. The following three chapters explore this interplay between, on the one hand, the occurrence of high-magnitude natural hazards, detailed case studies of which are considered in Chapter 3, the practical measures taken to mitigate and protect against their occurrence, in Chapter 4, and the religious and superstitious practices which were believed to provide additional protection, in Chapter 5.

---

147 Slack 1988: 433–439.

148 Anderson et al. 2016.

149 Behringer 1999.

150 As in the case of recurrent North Sea storm surge flooding over the late 13th and 14th centuries.

3 A litany of disasters

There are a near countless number of natural disaster events about which we know some details from the medieval period. What is rare, however, is for a sufficient amount of detailed information to survive to permit an approximate picture to be assembled of exactly what occurred, when it happened, and how the hazard impacted contemporary society. The majority of textual accounts that record occurrences of natural hazards are laconic – rarely stating more than the type of hazard, the date on which it occurred, sometimes just the year, and occasionally a vague remark concerning the impact of the event. Where more lengthy descriptions survive, it is often difficult to disentangle exactly what exactly is accurate reportage and what is a literary embroidery of the facts. Medieval writers were rarely eyewitnesses to the events they described and frequently exaggerated real world events or referenced well-known biblical tropes to give their writings greater contemporary relevance.

While most textual descriptions are low in detail, the vast majority of traces of disasters left in the material record provide only fleeting glimpses of the events that created them. By their nature, the physical impact of most meteorological hazards are either extremely ephemeral – as in the case of rainstorms, lightning strikes or windstorms – or liable to erasure shortly after the occurrence of a disaster – for example during the phases of site clearance and rebuilding that commonly take place in the aftermath of an extreme event. Despite these challenges, therefore, it is perhaps surprising that many physical traces of medieval disasters can still be found in the present-day at all. Since, on their own, textual and material sources of evidence offer only limited windows into the occurrence of disasters during the medieval period, it is only through the combination of a wide variety of source material that it is occasionally possible to observe, with some degree of certainty, how these events unfolded, the various ways in which contemporaries were affected and the different pathways they adopted in the aftermath in responding to disaster.

Of course, different types of natural disaster impact human communities in different ways, leaving behind different categories of evidence. Some of these material traces of disaster are more likely to survive over long periods of time than others. The extent to which the surviving evidence allows the impact of meteorological hazards to be reconstructed, as well as the limitations, are well demonstrated by two hazards which affected communities throughout Britain during the medieval period. Wind-blown sand – the encroachment of sand onto areas of human occupation as a result of storm activity – and lightning strikes both impacted medieval communities repeatedly and can be traced variously through archaeological and textual sources.

Wind-blown sand is the result of wind action on exposed expanses of sand causing the inland migration of bodies of sand – sometimes encroaching onto areas of human habitation or economic interest. The hazard usually impacts larger areas rather than specific point-locations. Wind blown-sand can be either a sudden-onset or 'creeping'

https://doi.org/10.1515/9783110719628-003

hazard since, while high-magnitude storms can provoke sudden and significant sand shifts that can entomb large areas, less extreme winds can result in more gradual sand migrations. Over the longer-term, this latter category can slowly bury areas of human settlement and agriculture. Of course, there was little that medieval populations could do in order to prevent the winds that blew sand across their lands but barriers could be erected in an attempt to slow its progress while the stability of sand environments was likely closely linked to medieval land-use decisions. Grazing cattle in dune areas, rabbiting and harvesting the plants that grew among the dunes – such as marram grass – were all activities that increased the risk of sand movement as these removed stabilising vegetation making dune environments more vulnerable to wind action. Certainly by the 15th and 16th centuries, this risk was recognised and measures were put in place in some areas to protect stabilising vegetation in these environments.[1]

These movements of sand on to areas of human settlement or activity can leave a considerable trace in the archaeological record – evident through layers of sand overlying archaeological remains. Documentary evidence also often attests to the problems the hazard posed for contemporary society – either rapid losses or, more commonly, gradual encroachment as sand was blown inland. Having experienced problems with encroaching sand for at least two centuries,[2] the Welsh town of Kenfig, Bridgend, seems to have been completely abandoned in the 15th or 16th centuries.[3] A number of comparable losses can be traced through archaeological and textual references elsewhere along the Welsh coast at Pennard,[4] Penmaen,[5] Merthyr Mawr[6] and Stackpole,[7] while in neighbouring Somerset wind blown-sand appears to have led to the abandonment of a medieval settlement at Berrow.[8] The impact of the hazard can also be traced at Scottish medieval settlements such as Eldbotle,[9] East Lothian, Forive[10] on the coast of Aberdeenshire and many sites in the Orkney and Shetland Islands.

Lightning strikes, on the other hand, typically affect almost a single point – rather than a large expanse of sand. They are produced when differential electric charges between storm clouds and the Earth's surface are equalised by a discharge. When this occurs over areas of human habitation, such as settlements, damage can occur at specific points on the ground – often the tallest object in the area of the storm.

1 Brown 2015: 143.

2 The earliest indication for this comes from the foundation of a new church away from the sand on higher ground in 1262 and documentary references to the issue in the 14th century. For full discussion see Brown (2015).

3 Wessex Archaeology 2012b: 24.

4 Lees, Sell 1983.

5 RCAHMW (The Royal Commission on Ancient and Historic Monuments in Wales) 1982.

6 Randall, Rees 1932: 119.

7 Benson et al. 1990.

8 Rippon 2000a: 152.

9 Hindmarch, Oram 2012.

10 Kirk 1957.

Trees and flammable structures are often set alight by the sudden surge of electrical energy. Lightning induced fires not only engulfed large medieval buildings but fires also sometimes spread throughout towns and cities – affecting a much larger area than the initial 'point' struck by lightning. Since no understanding of electric charge existed during the medieval period, no effective measures to prevent lightning damage – such as lightning conductors – were available to medieval populations though efforts to mitigate the risk of the secondary hazards they produced – urban fires – were sometimes put in place.

The tall towers and spires of cathedrals were frequent victims to lightning strikes over the course of the medieval period. The spire of London's St Paul's, for example, was struck repeatedly – in 1230, 1341, 1444 and 1561.[11] The central tower of Durham Cathedral was also struck by lightning in both 1429 and in 1459.[12] The first strike destabilized the Norman tower forcing episodes of repair which took place principally between 1433–1436.[13] Glasgow Cathedral too suffered a lightning induced fire shortly before 1406 and the rebuilding that followed in its aftermath can be traced within the standing fabric of the building today.[14] The cathedral at St Andrews seems to have suffered a very damaging fire in 1378 – lightning is one of the possible causes recorded by later chroniclers alongside a bird carrying a burning twig into its nest and a plumber leaving a hot iron in a crow's nest.[15] The fire seems to have precipitated the replacement of almost the entirety of the roofing, the rebuilding of one side of the central tower – involving careful engineering – the replacement of pillars in the transepts, the renewal of much of the nave and west front, and repairs and modifications to the east end.[16] This latter episode reveals an ever present problem in tracing the impact of disasters through the textual and material sources of evidence. While apparent correlations between documented disasters and the material evidence can often be found, it is difficult to verify that the physical evidence definitely relates to the documented occurrence of a natural hazard and not either a different occurrence of the same type of hazard or a separate cause entirely.

The remainder of this chapter, focuses on several case study events which – due to the relatively plentiful survival of both historical and material sources of evidence – can be reconstructed in more detail than is usually possible. These events are chosen for a number of reasons. The numerous, and high resolution, nature of the available sources allow a particularly deep exploration of the occurrence of the event(s) and their aftermaths. They also illustrate the types of impacts caused by particular hazards as well as the limitations of the surviving historical and material record. Although an

---

11 Saunders 2001: 16.
12 Cambridge 1992: 96, 116.
13 Cambridge 1992: 96.
14 Durkan 1975.
15 Batho, Husbands 1941: 340; Bellenden 1821: 455.
16 McRoberts 1976: 29–31.

argument could be made for preserving the comparability of the events analysed by choosing a specific type of hazard, region or type of impact, a number of different types of hazard have been selected allowing 'natural disasters' as a broad category to be examined. To this end the selected events cover different types of hazard, different areas of effect and different scales of effect, both geographically and temporally.

3.1 The storms and floods of 1287/88

During the winters of 1286/7 and 1287/8 a particularly severe series of storms caused dramatic flooding on the east coast of England as well as in the Low Countries. The sources which describe these floods invariably speak of *venti vehementia quam maris violentia*[17] which caused damage to buildings – *prostrata sunt aedificia*[18] – in addition to severe casualties amongst both people and livestock – *homines innumerabiles et jumenta pariter infinita submersit.*[19] Although such storms are a recurrent feature of the North Sea basin and many comparable examples are known from the medieval period,[20] the 1287/88 storms stand out in the historical record, and, as we shall see, there is a wide array of material evidence attesting to their occurrence. The analysis of these events that follows focuses on England while making reference to their impact in continental Europe.

3.1.1 Historical background

The documentary evidence relating to this series of storms is relatively rich with descriptions in at least 15 different extant chronicles. The information they provide, however, is difficult to interpret; many sources fail to give dates at all and where these are included years and seasons are sometimes unclear. Some events are only reported by a single source while equally, those details that are mentioned in several sources may reproduce errors copied from a single unreliable source. These are all common concerns with this type of source material,[21] but, despite these issues, useful information can still be extracted.

Many of the sources appear to discuss the same events – with the majority recording storms on 1st January 1287 and 4th February 1288 but there is also some disagreement. For example, the *Chronicon of Thomas Wykes* mentions a storm on 1st January 1286.[22]

17 Strong winds and violent seas. Ellis 1859: 268.
18 Ellis 1859: 268.
19 Luard 1869: 312.
20 E.g. Gottschalk 1971; Bailey 1991.
21 Gottschalk 1971: xiii; Brázdil et al. 2005: 373–374.
22 Luard 1869: 308.

This can be explained by the different calendar systems that were in use throughout the period meaning that there were up to 12 different dates from which the new year could be counted.[23] Wykes probably used the system whereby the new year was counted from 25th March rather than 1st January meaning that, by modern reckoning, his 1st January 1286 corresponds to 1st January 1287. Correction for these differences brings the majority of the dates given by the chroniclers into agreement (see Tables 3.1 and 3.2).

One of the most detailed accounts comes from John of Oxnead, a contemporary chronicler, who records four storms: one on the feast of the holy circumcision [1st January] 1287, one on 12th January 1287, one on 7 kalends January 1288 [December 26th 1287] and one on 4th February 1288.[24] The storms on 1st January, 4th February and to a lesser extent 26th December are mentioned to some degree by most of the other English sources (summarised in Table 3.1) including the *Annales de Wigornia*,[25] the *Annales de Dunstaplia*,[26] the *Flores Historiarum*,[27] Gervase of Canterbury,[28] the continuation of the *Chronicon ex Chronicis*,[29] the *Chronicon of Thomas Wykes*,[30] *Bartholomæi de Cotton*,[31] the *Hagnaby Abbey Chronicle*[32] and the *Chronicon Abbatie de Parco Lude*.[33]

On the other hand, the evidence from continental Europe shows only limited correlation with the English sources. The Belgian *Annales Floreffienses* records floods in Zeeland, Frisia and Holland at some point during 1287, providing no exact date, but no storms are mentioned in 1288.[34] The contemporary German churchman Alexander of Roes on the other hand, describes a single flood in 1288 in the same areas – Zeeland, Holland and Frisia.[35] This may be supported by the French *Annales Colmarienses Majores* which record strong winds in eastern France and terrible floods in Flanders responsible for 50,000[36] deaths on 2nd February and a thunderstorm on 4th February 1288.[37] Meanwhile, the anonymous continuation of the *Menkonis Chronicon*, from the

---

23 Cheney, Jones 2000: 8–9.
24 Ellis 1859: 268–271.
25 Luard 1866: 338.
26 Luard 1866: 338.
27 Luard 1890: 68.
28 Stubbs 1880: 293.
29 Thorpe 1849: 237–239.
30 Luard 1869: 308–312. Although Wykes also mentions another storm between 25th January–2nd February 1288. This date finds support in the *Annales Colmarienses Majores*.
31 Luard 1859: 167–168.
32 Owen 1986: 61–62.
33 Venables 1891: 19.
34 Pertz 1859: 628.
35 Grundmann, Heimpel 1949: 72–74.
36 As noted in chapter 1, this number is certainly an exaggeration as it seems doubtful the chronicler would have had access to any accurate figures if these were even produced. High figures such as these were merely used by medieval writers to illustrate that a very high number had died. See Ziegler 1969: 51–53.
37 Pertz 1861: 215.

Premonstratensian monastery of Bloemhof, present day Netherlands, describes in remarkable detail a flood on 14th December 1287 as well as storms at sea which wrecked many ships in the summer of 1288.[38] The Dutch rhyming chronicle of Melis Stoke[39] records the 1287 December flood on the 17th[40] while the February 1288 flood is dated to the 5th.[41] This is evidently a complex picture that is not easily reconciled but when viewed in tabular form (Table 3.2), we can be sure that extreme floods struck the Low Countries in December 1287 and February 1288 – though the precise dates remain debatable.

Therefore, the historical evidence suggests that between 1287 and 1288 storms were limited to 4 months,[42] with the most severe occurring in January 1287, December 1287 and February 1288. The 14th/15th August 1288 storm is only recorded by one British source,[43] which describes damage at Mablesthorpe, Lincolnshire, while the continental source in which a storm in August is mentioned only describes damage to shipping, failing to specify if any damage occurred on land.[44] This was probably, therefore, a less damaging event, although as severe conditions and structural damage are recorded in Lincolnshire, it is likely that similar conditions were felt at other locations along the east coast of England. As the evidence for this storm is sparse, however, it has been excluded from further analysis below.

The sources which record these events show a relatively high level of conformity (See Tables 3.1 and 3.2). GOTTSCHALK argues that the 1st January 1287 event should be regarded as a calendrical error and must in fact have occurred in January 1288.[45] If this were the case, the storm would fit into the same season as the other two, or more, events but this scenario is unsupported by the documentary evidence which always presents the January 1st storm separately and before the accounts of the December 1287 and February 1288 storms suggesting it took place in the preceding winter. Further support for this comes from the records of Canterbury Cathedral Priory which record a meeting at Snargate, Kent, in 1287 at which it was decided a new sea wall should be built to

38 PERTZ 1874: 565.

39 C.1255 – c.1305.

40 The same date given by Bartholomæi de Cotton.

41 BRILL 1885: 233–234.

42 The only other extreme weather events reported over this period are a drought in 1288 which is mentioned in some of the English chronicles and may also have affected Normandy as well as a hail storm which fell in Lincolnshire on 10th July [sixth of the Ides of July] 1288. For the drought in England see LUARD 1869: 495. For its possible occurrence in France see GOLB 1981: 168–169. For the hail storm in Lincolnshire see VENABLES 1891: 20.

43 OWEN 1986.

44 PERTZ 1874: 565.

45 GOTTSCHALK 1971: 261.

protect against flooding at Holewest.[46] This work was to begin on 10th February and would be finished by Easter.[47] Assuming the years have not been confused in all the chronicles as well as this document, this provides circumstantial evidence of flooding in January 1287.[48]

Only the events in December 1287 are difficult to reconcile – either this was a particularly stormy period with more than one event affecting different areas, which is perfectly possible,[49] or many of the authors were careless or ignorant as to the precise dates. The date given by *Bartholomæi de Cotton*, 17th December, matches that given by the Dutch chronicler Melis Stoke but Cotton's account of the event is very similar to the description of John of Oxnead, who gives the date as 26th December. Both chroniclers describe flooding at Hickling, Norfolk, and the deaths of *c.*200 which would support the interpretation that only one major flood occurred during this month.

The February 1288 storm appears to have had the greatest impact as it is reported by the most sources, both in Britain and the Low Countries. It should be remembered, however, that due to the occurrence of the preceding storms, which may have damaged existing sea defences, exacerbating the impact of future storms, it was not necessarily the event of greatest magnitude. As DE KRAKER points out in reference to flooding in the Netherlands, akin to communities hit by successive years of harvest failure, communities were rarely able to cope with extreme flood events occurring in quick succession as such rapid recurrence put their resilience strategies to "the highest test".[50]

An important point is that a correlation can be demonstrated between the dates presented above and the phases of the Moon. The position of the Earth in relation to the Sun and Moon are what control global tides. When the Sun and Moon are in alignment, at a full or new Moon, tide levels are at their maximum.[51] In relation to the major storms

---

46 TATTON-BROWN connects this place to an area within the parish of Dymchurch and suggests the wall in question was the precursor to a major sea wall still in existence in the 16th century which ran north from St Mary's Bay up to Palmarsh. TATTON-BROWN 1988: 108.

47 SMITH 1969: 170.

48 It seems unlikely that the dates are incorrect as this would mean the work was to begin on 9th February 1288, the Monday during the octave of the purification. This would have been only *c.*5 days after the major storm surge of 4th/5th February but it seems unlikely that such a fast response, which would have involved organising the meeting, agreeing on when and where the work was to commence and who would carry it out would have been possible in such a short timeframe. If the storm had occurred at the start of the previous January however, this would have allowed a little over a month for this organisation to have been put in place which seems more realistic.

49 GOTTSCHALK 1971: 262.

50 DE KRAKER 2015: 2680.

51 This is reversed when the Moon is in its first or third quarter when tides experience minimum or 'neap' levels.

Tab. 3.1: British historical sources recording the storms of 1287/88. Where the sources give specific dates these are shown. Where storms in particular months are recorded without a specific date, this is indicated by an 'X'. Created by the author.

British Sources					
Source	**Jan 1287**	**Dec 1287**	**Feb 1288**	**Aug 1288**	**Reference**
Chronica Johanis de Oxenedes	1st, 12th	26th	4th	–	Ellis 1859: 268–271.
Annales de Dunstaplia	X	–	X	–	Luard 1866: 338.
Chronicon Thomas Wykes	1st	–	25th Jan-2nd, X	–	Luard 1869: 308–312.
Flores Historiarum	–	26th, X	–	–	Luard 1890: 68.
Annales de Wigornia	1st	–	X	–	Luard 1869: 493–495.
Gervase of Canterbury	–	–	4th	–	Stubbs 1880: 293.
Bartholomæi de Cotton	1st	17th	–	–	Luard 1859: 167–168.
Hagnaby Abbey Chronicle	1st, 13th	–	4th	14th	Owen 1986: 61–62.
Chronicon Abbatie de Parco Lude	X	X	–	–	Venables 1891: 19.
Continuatio Chronici Florentii Wigorniensis	31st Dec 1286, 13th	X	3rd	–	Thorpe 1849: 237–239.
Chronica Buriensis	–	–	3rd	–	Arnold 1896: 35.

listed above, during 1287 full moons occurred on 1st January and 21st December.[52] The former exactly coincided with a documented extreme storm while the latter seems to have been a few days before or after the storm depending on which dates are taken as reliable. It seems likely therefore, that the December flood would have been less severe due to the tidal conditions. The following year, 1288, a new moon occurred on 3rd February while a full moon fell on 13th August, in both cases one day prior to a recorded storm. This suggests that the most devastating floods occurred when severe storms coincided with natural tidal maxima, a phenomenon which appears to have happened at least three times in the space of two years.

With the chronological occurrence of these events somewhat clarified, one of the main unanswered questions is the geographical areas that were affected by each storm. Only a minority of the sources name specific locations or regions in their accounts. Combined with the information discussed above, these data allow a map to be produced which illustrates the known areas that were affected by each storm event (Fig. 3.1).

52 Dates obtained from http://astropixels.com/ephemeris/phasescat/phases1201.html [Accessed 05 December 2020].

Tab. 3.2: Continental European historical sources recording the storms of 1287/88. Where the sources give specific dates these are shown. Where storms in particular months are recorded without a specific date, this is indicated by an 'X'. Created by the author.

Continental European Sources					
Source	**Jan 1287**	**Dec 1287**	**Feb 1288**	**Aug 1288**	**Reference**
Menkonis Chronicon	–	14th	–	15th	Pertz 1874: 565.
Annales de Floreffe	–	X	X	–	Pertz 1859: 628.
Annales Colmarienses Majores	–	–	2nd, 4th	–	Pertz 1861: 215.
Alexander von Roes	–	–	X	–	Grundmann, Heimpel 1949: 72–74.
Rijmkroniek van Melis Stoke	–	17th	5th	–	Brill 1885: 233–234.

As there is considerable disagreement on the specific dates, however, it is best to view this as a map of storm damage that occurred during the three months rather than maps of storm damage from three specific storm events. In addition, Fig. 3.1 has the caveat that it only shows areas where historical sources record the effect of the storms – which is unlikely to be a particularly comprehensive picture. Certainly, each storm would have had a negative impact for neighbouring coastal towns, for which either no documentation exists or such documentation is now lost or unknown.

What the available evidence does reveal is that the storms which occurred in January 1287, the worst of which seems to have occurred on the feast of the circumcision [1st January], appear to have affected England to a greater extent than the Low Countries. It is certainly possible, as a result of differing storm tracks, speeds, wind direction and tidal conditions, for the effects of storm surges to differ markedly.[53] This could explain why no damage was reported in the Low Countries in January 1287 where damage was presumably relatively minor. The following December, storms hit Frisia, and perhaps Holland and Zeeland,[54] in addition to England. The English sources suggest this storm (or storms) came later in December, either 17th or 26th, while in the Low Countries the storms are reported on the 14th or 17th. This divided picture is resolved during February 1288 when both the English and continental sources agree that a severe storm caused extensive flooding on the 4th/5th February. It is also possible that, thunderstorms and/or a less severe storm occurred in the days before this event, most likely the 2nd/3rd February. In addition, the two sources that mention a storm in August, the *Hagnaby Abbey Chronicle* and the *Menkonis Chronicle*, also agree on the date 14th/15th August (assuming the storm struck at night and continued into the following morning).

53 Gottschalk 1971: xiii; Gerritsen 2005: 1276; Wadey et al. 2015.
54 Brill 1885: 233–234.

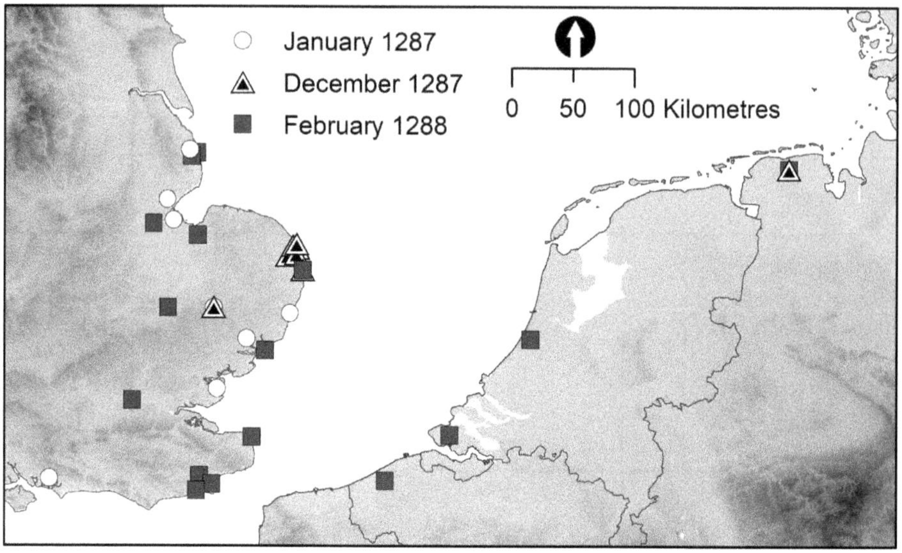

Fig. 3.1: The extreme weather recorded occurring in 1287 and 1288 as presented in tables 3.1 and 3.2. Created by the author.

The locations affected by flooding in these three, or more, events are all consistent with North Sea storm surges. These occur when extratropical cyclonic storm systems track across the North Atlantic, directing huge volumes of water into the North Sea basin, where the surrounding landmasses cause the water to pile up so that the sea level is significantly raised. As the sea is 'funneled' against the coasts of England and the Low Countries, with a narrow exit through the English Channel, the rise in sea level is particularly pronounced in these areas – precisely those that were affected in all three months described above. A modern analogue for this phenomenon is the 31st January/1st February 1953 storm which resulted in the deaths of over 2,000 people in the UK, the Netherlands, Belgium and Germany.[55] Its near co-incidence with a full moon, which had occurred 2 days prior, along with the comparable area of effect makes this a particularly comparable disaster to the storms of the late 13th century – and particularly the February 1288 event.

The historical sources provide a wealth of information concerning the impact of the 1287/88 storms on certain spheres. Chroniclers' accounts widely report high numbers of casualties drowned by the floods.[56] John of Oxnead, for example, describes how many climbed trees to escape the floodwaters but, unable to hang on as they waited for

55 Gerritsen 2005.

56 See, for example, Thorpe 1849: 237–239; Luard 1869: 311–312, 493, 495.

help to arrive, they fell back into the water.[57] Livestock and agricultural produce were also acutely affected. At Boston, Lincolnshire, for example, along with human victims, innumerable cattle were lost[58] while both the *Annales de Dunstaplia* and the *Hagnaby Abbey Chronicle* report the general loss of high numbers of sheep and cattle throughout the zone affected by the disaster.[59] At the Benedictine Abbey of St Benet at Holme, Norfolk, although the monastic community were forced to evacuate the Abbey precinct, two monks stayed behind to tend to the horses who would, no doubt, have otherwise perished.[60] Damage to structures is also frequently reported. In the January 1287 floods, the church of Mablesthorpe St Peter, Lincolnshire, was damaged, causing the chalice and pyx to be crushed underneath fallen masonry, while in the 1288 floods, which came the following February, the already damaged structure was "completely destroyed".[61] At Hickling, Norfolk, floodwaters rose a foot above the high altar of the church of the Augustinian priory before subsiding[62] and presumably other nearby buildings in the town were also damaged by the floodwaters. Structural damage is reported at Great Yarmouth as a result of the December 1287 floods when the stone walls of the cemetery were broken by the sea.[63] The historical record, therefore, provides a wealth of detail in certain areas regarding the impact of the storms and the floods they caused. As we will see, however, there is much that can be added to this picture by turning to the material record.

3.1.2 The archaeological evidence

The rich corpus of historical sources describing the 1287/88 storms, and the magnitude of flooding they describe, indicate that these storms inflicted severe destruction across the area of effect. As a result, as well as evidence for destruction across areas in the affected zones, the storms probably also left stratigraphic layers deposited during the storms themselves.[64] Of course, in the aftermath of the floods, most of the damage would have gradually been cleared and restored – ground might be levelled and repairs

---

57 Ellis 1859: 270–271.

58 Luard 1869: 308.

59 Luard 1866: 338; Owen 1986: 61–62.

60 Ellis 1859: 270–271.

61 Owen 1986: 61–62.

62 Ellis 1859: 270–271.

63 Luard 1859: 168.

64 Although in archaeological terms it is effectively impossible to distinguish between the, three or more, storm events, in theory it is possible that separate events could be distinguished archaeologically if distinct layers were formed by flood deposits. It is most likely that, as the storms occurred so close together in time, the later events would have incorporated and mixed sediments deposited by the earlier storms. Additionally, it is unlikely that absolute dating methods applied to a single layer could be resolved to a resolution that would allow the layer to be confidently ascribed to a single storm event.

would take place to remedy the damage caused by the storms. As a result, over time, the material evidence directly associated with these storms would diminish. In spite of this, it is still possible that damage on the scale that the documentary sources suggest may have left traces that can still be identified in the present day.

Notably, a number of geological and geomorphological sequences have been linked to the occurrence of the storms. For example, according to HEARNE et al. the shingle spit at Deal, Kent, was significantly extended by beach sediments deposited by the storms.[65] In Romney Marsh, in Kent and Sussex, the extent of flooding caused by the storms of the late 13th century can be traced through the later history of reclamation, place names and comparative soil analysis.[66] In a core from Winchelsea, East Sussex, DHOOP has identified a period of turbulent, stormy weather in the decades leading up to the 1287/88 storms, attesting to the occurrence of undocumented, lower magnitude events during this time, while the 1287/88 storms themselves appear to be represented by a slope-wash deposit of sandy/clayey sediments.[67] Late Holocene high energy deposits from Brittany, which are late 13th or early 14th century in date, have also been connected to this series of storms[68] though, as there is no known historical evidence for the impact of the 1287/88 storms in Brittany,[69] and other regional studies attribute comparable deposits to a different storm event,[70] it seems most likely that these deposits relate to a different storm.

Excavated deposits, such as an assemblage of pottery discovered in 1951 at Chapel St Leonards, Lincolnshire, have also been interpreted in relation to this series of storms – in this case as evidence of a site abandoned in the aftermath.[71] Similarly, at South Denes, Great Yarmouth, an assemblage of medieval pottery dated to the 13th century, discovered sealed by a layer of silt, was interpreted as an area of medieval activity disrupted by the storms.[72] Although documentary evidence for the impact of the storms at Hastings is lacking, the archaeological evidence in the form of a layer of clay encountered at a number of sites across the town has been interpreted as a deposit associated with the storms' occurrence.[73] This interpretation is strengthened by cores from the nearby deserted medieval village of Northeye, Sussex, which suggest the salt-works upon which the village's economy depended were severely affected by storms during

65 HEARNE et al. 1995: 243.

66 RIPPON 2000b: 196–197.

67 DHOOP 2016: 107–117, 206.

68 HASLETT, BRYANT 2007b: 217.

69 BRYANT and HASLETT suggest that the 1287/88 storms are candidate events because these floods affected northern France. The source for this information is unclear.

70 BAILIFF et al. 2014: 900.

71 THOMPSON 1953.

72 GREEN 1961: 21.

73 VAHEY 1989: 2–3; VAHEY 1991: 2.

this period.[74] A number of sites have also been investigated in New Romney, Kent, with evidence convincingly interpreted in relation to these storms[75] which is discussed in greater detail below. The wider Romney Marsh area (including 'Romney Marsh proper' and Walland Marsh; see Fig. 3.2) has been investigated in detail and a number of sites are known through both historical and archaeological evidence relating to the 1287/88 storms. At Broomhill, East Sussex, for example, GARDINER suggests that the lack of evidence found during the excavation of the church predating the late 13th century may indicate that the church was rebuilt at a new location after the occurrence of the storms which caused flooding throughout the majority of Walland Marsh.[76] The failure of the dikes protecting low lying reclaimed land in this area[77] resulted in widespread damage – including the destruction of a number of known structures such as the Court Lodge at Agney and the church at Midley.[78]

Some archaeological evidence that has not traditionally been tied to the 1287/88 storms could also be products of their occurrence. The bridge repair at Heigham Bridge, Norfolk, for example, provides a possible case of structural damage caused by the storms.[79] Excavated flood deposits at Wisbech, Norfolk, on the other hand, which date to *c.*1250–*c.*1350 and are 30cm at their deepest,[80] appear to have instigated a new phase of building[81] and accord well, both geographically and in terms of magnitude, with the 1287/88 storm series. Similarly, a flood level visible at more than 20 sites across King's Lynn, Norfolk,[82] where flooding is recorded in the February 1288 storm,[83] dates to the late 13th or early 14th centuries.[84] It may be possible to refine the dating of these flood deposits, thereby confirming or refuting whether they relate to the 1287/88 storm series, if samples could be dated using optically stimulated luminescence methods. Currently, however, it remains a possibility that such layers were deposited by the storm though this cannot be stated with certainty.

3.1.3 New Romney, Kent

At New Romney, Kent, there is a significant body of archaeological evidence relating to one, or more, of the 1287/88 storms. As a result of the introduction of developer

74 OXFORD ARCHAEOLOGY 2009: 11.
75 E.g. DRAPER, MEDDENS 2009: 59–69.
76 GARDINER 1988: 125; RIPPON 2000b: 194; GARDINER, HARTWELL 2006: 148–149; see Fig. 3.2.
77 GARDINER, HARTWELL 2006: 152.
78 TATTON-BROWN 1988: 108.
79 This is discussed in Chapter 4.4.1.
80 The surface of the flood level at Wisbech is *c.*4.5m above sea level.
81 HINMAN, POPESCU 2012: 24.
82 The surface of the flood level at King's Lynn is between 4.15m and 4.2m above sea level.
83 LUARD 1869: 495.
84 CLARKE, CARTER 1977: 63.

funded archaeology in the early 1990s, a wealth of well-recorded excavations have been undertaken across New Romney. This provides the opportunity to investigate a detailed case study of the effect of an extreme storm on the medieval population living in a single town. In order to assess the extent to which New Romney was affected by the 1287/88 storms, all of the available archaeological grey literature reports, held by Kent County Council,[85] relating to archaeological investigations in the town have been assessed. This allows a limited reconstruction of the impact of the 1287/88 storms and their aftermath. Following a brief review of the town's early history, the results obtained from this grey literature survey are presented below.

New Romney's underlying geology is made up of pebbly sand and shingle[86] which is encountered by excavations across the town.[87] The early history of the town is debated. Tatton-Brown suggests the settlement emerged as a re-foundation of a Saxon town located at present-day Old Romney, 3.2km to the west,[88] while Gardiner argues that New Romney evolved naturally from an existing nucleus of settlement.[89] Whichever the case, the town expanded rapidly from relative obscurity during the 12th century, becoming an important maritime trading centre, and one of the Cinque Ports, able to benefit from its fortunate location within easy sailing distance of some of north western Europe's largest urban centres.[90] This expansion was short-lived however. During the 12th century, the River Rother, which up to that point had drained out into the North Sea at New Romney, shifted south to drain out near Rye, Sussex. At least since the time of William Camden, this change in the course of the Rother has often been interpreted as a result of the 1287/88 storms themselves[91] yet the available historical and palaeoenvironmental evidence demonstrates this change had already occurred by the 12th century.[92] To remedy this alteration in the Rother's drainage regime, a major water channel, the Rhee Wall,[93] was dug in sections from Appledore to Old Romney in an attempt, by diverting water from the Rother, to flush out silt clogging up the tidal creek upon which the harbour at New Romney was located. Following a phase of storms in the early 1250s, described by the chronicler Matthew Paris,[94] this feature was extended to New Romney itself but it did not prove effective in the long term and was unable to prevent the development of problems caused by siltation.[95] At least in New

---

85 As of 27/02/2017 this included 55 reports spanning 1993–2016.

86 Green 1960. 22.

87 E.g. Wilson, Linklater 2002: 14; Linklater 2003.

88 Tatton-Brown 1988: 107–108. see Fig. 3.2.

89 Gardiner 1994: 344–345.

90 Gardiner 2000: 86; Martin, Martin 2004: 7–19; Campbell 2010a: 28.

91 Camden 1610: 350–351.

92 Rippon 2000b: 191.

93 See Fig. 3.2.

94 Luard 1880: 395, 453; Luard 1890: 402.

95 The effectiveness of such a feature would have been highly dependent on the velocity of the water flowing through the tidal creek at New Romney – without a sufficient gradient to produce fast flowing

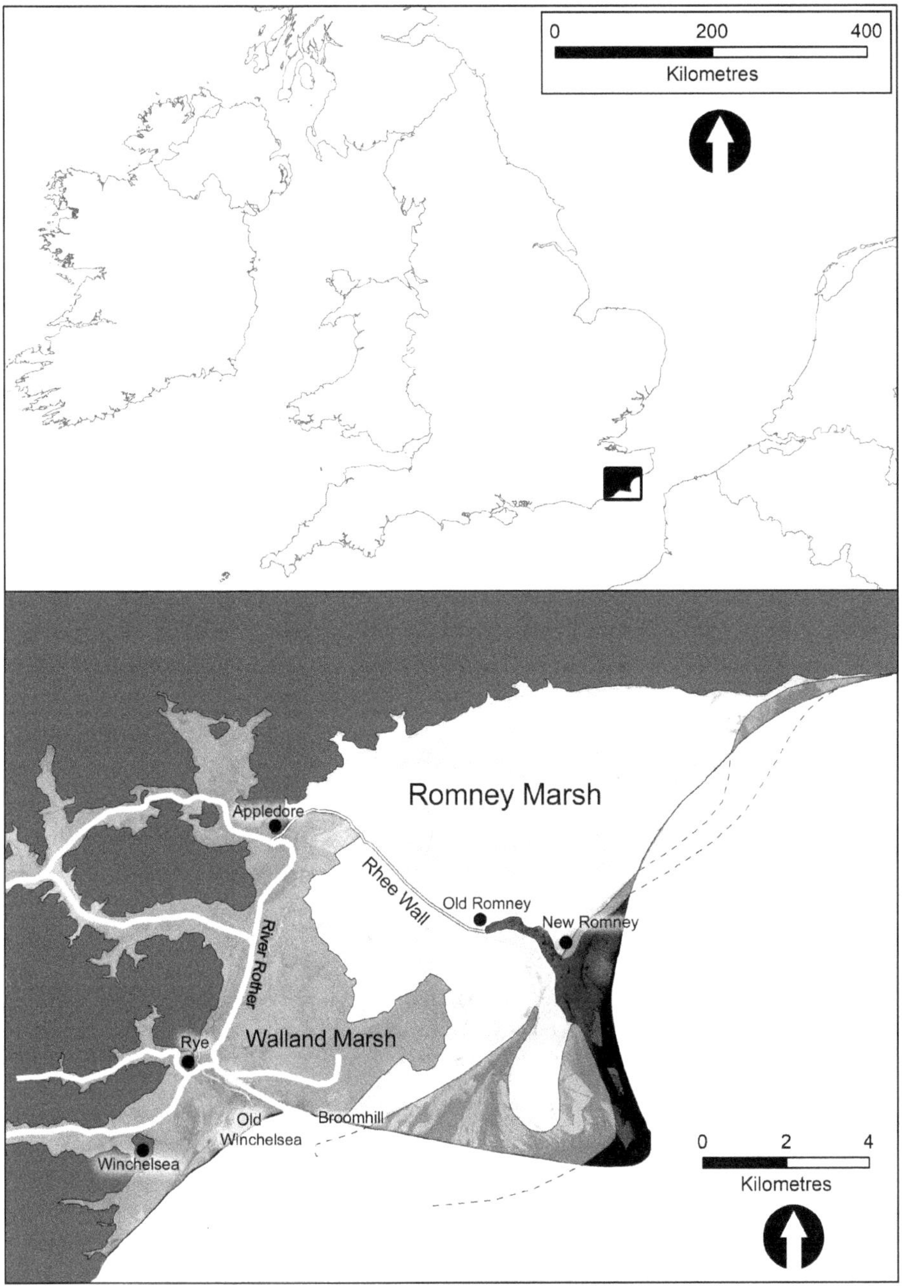

Fig. 3.2: New Romney and Romney Marsh at the time of the storms of 1287/88 overlain against present-day LiDAR data. Postulated coastlines in the late 13th century are indicated by the dotted line. Created by the author after Rippon 2000b: 193.

Romney, therefore, the storms of 1287/88 occurred in a context of decline as a result of pre-existing environmental conditions.

The only chronicler to record the effect of the 1287/88 storms within Romney Marsh itself was Gervase of Canterbury who states that:

> On the day before the nones of February [4th February] the sea swelled up so high in Thanet, and in Romney Marsh and in all other adjacent places, that all walls were broken and almost all the land was covered from the Great Wall of Appledore to Winchelsea to the south west.[96]

It remains a possibility that damage also occurred in this area in one, or all, of the earlier storms in 1287. The identification of the 1288 event as the storm of greatest intensity in this particular area, though, is strengthened by the fact that a charter of unknown provenance from nearby Rye, Kent, referenced by the 17th century antiquary Samuel Jeake states that:

> In the year MCCLXXXVII,[97] on the night of St Agatha the Virgin [5th February], the town of Winchelsea was flooded as well as all the land between Cliffsend and Hythe.[98]

The fact that the text of this document provides fresh information – Cliffsend, Hythe and St Agatha's Day all go unmentioned in the English sources discussed in Section 3.1.1 – as well as the fact that it agrees with the other available sources suggests, since it is unlikely to have been copied from any other known source, that the document is genuine or at least preserves a communal memory of an event, only written down at a later date. Taken together then, the historical evidence suggests that if only one of the 1287/88 storms struck New Romney, this was the 4th/5th February 1288 storm. The meeting at Snargate to discuss flooding at Holewest in 1287,[99] discussed above, however, may provide evidence that Romney Marsh was also affected by flooding in the January 1287 storm although, if this was the case, the damage does not appear to

---

water, therefore, it seems unlikely that the Rhee Wall could have had much impact on the silting up of the harbour and, if the water carried by the Rhee Wall flowed slowly enough, it could even have exacerbated the problem by increasing sediment deposition in the tidal creek. See RIPPON 2000b: 197–198.

96 STUBBS 1880: 293. Translation by the author. Original latin text: *Pridie nonas Februarii mare ita erexit se in Thaneto, et juxta, et in marisco de Romenal et omnibus locis adjacentibus, quod omnes wallias dirupit, et fere omnes terras operuit a magna wallia de Apuldre usque Winchelese versus austrum et versus occidentem*. Note that this text, which is considered to be original rather than a later copy from an earlier document STUBBS 1880: xxxiv., was very likely the "ancient French Chronicle, sometime belonging to the Church of Canterbury" cited by SOMNER 1693: 57–58.

97 This date almost certainly relates to 1288 by modern reckoning due to the medieval New Year commonly being counted from 25th March rather than 1st January.

98 JEAKE 1728: 105. Translation adapted by the author. Original latin text: *M.D. quod anno Domino Millesimo CCLXXXVII. In vigilia sanct. Agathæ virginis, submersa fuit villa de Wynchelsee, & omes terræ inter Climesden usq. le Vochere de Hethe.*

99 SMITH 1969: 170.

have been as severe as in the subsequent February 1288 storm. As discussed above, this storm caused severe damage and destruction across the wider area and especially in Walland Marsh.[100]

In order to gauge the impact of the storms across the unit of an entire town, all of the 'grey literature' held by Kent County Council[101] relating to excavations in the town was assessed. This allowed the locations of all excavations, test pits, boreholes and watching briefs to be plotted in a GIS environment. The result of this activity produces a plot containing records of 351 archaeological interventions (see Fig. 3.3). To this spatial data was added any stratigraphic information noted in the reports of relevance to the impact of the storms on the town. This has been broken down into six categories of evidence:

1) Evidence for destruction sealed by stratigraphic layers identified as material deposited by 13th century storm surges. This category relates to in-situ remains of the pre-storm town which were heavily damaged and sealed by sediments deposited by the storm.
2) Layers identified as medieval storm deposits. This category covers areas where no archaeological pre-storm material was encountered below the storm deposited layer or where the excavations did not penetrate beneath this layer.
3) Contexts with evidence for abandonment or decline contemporary with the storms of 1287/88. This category reveals areas which may have been abandoned, fell out of use or ceased to be maintained in the aftermath of the storm. This could relate to depopulation and/or declining economic opportunities caused by the storm.
4) Stratigraphic layers identified as marine which are now located in terrestrial locations which assist in defining the town's coastline at the time of the storm.
5) Contexts containing archaeological material which can be dated to between *c.*1250–*c.*1350. These are mainly dated through ceramics which provide a date range of about 75 years meaning there is some uncertainty as to whether these locations were occupied before or after 1287/88.
6) No contemporary archaeological evidence. These contexts may have been truncated by later development or sit outside the zone of medieval occupation and do not reveal anything about the effects of the storm on the town.

This 'grey literature' survey provides another method of gauging the impacts of the storms of 1287/88. The occurrence of severe storms in the town, however, was not without precedent, and several of the excavations revealed evidence for earlier flood events – presumably those recorded by Matthew Paris.[102] At the Southlands School site,

---

100 Tatton-Brown 1988.

101 This analysis included all of the archaeological grey literature relating to New Romney held by Kent County Council up until 12/08/2016. I am grateful to Paul Cuming for assisting me in accessing these data.

102 Luard 1876: 379; Luard 1880: 395, 453; Luard 1890: 219, 402.

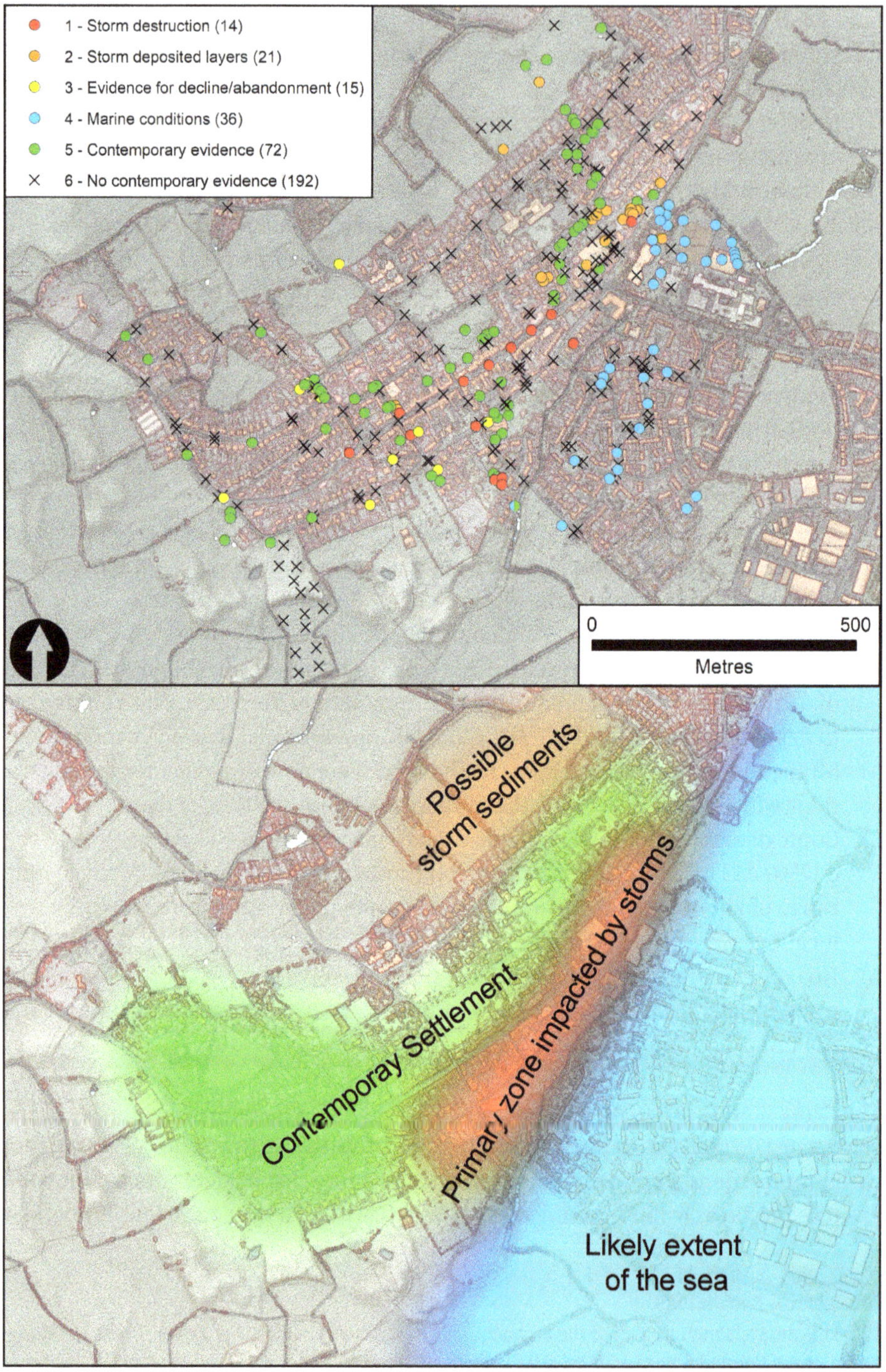

Fig. 3.3: Above: The results of a survey of the 'grey literature' relating to archaeological excavations in New Romney, Kent. The classification of this evidence is described in more detail in Section 3.1.3. Created by the author.

off Dymchurch Road – along what would have been the medieval waterfront, the early 13th century beach front was sealed by up to 50cm of storm deposited sand. This layer contained a silver long-cross penny which dated to the reign of Henry III (1247–1272).[103] This accords well with one of the inundations recorded by Matthew Paris in the 1250s, however, GOTTSCHALK argues that these must in fact have occurred in 1248 or 1249 due to the fact that no storm surges of comparable magnitude are recorded in the Low Countries in the early 1250s.[104] In either case, the coin supports the dating although it must have been relatively recently minted at the time of deposition.

Focussing on the storms of 1287/88, archaeological evidence across the town has been widely attributed to their occurrence. Work along the High Street in advance of the installation of a gas main in 1995, for example, uncovered contexts containing evidence for early 13th century activity – dated by ceramics – sealed by layers[105] of silty sand with a high gravel content which were interpreted as storm deposited.[106] As is the case with many of the excavations undertaken across New Romney, there is some doubt as to whether these layers were deposited as a result of storm action or not.[107] In the absence of geomorphological analysis, which may have allowed such contexts to be more firmly attributed, these interpretations should be appraised critically.

Nevertheless, many of the archaeological sequences obtained from sites across the town appear to correlate closely with the historical evidence. Excavations off Church Road, for example, appear to have encountered the banks of the medieval river mouth. Geoarchaeological analysis in this area revealed two phases of erosion, evidenced by two sharp contacts between stratigraphic units, which could correspond to the 13th century storms. These were both followed by periods of reduced energy sediment deposition[108] which accords well with the hypothesised gradual siltation of the river mouth over the course of the 13th century. Similarly, outside the church of St Nicholas, the east end of which is thought to have been largely swept away in the 1287/88 floods,[109] excavation for a manhole revealed two earlier phases of road surface separated by 20–40cm of 'storm deposit'.[110]

The highest volume of evidence for the impact of the storms of 1287/88 comes from the area presumed to have been the shoreline during the 13th century. This roughly corresponds to Church Road, which becomes Dymchurch Road to the north. At the

103 DRAPER, MEDDENS 2009: 60. The coin was initially thought to post-date 1250 due to its heavily worn condition prior to post-excavation analysis. For this initial interpretation see MEDDENS, WOOLDRIDGE 2002: 17.
104 GOTTSCHALK 1971: 190, 199.
105 The contexts interpreted as the result of the storms of 1287/88 are contexts 10, 21 and 29.
106 HERDMAN, JARMAN 1996: 15–16.
107 HERDMAN, JARMAN 1996: 15.
108 PRIESTLEY-BELL 1999: 11.
109 TATTON-BROWN 1987: 344.
110 CANTERBURY ARCHAEOLOGICAL TRUST 2010: 6.1.3.

Old School House, located *c.*100m seaward of St Nicholas' Church, which would have been a coastal location in the medieval period, occupation layers, evidenced by early-mid 13th century pottery and detritus such as animal and fish bones, were sealed by high energy gravel deposits.[111] In this case, the excavators suggest the important point that the storm(s) may have initially caused erosion prior to the deposition of sediments. In certain circumstances, this could have resulted in the immediately pre-storm remains being heavily disrupted or entirely removed which would have erased evidence for any earlier storms which had recently occurred. Further damage can be seen at a site on the seaward side of Church Road where a timber structure was destroyed and overlain by 40cm of coarse yellow sand and rounded beach pebbles[112] (see Fig. 3.4). This was interpreted by the excavators as "deposited by severe marine action".[113] The flood layer which sealed this structure was overlain by a later stone building which, at some point, also became inundated by the sea – a fact demonstrated by the presence of a marine deposited layer.[114] Unfortunately, no dateable material was recovered relating to the earlier timber structure although unstratified 13th century pottery from the site is indicative of activity contemporary with the storms.[115] The later inundation, which affected the masonry structure, sealed a pit containing late 15th–early 16th century pottery, demonstrating that this was a significantly later flood.[116] A possible interpretation, therefore, is that the earlier timber structure was destroyed as a result of the 1287/88 event.[117]

A similar narrative was revealed at the Southlands School site, less than 300m north-east of the Church Road site, where 'Building 2' was severely truncated on its south-eastern half by *c.*50cm of storm deposits – attesting to the storm's erosive power (see Fig. 3.5). The structure measured 7x11m and was probably timber framed with a floor constructed of either timber or trampled earth and can be roughly dated to the mid 13th century by ceramics and other artefactual evidence.[118] At some point a small extension, measuring 2x2.3m was added to the structure's north-western side where a contemporary area of burning and a number of hearths were also encountered. The building's coastal location, as well as the recovery of finds such as lead fishing weights and a bone net needle, suggest a function related to fishing and seafaring.[119] The impact

111 Thomason, Stafford 2001: 7–11.
112 Context 28.
113 Wilson, Linklater 2002: 9–10.
114 Context 17.
115 Wilson, Linklater 2002: 15.
116 Wilson, Linklater 2002: 12.
117 The thickness of the layer, context 28, representing the earlier inundation presumed to relate to the 1287/88 event, preserved in section 5 is *c.* 40cm which is roughly double the thickness of the later 15th or early 16th century inundation attested by context 17. This suggests that the earlier flood was significantly more severe, mobilizing and depositing a significantly higher volume of sediment.
118 Meddens, Wooldridge 2002: 21.
119 Meddens, Wooldridge 2002: 21.

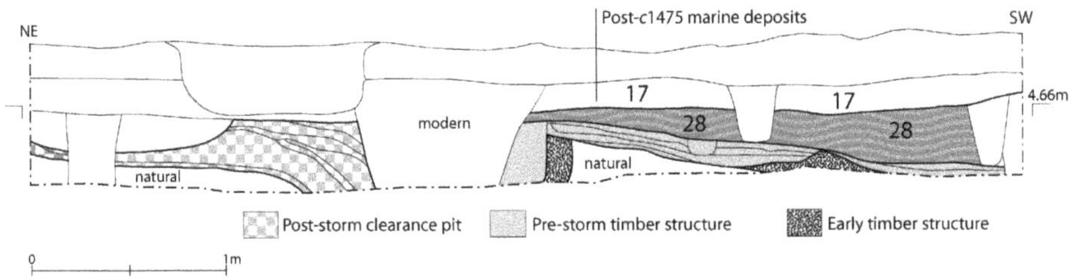

Fig. 3.4: Section 5 from the Church Road site, New Romney. A pre-storm timber structure was overlain by a 40cm layer of storm deposited material (context 28). Although limited dating evidence was recovered, this layer likely relates to the storms of 1287/88. Redrawn by Alejandra Gutiérrez after Wilson, Linklater 2002: Fig. 11.

of the storm(s) is evidenced by the absence of a large portion of the south-eastern half of the structure which was, at some point, entirely eroded and destroyed by the powerful action of sea and storm (see Fig. 3.5). Notably, the excavators suggest that this portion of the building collapsed after the storm had swept the sands upon which it was built out from underneath the building.[120] The void caused by this erosional activity was in-filled with layers of sand which incorporated pottery dated from *c.*1150–*c.*1450 and two coins dating to 1180–1247 and 1100–1135 respectively.[121] The early date of the latter of these coins suggests that the force of the storm was possibly disturbing and reworking material contained in earlier contexts. In the aftermath, the entire building was sealed by a layer of sandy gravel and silt – interpreted by the excavators as an attempt by the townspeople to repair the storm damage and stabilise the shoreline.[122] As well as loosely dated 12th–15th century pottery, a silver farthing of Edward I was recovered from this layer which provides a *terminus post quem* of 1272[123] – thus excluding the earlier storms of the 1250s as candidates and securing the identification of the 1287/88 storms – and most probably the 4th February 1288 event – as the most likely event(s) to have brought about the destruction of 'Building 2'.

The picture obtained from the archaeological evidence in New Romney is that the floods appear to have affected a significant swathe of the town (see Fig. 3.3). The clearest examples of storm damage, identifiable through the archaeological record, are found along the medieval shoreline where, presumably, structures would have faced the full force of the storm. The evidence also points to the fact that the effect of the storm penetrated beyond Church Road and even the High Street – as the 1995 gas main excavations make clear. This picture supports the conclusions of Parkin who argues

120 Meddens, Wooldridge 2002: 23.
121 Meddens, Wooldridge 2002: 23.
122 Meddens, Wooldridge 2002: 23.
123 Meddens, Wooldridge 2002: 23.

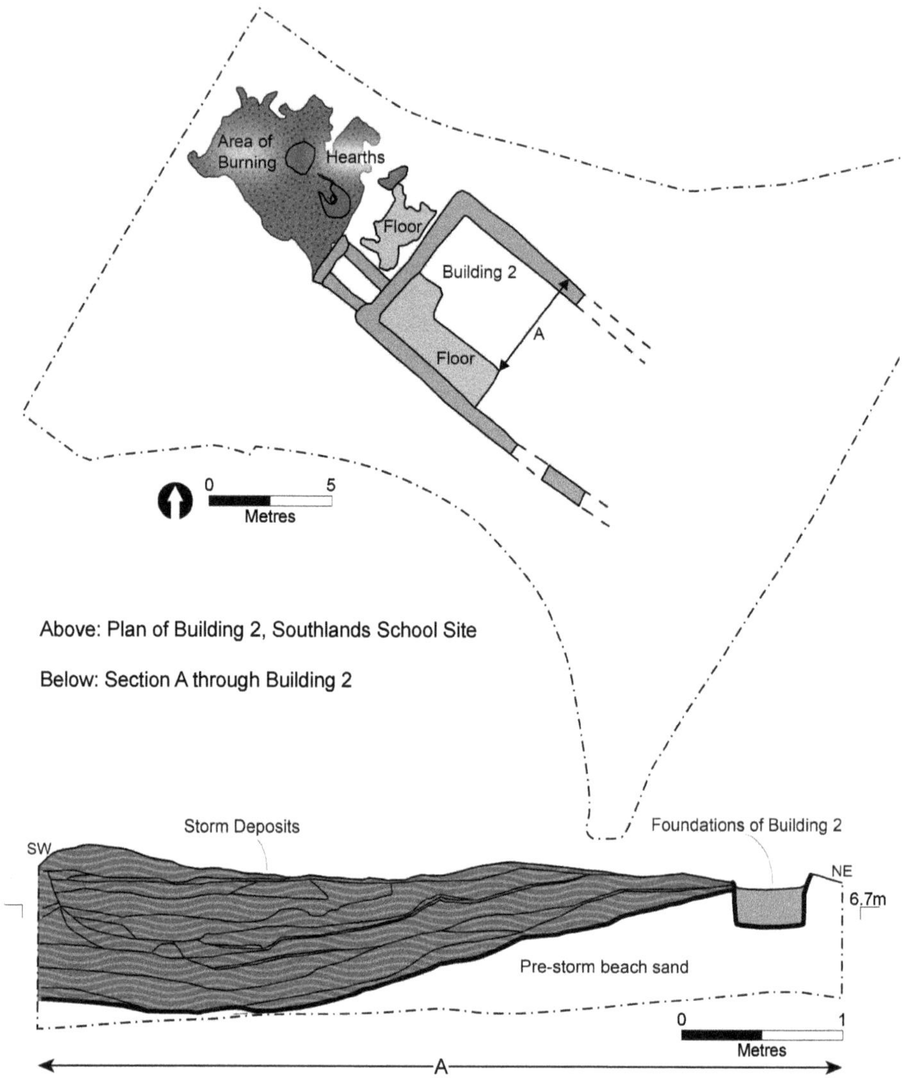

Fig. 3.5: Building 2, Southlands School site, New Romney. Building 2 was severely truncated to the south-east by the storms of 1287/88. Only evidence of the north-western half of the structure was detected due to the storm erosion and deposition visible in a section through the structure – the location of which is indicated by line A. Redrawn by the author after MEDDENS, WOOLDRIDGE 2002: 22 and DRAPER, MEDDENS 2009: 63, 65.

that the floor levels of many of the oldest standing buildings of the town, which are below the current ground level, relate to the original land level of the town prior to the deposition of sediments by the late 13th century storms.[124] This is especially the case in buildings along the High Street, and adjoining roads,[125] in broadly the same area indicated by the excavations in Figure 3.3. The furthest inland archaeological intervention with evidence for a storm deposited layer comes from a field abutting Rolfe Lane on the town's western fringe.[126] Assuming this interpretation is reliable, it presents the possibility that the town was either entirely overflown by floodwaters or, perhaps more likely, encircled as the floodwaters broke through neighbouring low-lying areas. Clearly, the historical and archaeological evidence relating to the 1287/88 storms are complementary with the material evidence largely corroborating the historical sources. At the local level the storm(s) had a catastrophic impact, especially on buildings fronting the seafront which must have led to a serious disruption in town life for some time after the occurrence of the disaster. The evidence which allows a limited reconstruction of what took place in the aftermath of this episode of catastrophic flooding is discussed in more detail in Chapter 4.

3.2 Flooding in Marshland, Norfolk, 1287–1349

Another area where flooding occurred in 1287/88 was Marshland, Norfolk,[127] – a group of parishes situated in an area of low-lying wetland which had been gradually enclosed against inundation from marine and freshwater from the late Saxon period.[128] This section investigates the numerous flood disasters which occurred over the 13th and 14th centuries during the reigns of the three Edwards[129] between 1287–1349 and their impact on the local populations in this area. This case study has been selected based on the availability and survival of source material which include extant historical maps – of which the earliest examples date to *c.*1582[130] and *c.*1610[131] – which show the

---

124 Parkin 1973.

125 Parkin 1973: 122, Fig. 2.

126 Margetts 2010: 57, 60.

127 Maxwell Lyte 1893: 298. At least, a Royal Commission *de Walliis et fossatis* (on walls and ditches) was instigated in 1288 to address a complaint that the lands of Robert de Scales in Marshland had been inundated as a result of a defect in the repair of flood defences on lands belonging to William de Carleton and William de Middilton.

128 Silvester 1988: 160; Rippon 2002a: 65; Rippon 2013: 335–339.

129 Edward I (1272–1307), Edward II (1307–1327) and Edward III (1327–1377).

130 The *c.*1582 map is held by the British Library under shelfmark: Add. MS 71126 f.1 and is available at: http://www.bl.uk/ [Accessed 04 February 2018].

131 The *c.*1610 map is believed to be a copy of an earlier map from *c.*1591 and is held by the British Library under shelfmark: Cotton Augustus I.I f.79 and is available at: http://www.bl.uk/ [Accessed 04 February 2018].

layout of the late medieval field system complete with local toponyms. In addition, the rich documentary record relating to flooding and sea defence in the area provides well dated descriptions of flood events, the damage they caused and the steps taken by society in their aftermath. The 17th century antiquarian William Dugdale's '*The history of imbanking and drayning of divers fenns and marshes*' provides an invaluable collection of primary sources which record this type of information – some of which have subsequently been lost or are no longer extant.[132] The region has also been the subject of extensive archaeological research, including detailed surveys of the parishes listed above.[133] This constellation of high-resolution source material, combined with modern methods such as GIS analysis and LiDAR data, permits a detailed exploration of the impact and subsequent responses provoked by severe flooding on medieval society.

Marshland usually refers to the west Norfolk parishes of Clenchwarton, Tilney, Terrington, Walpole, West Walton, Walsoken and Emneth.[134] Wiggenhall is sometimes also considered a Marshland parish, as is the case here, and it is frequently included alongside the parishes listed above in many of the contemporary historical sources. This area of Norfolk is formed of flat, low-lying, siltlands which formed a region with important stock rearing and salt-manufacture industries during the Roman and Saxon periods.[135] Although definitive historical evidence for drainage and embanking comes only after the Norman Conquest, a number of strands of evidence suggest that this process began somewhat earlier. An intensification and shift in the settlement pattern from the inland margins of the intertidal zone to the coastal marshes took place around the 8th century. This was accompanied by a shift in the character of settlements, from what had most likely been only seasonally occupied locations[136] to permanently occupied nucleated settlements, suggesting that early medieval populations actively decided to move out into these wetland locations forcing them to adapt to the difficulties associated with such locations.[137] Palaeoenvironmental sequences from Fenland Saxon sites indicate that the area remained intertidal – meaning unembanked – until the 10th century when the sequences demonstrate a switch from tidal to freshwater conditions – presumably related to the construction of sea walls preventing marine inundation of the area.[138] Confirmatory evidence for this process comes from the trio of villages, known to have been in existence by the 10th century, on the western edge of Marshland with names

132 SILVESTER 1988: 5.
133 SILVESTER 1988.
134 SILVESTER 1988: 4; see Fig. 3.6.
135 SILVESTER 1988: 156–160.
136 Exploiting the rich grazing available in the marshes during the summer months.
137 RIPPON 2009: 43–44.
138 CROWSON et al. 2005: 168–170.

including the 'wal' prefix: Walsoken, Walton and Walpole, which may relate to the presence of nearby sea walls at an early date.[139]

RIPPON has suggested that the morphology of the field boundaries visible in some of the Marshland parishes, notably Walpole St Peter and West Walton, are indicative of piecemeal reclamation – with core areas enclosed against tidal flooding prior to the construction of a cohesive sea wall along the entirety of the coastline.[140] From the beginnings of the embanking process in the 10th century, progress probably continued gradually as and when new land was required, sufficient manpower and capital were available and favourable market conditions existed, with most of the medieval field systems discernible on later maps in place by the close of the 13th century.[141] The banks and ditches which drained this field system, preventing the ingress of both sea- and fresh- water, emerged in concert with this field system and are visible both on historical maps and in the modern landscape. The agricultural exploitation of the fertile soils in this wetland zone, permitted through the creation of an infrastructure of flood embankments and drainage channels, permitted the Marshland parishes to prosper, achieving a high level of economic prosperity by the 14th century, with an average taxable wealth per square mile across the area of over £30 by 1334.[142] This prosperity, however, was at constant risk – any failure in the dikes or the drainage infrastructure had the potential to flood large areas of arable cultivation or human settlement and cause considerable losses.

3.2.1 The medieval landscape

To investigate the medieval landscape, it is helpful to exclude areas of post-medieval and modern reclamation which in Marshland, north towards the sea, have been extensive.[143] The late medieval field system depicted by the maps of *c.*1582 and *c.*1610 more or less represents the farthest extent of medieval arable agriculture in the area. This is clearly demonstrated by LiDAR data covering the region. Areas of artificially embanked wetland typically experience a decrease in land level, relative to neighbouring unenclosed areas, which can allow relative dating of discrete landscape units.[144] This is due to a combination of factors including increased sedimentation on the seaward side relative to the landward side and a decrease in land level on the landward side as a result of sediment compaction.[145] This picture is visible in Marshland, with the

139 BELOE 1895: 315; SILVESTER 1988: 160.
140 RIPPON 2002a: 64–65.
141 SILVESTER 1988: 164.
142 GLASSCOCK 1975: xxvii.
143 ALLEN 1997: 24.
144 SHENNAN 1992: 79–80; ALLEN 1997: 20.
145 GARDINER 2002: 108–109.

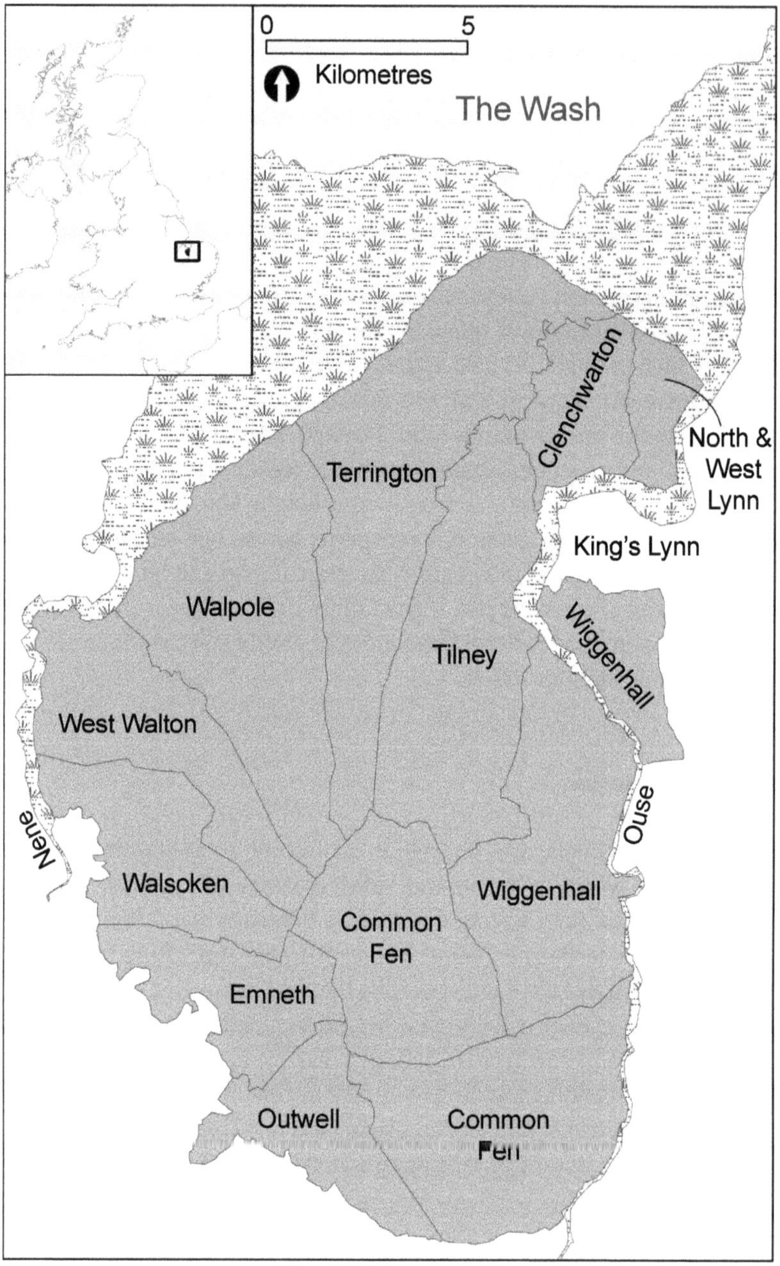

Fig. 3.6: Map of Marshland, Norfolk, with the medieval parishes marked. Parish boundaries after SOUTHALL, BURTON 2004. Created by the author.

area of medieval agriculture – more or less analogous to the zone within the parish boundaries in Figure 3.6 – closely corresponding to an area of topographically lower ground with a sudden rise immediately outside these bounds. Almost all of this area, therefore, excluding only the most northerly areas of the parishes of Terrington and Clenchwarton, were embanked during the medieval period.

3.2.2 Edwardian floods (1287–1349)

Dugdale's compilation of documentary evidence provides a substantial body of evidence relating to the impact of flooding in Marshland during the reigns of the Edwardian Kings between 1287–1349. The first of these incidents occurred during Edward I's 16th regnal year, 1287–1288 and was presumably a result of the storms of 1287/88 discussed in Section 3.1.[146] In this case, breaches occurred at Islington, a hamlet in the modern day parish of Tilney St Lawrence, and 'the Hawe'. The latter toponym does not appear on either the *c.*1582 or *c.*1610 maps. Etymologically, *hawe* describes an enclosed area[147] which could correspond to a parcel of land defended by flood embankments. A clue to its location may come from the mention of land called 'Howe' in the 15th century *Inquisition Post Mortem* of Robert Scales – whose ancestor held 'the Hawe' and Islington in 1287/88.[148] This document lists a "manor called Howe by King's Lynn"[149] which can be linked to the manor of Scales Hooe shown on an 18th century map, located on the banks of the Ouse to the immediate south of King's Lynn and defended against the sea by a, now vanished, dike visible on early aerial photographs.[150] The *hawe* place-name element, and its variations, can also be seen nearby at an area beyond the Clenchwarton sea wall shown on the *c.*1610 map named 'Popeshowe' and, although very faint, this can also be seen in the *c.*1582 map. Both locations containing the *hawe* place-name element, or a variant, appear to have been areas which were highly vulnerable to flooding – exposed to both the full force of the sea and riverine flooding from the Ouse. Indeed, the 'Hawe' belonging to the Scales family was again reported to have been flooded in 1369/70.[151] It is unclear to what extent such areas would have been permanently occupied. The 'Scales Hooe' is now part of King's Lynn, and has

146 Dugdale 1662: 245; Maxwell Lyte 1893: 298.
147 Hanks et al. 2016: 1225.
148 Dugdale 1662: 245.
149 Kirby, Stevenson 2002: 75.
150 Norfolk Historic Environment Record No. 21808. Available at http://www.heritage.norfolk.gov.uk/record-details?MNF21808-Scales-Howe-Bank-medieval-sea-defence-bank [Accessed 01 February 2021].
151 Dawes, Chapman 1938: 400.

seen heavy modern development,[152] while, although not a main focus of the fieldwalking undertaken by SILVESTER, 'Popeshowe' and its adjacent areas have not revealed evidence for anything beyond low level industrial activity such as salt-production.[153]

As can be seen, the descriptions of flood events in the medieval sources often provide locational information which, using historic maps (e.g. those of *c.*1582, *c.*1610, Tithe Maps and 1st Edition Ordnance Survey), can often be linked to a precise or, at least, rough location within the landscape. As in the latter example, this often comes with some difficulties or uncertainties and reconciling certain toponyms mentioned in medieval sources with real-world locations is not always possible. Despite these difficulties, as far as possible, the flood events which occurred in Marshland between 1287–1349 (see Fig. 3.7), after which Dugdale's sources become less detailed, have been examined in order to assess their landscape impact. This produces a database of 'point-data' recording flood damage at particular locations in specific years. When plotted by decade, as in Figure 3.8, some patterning is evident both spatially – with a number of areas registering as weak points which suffered damage repeatedly – as well as temporally – with relatively consistent low levels of damage evident across the period. The decade of the 1330s is the obvious outlier with a high number of reported impacts all along the Sea Wall that protected the Marshland parishes from marine inundation. This is unlikely to be a complete dataset of all the flood events which occurred in the area over this period, with studies both nearby – in eastern England[154] – and further afield – in the Netherlands[155] – attesting to a higher frequency of significant flood events than is suggested by Figure 3.7. It is likely, therefore, that the evidence presented here represents only the inundations that had a severe impact locally or, at least, those flood events which generated a high volume of documentary evidence as a result of disputes between landowners and the escalation of issues to higher authorities. Presumably, more minor flood events were relatively routine and occurred on close to an annual basis.

Prior to the 1330s the floods reported in Marshland appear to have been relatively minor events. Indeed, in the 1290s, the only reports of damage to the flood defences are as a result of misdeeds by members of the community rather than extreme weather events. In 1293, therefore, certain "persons … perforated the dyke called Pokediche"[156] – the main dyke which kept inland freshwaters from flowing through the Marshland parishes, diverting them along drainage channels to the nearby town of Outwell where they flowed out into the Nene. The damage to the 'pokediche', or podike,

---

152 Although a desk based assessment of the area in advance of a proposed Paper Mill development in 2007 noted the potential for medieval remains related to the Scales Hawe manor, the proposed development was judged unlikely to impinge on these deposits. See MEAGER 2007.

153 SILVESTER 1988: 18–22; 26–27.

154 BAILEY 1991: 190–191.

155 GOTTSCHALK 1971: 215–465.

156 DUGDALE 1662: 246; MAXWELL LYTE 1895b: 24.

Fig. 3.7: Flooding in medieval Marshland, Norfolk, between 1287–1349 as revealed by the sources amassed by DUGDALE 1662: 245–260. Compiled by the author.

in this case resulted in the flooding of "land and pasture"[157] but does not seem to have caused significant damage. More disastrous flood events seem to have begun to occur from the early 1330s. On 7th January 1330,[158] for example, a bank on the west bank of the Ouse in Wiggenhall was breached, reportedly causing the inundation of 1000 acres of corn. Although this figure is unlikely to be reliable, it is not an entirely fanciful estimate if the breach allowed the river to flow into a large area of low-lying farmland. Although the stretch along which this bank must have run is easily identifiable, the precise location of the breach is unknown – the only toponym provided "Burtys bithe", does not appear on historic or modern maps. The vulnerability of this area, however, is demonstrated by the fact that it was flooded once more on 29th November 1334 and again in January 1338.[159] The 1334 inundation also breached the sea bank at Ristoft field in the parish of Walpole leading to the flooding of 200 acres[160] while in the same year, further flooding was caused by a drainage problem in Walsoken which had emerged the year before, but had evidently not been rectified.[161]

The flood with the highest volume of historical data to have affected the region occurred in January 1338. A number of sources refer to a similar flood in 1337,[162] while Dugdale's

157 MAXWELL LYTE 1895b.

158 The morrow after the Epiphany, in the third year of the then King [Edward III]. DUGDALE 1662: 257.

159 DUGDALE 1662: 257.

160 DUGDALE 1662: 255.

161 DUGDALE 1662: 256.

162 E.g. TNA: SC 8/152/7565, TNA: SC 8/170/8456 and TNA: SC 8/78/3856; PEVSNER 1962: 745; WATT, COLSTON 2000: 738.

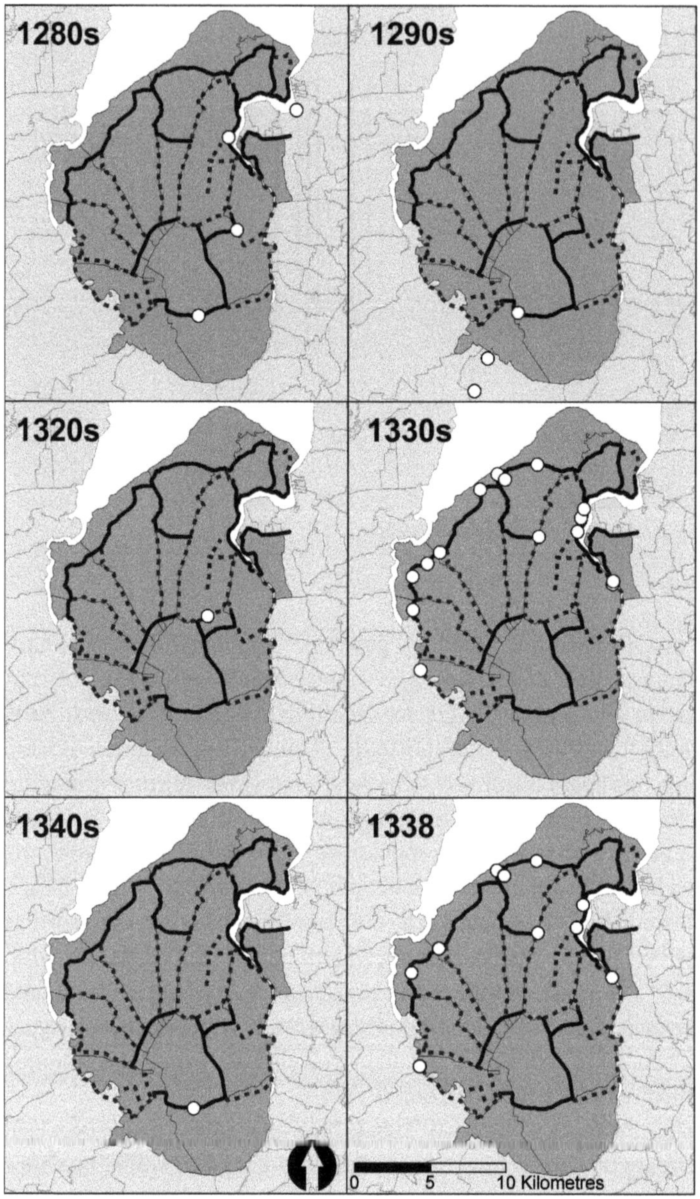

Fig. 3.8: The impact of flooding in Marshland by decade between 1287–1349. As the locations referred to in the historical evidence are often vague, the precise locations shown in the maps have been assigned with a degree of uncertainty although always with supporting evidence. Flood banks marked by a solid line are those marked on Silvester 1988.'s medieval period parish maps while those indicated by a dotted line are dikes shown on Hayward's *c.*1610 map. A 'swarm' of flood damage during the 1330s relative to the other decades is visible and, notably, the majority seems to have occurred in a single flood event in January 1338. Created by the author.

sources state that this event struck the Fens (including the coasts of Lincolnshire, Cambridgeshire and Norfolk) in the 11th year of the reign of Edward III[163] (1337–8). The majority of Edward III's 11th regnal year, therefore, corresponds to 1337 but as the sources which give a calendrical date agree the floods occurred on 12th and 13th of January, these would have in fact taken place in 1338 by modern reckoning. Supporting evidence for this event can be found in the rolls of Edward III; the Close Rolls record the inundation "by sea storms" of Tilney, Walpole, Walsoken, West Walton and Emneth on "Monday before Hilary" [12th January 1338] while Wiggenhall and Terrington were flooded the following day [13th January 1338].[164] Dugdale's sources do not specifically mention the date given above but a passage does refer to a flood on "the Eve of Hijlarie then last past" which, if 1338 is the year referenced, would correspond to 12th January.[165] The Fine Rolls covering the period also reference the fact that the residents of the seven towns listed above were affected by flooding in 1338.[166] Several secondary sources report supporting evidence for a flood in the area at this time,[167] particularly mentioning severe damage to the church of Walpole St Peter[168] – an occurrence which is also referenced by DUGDALE.[169] The structural evidence of the standing church does fit the narrative that the original church was damaged or destroyed in the early-mid 14th century. The tower is the only surviving part of the original structure with a later nave constructed in the transitional perpendicular style by about 1360 and a later perpendicular chancel added in around 1425 (see Fig. 3.9). Although the material evidence is scarce, it appears that a severe flood occurred in January 1338. While it is possible that a similar event occurred during the previous year, it is more likely that a 1337 flood, referenced in subsequent literature, has been confused with the 1338 event due to calendrical differences.

The documentary sources which reference the impact of the 1338 flood particularly stress the fact that the sown corn, as well as many goods and chattels, had been destroyed or lost.[170] More detailed evidence for the 1338 inundation comes chiefly from the sources brought together by DUGDALE.[171] These provide acreages of land lost to

163 The 11th regnal year of Edward III lasted from 25th January 1337–24th January 1338.
164 MAXWELL LYTE 1900a: 293.
165 DUGDALE 1662: 257.
166 MAXWELL LYTE 1915: 62.
167 PEVSNER 1962: 745; WATT, COLSTON 2000: 738. No references are given for this information in either of these works.
168 This claim is repeated in the church's guidebook with the additional assertion that this was in some way the result of "a tidal wave from a volcanic eruption off Iceland". The evidence on which this claim is based is unclear; no major Icelandic eruption is known from 1337/8 and a hypothetical 'tidal wave' generated off Iceland would need to be of an extremely high-magnitude to cause inundation in The Wash. ANON n. d.: 3.
169 DUGDALE 1662: 255.
170 MAXWELL LYTE 1900a: 293.
171 DUGDALE 1662.

Fig. 3.9: The west front of the parish church of Walpole St Peter, Norfolk. Clear distinctions between architectural phases are visible with the original tower, constructed in *c.*1300, the later nave dating to the 1360s and the later south porch (*c.*1425) on the right-hand side. Photograph by the author.

flooding for each parish,[172] and the financial costs of repairs, as well as qualitative information such as the nature of flooding at particular locations. While the acreages lost to flooding must be regarded with care, they provide a rough picture of the relative impact of the floods in the different parishes (see Fig. 3.10 and Table 3.3). In a number of instances, this qualitative information allows specific affected areas to be identified. We are told, for example, that 'Rushemershe' and 'Newemershe', belonging to the parish of Terrington, were inundated.[173] While the location of the latter is unknown, the location of the former can be inferred from the location of 'Rushgate' shown on the 1610 map of the area set into the sea walls running along the northern edge of the parish. These two areas, therefore, were probably areas of salt marsh beyond the sea walls and as such are unlikely to have contained much human settlement beyond functional and productive structures such as small outbuildings and, perhaps, structures related to salt-production.

172 As these documents relate to a petition for the reassessment of taxes, more attention may have been paid to the accuracy of the figures than most contemporary chroniclers recording similar circumstances but these figures cannot be taken at face value as medieval information of this type is renowned for its inaccuracy. For an example see ZIEGLER 1969: 51–53.
173 DUGDALE 1662: 259.

Tab. 3.3: Repairs and maintenance costs to the flood defences in Marshland following the 1338 flood. Created by the author after data derived from DUGDALE 1662: 258–259.

Parish	Affected Acreage	Costs	Details
Walpole	140	£162	Repairs to sea bank, 'Pokediche' and other works
Terrington	1200	£102 18s 24d	Lost income from inundated lands and repairs to the Sea-bank and the 'Bank of Pokedich'
Tilney	10	£236 13s 4d	Damage to the town by flooding, Repairs to 'Pokediche' and 'Blakediche' other works and maintenance
Wiggenhall	100	£125	Repairs to sea banks and the 'pokediche'
Emneth	100	£30	Repairs to the 'pokediche'
Walsoken	200	£50	Repairs to the sea bank, the 'pokedike' and a sewer 'containing five miles in length'
West Walton	420	£90	Repairs to sea-banks, the 'Pokediche' and other 'gutters and sewers'

Evidently, the existing flood defences were insufficient to protect the Marshland parishes in January 1338. Non-specific breaches in the sea banks which embanked the area are frequently reported in the documentary account.[174] When sea banks were insufficient to withstand marine or fluvial forces, the resulting breaches could range from relatively minor holes, which typically occurred at points which were particularly vulnerable, or had been weakened for some reason to, in the most extreme conditions, the removal of long stretches of embankment which could be over 100m in length.[175]

Small scale breaches were usually repaired with an encircling repair, surrounding the scour hole created by the water pouring through the hole in the embankment[176] while the loss of longer sections was more likely to be repaired by a more significant realignment of the sea wall. This often forced the construction of a new defence connected to the limits of the breach. Often these were removed some distance inland from the original location in order to provide increased protection.[177] Over a long period of time, therefore, intermittent breaches, and the repairs and adjustments to the sea walls these forced, generated distinctive patterns in the landscape which often remain recognisable today. Archaeologically, breaches of medieval date can be found in the parish of Terrington St Clement in the sea bank running toward Clenchwarton[178] where two 'horse shoe' shaped features indicate encircling repairs constructed around failures in the sea bank. Although it is difficult to precisely date these features, the fact

174 DUGDALE 1662: 254–259.

175 GARDINER, HARTWELL 2006: 138.

176 This is discussed in greater detail in Chapter 4.5.

177 GARDINER, HARTWELL 2006: 139.

178 SILVESTER 1988: 41.

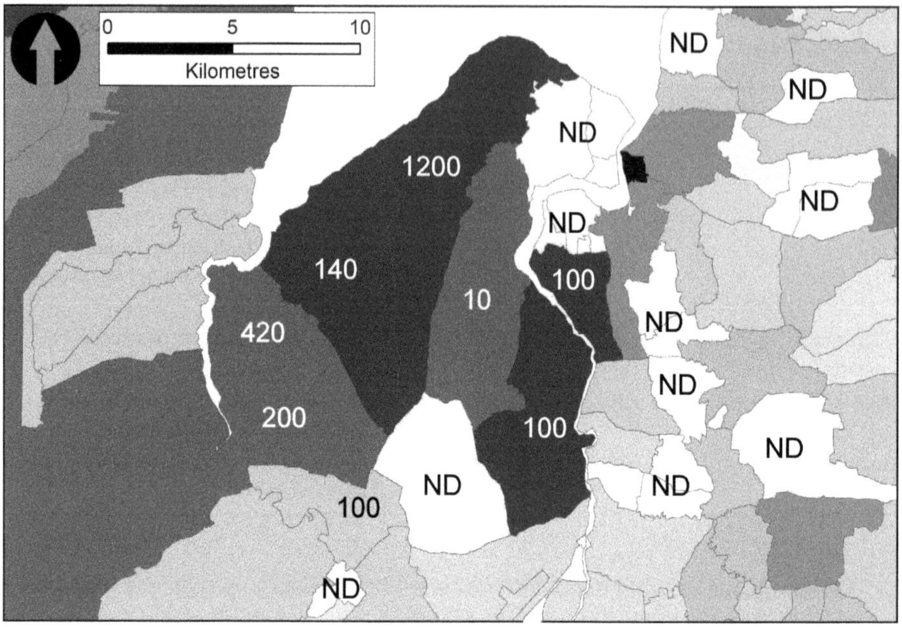

Fig. 3.10: Acreages affected by flooding in the parishes of Marshland overlain against taxable wealth in 1334 (darker=more, lighter=less) from the Lay Subsidy of 1334 ('ND' denotes parishes which are unmentioned). Compiled by the author, data from DUGDALE 1662; GLASSCOCK 1975: 197.

that they appear on the *c.*1582 map of Marshland provides a *terminus ante quem* for their breaching and subsequent repair. SILVESTER ascribes them to the 13th or 14th centuries[179] and although it is impossible to confirm with certainty when these breaches occurred, their location is consistent with the repair pattern described after the 1338 inundations.[180] A further possible medieval breach may be indicated by the toponym 'the breach' shown on the *c.*1610 map in the sea bank between Terrington St Clement and Walpole Cross Keys. In addition, the re-alignment of the sea bank in the parish of Walpole St Andrew, evidenced by the continuity of field boundaries underneath its current path, indicates an adjustment necessitated by the loss of a significant stretch of sea wall to flooding – although when exactly this may have occurred is unknown.[181]

Some effects of the 1338 floods are, however, more difficult to identify. Reference is made to a sluice known as 'Scales Gole', which was presumably located on land belong-

179 SILVESTER 1988: 41.

180 DUGDALE 1662: 259.

181 SILVESTER 1988: 41; RIPPON 2001: 20–22.

ing to the already mentioned Scales family,[182] throughout the material amalgamated by Dugdale. The various references to the sluice place it in the parish of Tilney but the fact that it does not appear on either of the early maps of Marshland or on the first edition OS map may be explained by the fact that, in the 1338 floods, Scales Gole was "destroyed by the Sea-tides; by reason whereof, a great part of the said Town was drowned and made yearly much worse".[183] Its rough location, however, can be inferred through further information such as the fact that, in the same flood, land in Terrington Parish was flooded as a result of the loss of Scales Gole, which, in turn, had caused damage to a sluice "called Oxhowe" which channelled freshwater downstream to Scales Gole where it presumably drained into the Ouse.[184] 'Oxhowe' may relate to 'Oxhow Clough', which is marked on the first edition OS map at the boundary between the modern parishes of Terrington St Clement, Terrington St John and Tilney St Lawrence where several drainage ditches run together (see Fig. 3.11). Today the waters which meet here flow to the south before being diverted towards the Ouse. As the Scales family held land in the parish of Tilney it is most likely that Scales Gole was located in this parish though either it was destroyed or fell out of use during the medieval period or it was, at some point, renamed.

As seen above, many of the locations in Marshland which experienced damage during the floods of 1338 can be located and explored through a variety of sources. Together with the historical descriptions of the inundation, the spatial data marks out the 1338 flood as a particularly damaging event at the local level of the Marshland parishes. In light of the record of flood events in Marshland from the 1280s, however, the community was clearly experienced in dealing with these disasters and this no doubt affected their responses to the conditions brought on by these floods. The social organisation and structural responses which allowed them to cope with these events are further explored in Chapter 4.

3.3 The St Maur's Day windstorm of 1362

On 15th January 1362 an extreme windstorm tracked across England, from the south west through central southern England to East Anglia, crossing the North Sea to arrive at the coasts of Denmark and northern Germany the following morning. Chroniclers' descriptions of the damage caused by the storm in England are evocative of the fear and alarm the event must have sparked. *The Brut Chronicle*, for example, describes how the storm "blew down to [the] ground high houses and other strong things and

182 Kirby, Stevenson 2002: 75. Reference is particularly made to Robert de Scales who owned land in Islington, Tilney Parish, and the Hawe, near King's Lynn. For example see Dugdale 1662: 245, 251, 253, 255.

183 Dugdale 1662: 259.

184 Dugdale 1662: 259.

Fig. 3.11: Oxhow Clough in the present day. Today this confluence of ditches drains water from the surrounding fields south to the fen. This has probably changed since the medieval period as, in 1338, it seems to have been connected to a sluice in the parish of Tilney. Photograph by the author.

all other strong works … were so shake[n] therewith that they … shall be evermore the feebler and weaker".[185] The historical sources, discussed in more detail below, reveal that the 1362 windstorm resulted in widespread damage across the area of effect, uprooting trees and bringing buildings crashing down. In these respects, and given the geographical coverage of the storm, it is somewhat comparable to the more recent 15th–16th October 1987 storm, known as the 'Great Storm' of 1987, which caused similar levels of destruction across a broadly comparable area. The peak wind-speeds recorded in 1987 were 196 km/h. While such speeds are not exceptional for the British Isles as a whole, south west of an imaginary line between Norwich and Southampton, wind speeds of this magnitude have an estimated recurrence interval of 200 years.[186] Rowe has identified the storms of 1362, 1662, 1703 and 1987 as the events which best fit this pattern,[187] although each event followed a slightly different track and, although instrumental readings are lacking for the three pre-modern storms, with differing wind speeds – resulting in variations in the levels of damage and areas affected across the British Isles.

185 Brie 1906: 315. Spellings modernised by the author.
186 Burt, Mansfield 1988: 101–103.
187 Rowe 1988.

3.3.1 Historical background

The 1362 storm is primarily known through the descriptions of chroniclers. There is, however, some difficulty in assessing to what extent their descriptions are contemporaneous. While some may have been composed soon after the event, others could have been recollected – or copied from other sources – decades later. Ranulph Higden's *Polychronicon* provides one such example. The core of this historical work comes to an end in the year 1348 but an anonymous continuation extends the text to 1362, including a succinct record of the storm.[188] This continuation was most likely composed during the late 14th or early 15th centuries but as the authorship is unknown it is difficult to gauge its reliability.

Although not always a good indicator of reliability, the vast majority of sources agree on the date of the storm's occurrence, its timing and direction. Most state it struck on the evening of St Maur's Day (15th January) 1361,[189] a date which is also found in a contemporary legal document which references the storm.[190] This date corresponds to 15th January 1362 by modern reckoning because, as discussed in relation to some of the other disasters considered in this chapter, during the medieval period the new year was commonly counted from Lady Day (25th March) rather than 1st January.[191] Outliers include *Knighton's Chronicle* which dates the storm to St Anthony's Day (17th January),[192] probably a simple lapse of memory, and the Irish *Annals of the Four Masters* which gives the year as 1363.[193] Errors of misdating, usually to one year before or after, are very common with this category of evidence[194] and whilst the Irish source conceivably documents a different storm in the following year, further evidence demonstrating that the St Maur's Day storm certainly affected Ireland[195] suggests this is not the case.

The timing of the event can be narrowed with some precision as a number of sources indicate the storm struck at evensong or vespers. These evening prayers were usually conducted at around 6pm giving an approximate time for the arrival of the storm.[196] The majority of continental sources, on the other hand, document the storm occurring on St Marcellus Day (16 January).[197] After passing over England on the evening of 15th January, therefore, the storm must have arrived at the North Sea coasts of the Low Countries, Germany and Denmark the following morning. This accords well with a chronology based on the speeds and progression of comparable modern storms such as the Great

188 Lumby 1882: 360.
189 Venables 1891: 40–41; Brie 1906: 315; Tait 1914: 150.
190 TNA: JUST 2/18/58.
191 Pfister 1996: 96.
192 Martin 1995: 185.
193 O'Donovan 1856: 625.
194 Pfister 1996: 96.
195 Gilbert 1884: 396.
196 Hingeston 1858: 221; Haydon 1863: 229; Schmidt 2011: 176–177.
197 Weikinn 1958: 232–235; Gottschalk 1971: 371–376.

Storm of 15th–16th October 1987, or the St Jude's Storm of 28th October 2013. Had these storms arrived over England at 6pm, they would have reached the German and Danish coasts around 5am the following morning[198] (see Fig. 3.12). Although a number of chroniclers state that the storm continued for seven days after St Maur's Day,[199] this is almost certainly an exaggeration – although the weather may have remained inclement during this period. All of the chroniclers who discuss the direction also agree that the storm came from the south or south-west.[200] The evidence therefore supports the identification of the St Maur's Day event as a high-magnitude extratropical windstorm which tracked from west to east, from south-west England across to the east coast, on the night of 15th January 1362 before proceeding across to the North Sea coasts of continental Europe early the following morning.

The documentary evidence widely attests to the storm's impact in England. Most of the chronicles provide qualitative statements describing, for example, damage to structures and felled trees. A typical example is the *Chronicon Angliae Petriburgense* which describes *domos et molendina innumera prostravit, arbores et integras silvas in multis locis a fundamentis evulsit.*[201] In some cases, local details such as damage to specific prominent buildings are included such as at the Dominican Friary in Dublin,[202] the Augustinian Friary in London, the bell towers of Bury St Edmunds, Suffolk, and Norwich, Norfolk,[203] as well as the gatehouse of the Benedictine Abbey of St Albans, Hertfordshire.[204] Some chroniclers provide anecdotal accounts of local occurrences. For example, at St Augustine's Abbey in Canterbury, Kent, a chaplain was killed after seeking shelter from the storm when a roof beam of the chapel of St Pancras was blown down into the nave.[205] Similarly, in London, an Augustinian friar was reportedly blown through a window by a particularly strong gust.[206] Such incidents, together with the recorded structural damage, would correspond to a storm of force 11–12 on the Beaufort Scale. This, together with the use of terminology describing the storm as "never … seen or heard before … in England"[207] or "as it was thought had never been seen in

198 This conclusion has been drawn through comparison with data from www.europeanwindstorms.org [Accessed 05 May 2016], Copyright Met Office, University of Reading and University of Exeter. Licensed under Creative Commons CC BY 4.0 licence: http://creativecommons.org/licenses/by/4.0/deed.en_GB.

199 Brie 1906: 315; Tait 1914: 151.

200 Hog 1846: 196; Venables 1891: 41; Tait 1914: 150.

201 Giles 1845: 172.

202 Gilbert 1884: 396.

203 Galbraith 1927: 50.

204 Riley 1869: 387.

205 Davis 1934: 564.

206 Scott-Stokes, Given-Wilson 2008: 119.

207 Scott-Stokes, Given-Wilson 2008: 119.

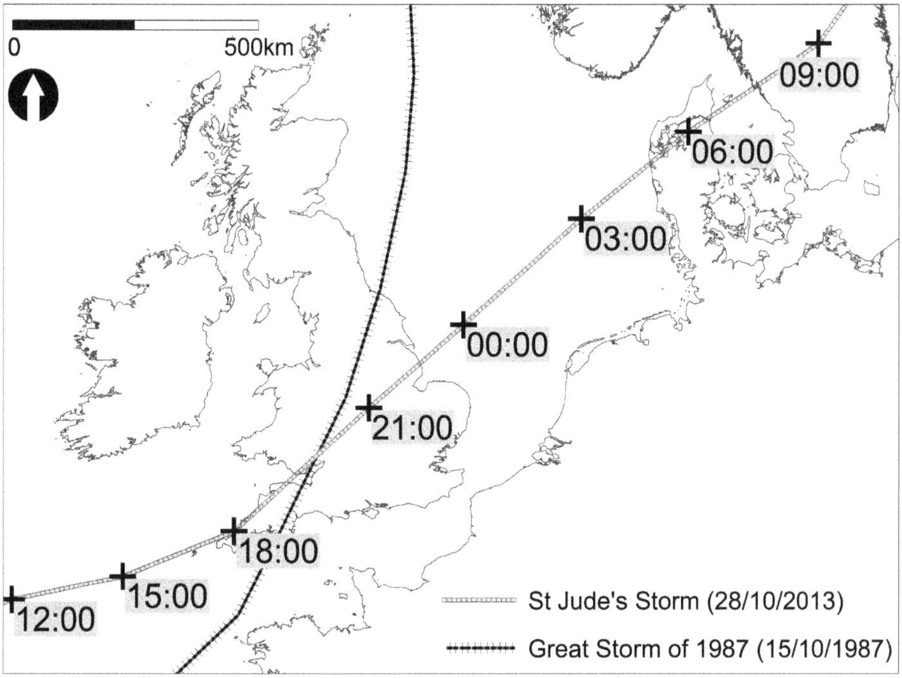

Fig. 3.12: Reconstruction of the timing of the storm of 15th January 1362. Shown are the tracks of two comparable modern storms, the Great Storm of 15th October 1987 and the St Jude's Storm of 28th October 2013. The timings shown indicate the location of the eye of the 1362 storm based on the progression of these modern storms but based on an arrival over England at 6pm as reported by a number of historical sources. Created by the author. Modern storm data from www.europeanwindstorms.org [Accessed 05 May 2016], Copyright Met Office, University of Reading and University of Exeter. Creative Commons CC BY 4.0 licence: http://creativecommons.org/licenses/by/4.0/deed.en_GB.

earlier times",[208] leaves no doubt that this event was beyond what was considered 'normal' during a winter storm season.

Beyond written chronicles, a number of other types of document provide evidence for the damage wrought by the storm. Many of these sources, however, only date the storm to the year 1362 so it must be assumed that they deal with the 15th January event rather than a separate storm. The coroner's rolls, for example, provide further evidence of fatalities, describing how two parishioners were killed inside the church at Longstanton, Cambridgeshire, when a tree was blown against the building, causing masonry to fall down upon them.[209] Manorial accounts such as those from Thaxted,

208 MARTIN 1995: 185.
209 TNA: JUST 2/18/58.

Essex, where two windmills and a grange were heavily damaged,[210] highlight the damage faced in affected rural areas. A particularly useful source is the *Register of the Black Prince* which covers the administration of the estates held by the heir to the throne, Prince Edward.[211] These provide details of felled trees in the Prince's parks as well as damage to housing, mills, manors and infrastructure across his estates.[212]

The widespread damage at royal parks, which contained swathes of woodland, documented in the Black Prince's *Register* are symptomatic of the high numbers of trees blown down across England. For example, in addition to damage to structures and park paling (fencing) *per tempestatem venti*[213] a high number of trees were felled by the wind at the royal palace and deer park at Clarendon, Wiltshire.[214] Interestingly, Clarendon was again a victim to storm damage in October 1987, with extensive damage caused across the site.[215] For comparison, the 'Great Storm' of 1987 toppled around 15 million trees in England – although as the 1987 storm struck in mid October, when the leaves were still largely on the trees, the proportion blown down in 1987 may have been higher relative to January 1362.[216] One characteristic of the damage in 1987, which is likely to have also occurred in 1362, was the almost total felling of 'swathes' of woodland, clusters of trees ranging from *c.*10 trees covering 0.025 ha to hundreds spread over areas up to 5 ha.[217] A wide range of documentary sources attest to the felling of trees at other locations, illustrative of the huge number of trees affected across a wide area. For example, felled trees are reported at Eaton, Norfolk[218] while at Alveston, Warwickshire, one aubel[219] as well as four elms were toppled by the wind. The fact that one of these elms was valued at 13s 4d (compared to the 2s value assigned to one of the others) indicates that this must have been a very large, old tree and perhaps a local landmark.[220] The records of the repairs at Wallingford Castle in the storm's aftermath provide further evidence for the toppling of woodland resources by the wind, with 1,700 beech trees felled at nearby Watlington, Oxfordshire. Some of this timber appears to have been put to use in the repairs of the castle while the remainder was sold for £28 6s 8d.[221]

Documentary evidence, in a number of different forms, records damage affecting many structures across a wide area which can be linked to the St Maur's Day windstorm.

210 Newton 1960: 71, 75.
211 b. 1330 d. 1376.
212 Dawes 1931: 188–189; Dawes 1933: 416, 420, 426, 429, 431.
213 TNA: E 101/460/2.
214 Maxwell Lyte 1912: 117, 183.
215 Beaumont James, Gerrard 2007: 177.
216 Quine 1988: 115.
217 Allen 1992: 340.
218 Pribyl 2017: 241.
219 A term for a white poplar (*Populus alba*).
220 WCL: E20. I am grateful to Chris Dyer for this information.
221 Keats-Rohan 2015: 94.

At high status structures such as Wallingford Castle, Oxfordshire, the damage and repairs are described in considerable detail by the surviving documentary evidence. Within the castle, the *Register of the Black Prince* specifically mentions the storm causing damage to the roofing of the chapel of St Nicholas, as well as housing belonging to the chaplain and clerks.[222] A campaign of repairs followed between 1363 and 1367 although some of these addressed issues which pre-dated the storm[223] and may have also sought to make the castle more habitable as it was to become the main residence of the Black Prince's new wife, Joan of Kent.[224] It is therefore likely that, although the storm provided a catalyst for repairs, high winds were not responsible for all of the works undertaken over these years. Some of the damage most likely to have been caused by the storm itself include the ironwork and glass in a window of the Tower chapel[225] while the re-roofing by, one John Tyler, of the wardrobe, Queen's chamber, another chamber and a hall[226] likely attests to significant damage to the castle's roofing by the storm. Low status structures, on the other hand, attracted considerably less detail in the documentary record – where they were recorded at all. A typical example comes from the records of Worcester Cathedral Priory which records that the house of one Nicholas Robines "was prostrated by the wind".[227] There is, therefore, a significant bias in the historical record towards high status structures and the affairs of the elite more generally.

In a number of cases, although evidence exists documenting repairs in 1362 which could relate to storm damage, this connection cannot be made with certainty. This can be seen when, in mid-February 1362, the Black Prince ordered his steward in Cornwall, John Dabernoun, to tour his Cornish estates to assess the damage.[228] As no specific locations are named, however, no locations can be linked to the storm with certainty. Later in the year though, repairs are mentioned at a number of sites, possibly indicating affected locations. On 20th September 1362, for example, the Prince ordered that the carpenters working at Liskeard, Cornwall, were to be paid 60 shillings as a gift[229] which may indicate ongoing repairs were taking place. Another uncertain but possible location where damage may have occurred is indicated by an entry from 6th October 1362 when the Prince ordered repairs to "the Chapel of the Trinity in the park of Lostwithiel".[230]

222 Dawes 1933: 426.
223 Keats-Rohan 2015: 84.
224 Keats-Rohan 2015: 64.
225 Keats-Rohan 2015: 94.
226 Keats-Rohan 2015: 86.
227 WCL: E20. I am grateful to Chris Dyer for this information.
228 Dawes 1931: 189.
229 Dawes 1931: 193–194.
230 Dawes 1931: 194.

A number of conspicuous repairs to bell towers in the region affected by the storm shortly after its occurrence provide some further likely if still unconfirmed cases of storm damage. One such case is the donation of £100 towards the construction of a new bell tower at the Augustinian friary of Clare, Suffolk, in 1363[231] while, similarly, a bequest in 1364 to fund the repair of the belfry at St Nicholas, Hinxworth, Hertfordshire, provides another probable candidate.[232] Less certain are the repairs ordered to the Tower of London on 10th June 1362,[233] specifically to one of the towers in which the chancery rolls were kept, perhaps indicating that damage occurred here in the storm, although it is impossible to be certain. Other reports of damage and repair in southern England in that year can probably be discounted or, at least, qualified. For example, on 12th February 1362 Edward III commanded 200 masons to be sent to work at Windsor Castle,[234] but there was an ongoing building programme and the shortage of labour resulting from the 1361 plague outbreak, rather than storm damage, may have been the primary driver for this command.[235] Similarly, although extensive damage was reported at St Briavel's Castle, Gloucestershire, on 20th May 1362,[236] the damage was described as occurring "when Robert de Sapy was sensechal" which would have been decades before the storm, either 1322–1325 or 1330–1335.[237]

Although in many cases the historical record recording the storm of 1362 is highly detailed it also has many caveats. The historical record is richest for high status and ecclesiastical buildings and estates which were better documented and have survived in greater numbers into the present. Certain spheres that such a high magnitude storm might reasonably have been expected to affect are conspicuously absent from the record. The storm's effect on shipping for example is attested by very few known sources. The only vessel known to have been affected, the *Tarrit*, was forced to shore at Plymouth by the storm and subsequently became the subject of a legal dispute after being plundered by the locals.[238] The vague statement from the Irish *Annals of the Four Masters* that the storm "also sank many ships and boats"[239] demonstrates that although no further specific cases are known in detail, the case of the *Tarrit* cannot have been an isolated incident. Certainly, when another great storm occurred in 1703, high numbers of both naval and civilian shipping were lost at sea.[240]

231 Harper-Bill 1991. 85.
232 Pevsner 1977: 196.
233 Maxwell Lyte 1909a: 336.
234 Maxwell Lyte 1912: 164.
235 Tighe, Davis 1858: 173–174.
236 Stamp 1937: 185–186.
237 Baggs, Jurica 1996: 414. Although note that a severely damaged castle would have been particularly vulnerable to what appears to have been the period's highest magnitude storm.
238 TNA: SC 8/247/12320. Hardy 1869: 420.
239 O'Donovan 1856: 625.
240 Defoe 1704: 165–224.

The evidence presented above goes some way to reconstructing the occurrence of the 1362 windstorm. It is briefly necessary, however, to clarify some points suggested by a number of secondary sources in relation to the windstorm. Firstly, it has been suggested that the storm was the result of, or at least exacerbated by, the Icelandic volcano Öraefajökull which erupted in 1362.[241] This can be swiftly debunked, however, as the windstorm struck England on 15th January while the sources which record the eruption suggest that Öraefajökull lay dormant until early June[242] – long after the storm had come and gone. Furthermore, in a number of cases the windstorm of 1362 has been associated with flooding around the shores of the British Isles yet there is no concrete evidence to support such an assertion. The only chronicle which mentions flooding in association with the windstorm is the continuation of Ranulph Higden's *Polychronicon* which states: *inundationes aquarum protinus sequebantur*.[243] This description, however, is very general and fails to specify any affected locations. Nevertheless, certain sites have traditionally been connected to flooding caused by the storm of 1362. For example, the loss of the port of Ravenser Odd on Spurn Point, East Yorkshire, has been attributed by some to the 1362 windstorm.[244] The available documentary evidence, however, suggests that the town was lost prior to the occurrence of this storm.[245] Severe flooding, presumably as a result of water driven across the North Sea overnight on the 15th/16th January 1362, is, however, attested from the German, Frisian and Danish coasts, in both the documentary record as well as the archaeological record.[246] As well as traces of former settlement identified through remote sensing,[247] a discrete high-energy layer has been detected through geo-archaeological investigations around Hallig Südfall, North Frisia, Germany. This layer is dated through a variety of means – it is association with a mollusc which yielded a calibrated radiocarbon age of AD 1312–1445 as well as a piece of wood dated to AD 1228–1269 thought to relate to the enclosure of coastal marshland in the preceding century – which provide a secure *terminus post quem* for the deposition of the storm layer consistent with the historically attested flooding which the 1362 windstorm caused in that area.[248]

241 Beaumont James, Robinson 1988: 39.

242 Thorarinsson 1958: 29.

243 Lumby 1882: 360. Inundations of water immediately followed. Translation by the author.

244 See for example Murphy 2009: 35.

245 Maxwell Lyte 1909a: 453; de Boer 1964: 83.

246 Hybel, Poulsen 2007: 47–48; Mauelshagen 2012: 63–65.

247 Gade et al. 2017. Note, however, that these remains cannot be accurately dated without ground-truthing and, as another severe storm surge affected the region in the 17th century, there is significant potential for misidentifying later remains as those of medieval settlement.

248 Hadler et al. 2018: 51.

3.3.2 The material evidence

As the storm layer detected on the German coasts attests, the storm did leave a physical signature and it is also possible to detect traces across the area of effect in the British Isles. Structural evidence is perhaps the most frequently documented impact of the storm and as such, it is a reasonable assertion that standing buildings dating from the time of the storm may preserve evidence for its occurrence. There are many of these cases for which no obvious supporting material evidence can be found. At Rochester Castle, Kent, for example, although extensive repairs motivated by storm damage took place between 1367 and 1370, the castle had been in a state of disrepair and neglect since a siege in 1264.[249] Thus, although documentary evidence does describe the damage suffered in the storm[250] and the 14th century repairs do seem to have particularly focussed on the east curtain wall and the two towers in this area,[251] disentangling in detail the damage which occurred in 1362 – as opposed to existing issues of disrepair and neglect – proves impossible. Similarly, at Portchester Castle, Hampshire, although extensive works were instigated on 20th January 1362, a major part of which were roofing repairs employing plumbers and tilers, this cannot be correlated with the surviving archaeological evidence.[252]

In a number of cases where damage is documented at specific structures, however, it is possible to trace this through the archaeological record, mainly through standing building evidence. A good example is Norwich Cathedral, where the Romanesque arches on the ground and first floors are superseded by a later Gothic clerestory.[253] Although later remodelling has taken place, this stylistic disjuncture (see Fig. 3.13) must be a direct result of the storm which, in blowing from the east, caused the spire to fall into the presbytery, destroying the roof and upper stories in this area – a detail corroborated by the written evidence which records severe damage to the presbytery.[254] Similarly, the gatehouse at St Albans, Hertfordshire, which was built in the aftermath of the storm on the site of an earlier gatehouse and almonry which had been heavily damaged,[255] contains structural elements which predate the current structure. A plausible scenario, therefore, is that these fragments belonged to earlier structures destroyed or damaged by the storm, which were later re-used in the construction of the new gatehouse.[256]

---

249 Allen Brown 1969: 18–19.
250 Stamp 1937: 282.
251 Allen Brown 1969: 29.
252 Cunliffe, Munby 1985: 145, 302.
253 Woodman 1996: 179, 192.
254 Wharton 1691: 415.
255 Niblett, Thompson 2005: 254.
256 RCHME 1982: 31. This is discussed in greater detail in chapter 4.4.2.

Fig. 3.13: View of Norwich Cathedral's clerestory showing the stylistic disjuncture between the lower Romanesque arches supplanted by later Gothic arches above – a tangible result of the storm of 1362 and the repairs required in its aftermath. Photograph by the author.

Fig. 3.14: Exterior view of Ashwell St Mary, Ashwell, Hertfordshire. The evidence suggests that masons were working on the building when the storm occurred or shortly after. One possibility, therefore, is that the unfinished building was damaged by the storm. Photograph by the author.

One building which goes unmentioned by the documentary record in connection with the storm can be linked through a combination of alternative strands of evidence. St Mary's Church, Ashwell (see Fig. 3.14), Hertfordshire, contains a graffito at the base of the tower (see Fig. 6.3) which mentions the St Maur's Day storm.[257] The graffito is *c.*2m above ground level and would have been challenging to carve from the current floor level. This observation, together with other nearby graffiti which seems to record workers' wages,[258] suggests that soon after the storm, a number of masons were at work in the tower with scaffolding in place. The chancel was completed in 1368 whilst work on the tower continued until 1381[259] with structural timbers from the tower tree-

257 This is discussed in greater detail in Chapter 6.2.

258 Champion 2015a: 209.

259 Pevsner 1977: 74–75.

ring dated to 1365–1376.[260] It is therefore possible, although unconfirmed, that the unfinished building was damaged in the St Maur's Day storm, necessitating repairs which would have been made over the following decade as the building was completed. In any case, one of the masons working on the structure chose to commemorate what must have appeared a troubling and noteworthy event through the graffito inscription.

Dendrochronological dating of timbers from contemporary structures may indicate a number of additional cases of storm damage. The tower of the church of St Peter and St Mary, Stowmarket, Suffolk, for example is constructed around an internal timber framework felled in one phase during 1362/3.[261] That this is likely to have been a repair rather than a fresh construction is indicated by a will from 1453 which describes it as "the new tower".[262] Similarly, at St Patrick's Cathedral, Dublin, only 750m from the documented storm damage at the Dominican Friary,[263] a timber has been dated to winter 1361/1362, the season of the storm.[264] In this case the damage itself has been interpreted as the result of a fire in the 1350s but the timber could have come from a tree felled by the storm. Another possible candidate is a low-status house from Long Wittenham, Oxfordshire, the timbers of which date to *c.*1363.[265] While this close dating alone is inconclusive, documentary evidence does record damage to the local parish church of St Mary in the St Maur's Day storm[266] strengthening the possible identification of the house as a structure which either required repair or reconstruction as a result of storm damage.

Through the collation of both the documentary and material evidence, a picture of the damage across the area of effect can be reconstructed (see Fig. 3.15). Although the available evidence does not permit many sites to be definitively attributed to the St Maur's Day storm, the evidence certainly attests to extensive damage throughout eastern England. According to a kernel density plot of these data (Fig. 3.15: 3), London and its surroundings register as the epicentre of the damage – and, indeed, urban areas must have focussed the damage with their higher densities of population and structures. It must be remembered, however, that only the density of known, documented, damage appears in these maps – rather than the density of the total damage which occurred – and damage was certainly more likely to be recorded in areas of high population. This may explain why there is no known data from south Wales and little evidence from Cornwall and Devon – even though the storm almost certainly affected these areas. The known area of effect, and particularly the areas where damage certainly occurred, on the other hand, covered some of England's most populous counties

260 Miles et al. 2003: 110–111.
261 Howard 1994: 38.
262 Pevsner 1974: 443.
263 Gilbert 1884: 396.
264 Brown 2010: 120.
265 Alcock 1989: 43–44.
266 RCHM 1874: 128.

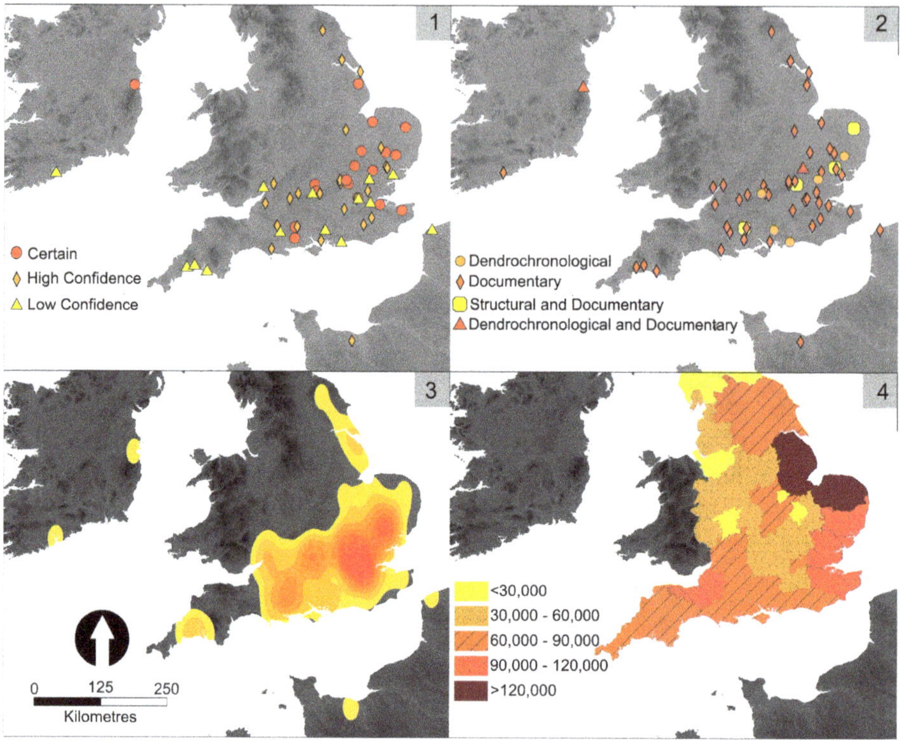

Fig. 3.15: Four maps exploring the impact of the 1362 windstorm in the British Isles. Clockwise from top left: (1) Confidence rating of data. (2) Different data types. (3) Kernel density plot of known damage created in ArcGIS 10.4. (4) Contemporary spread of the English population at the county level as derived from the poll tax return of 1377, after BROADBERRY et al. 2015: 25–26, only 15 years after the storm. Created by the author.

(Fig. 3.15: 4), especially Norfolk, Suffolk, Essex and Kent. Taken together the evidence augments that derived solely from the written evidence – that the storm originated from the south west and particularly severely affected the south east.

3.4 Landslides in medieval northern Europe

As discussed in Chapter 1, landslides or mass movements are common occurrences in some parts of northern Europe today. The triggering of such events bears a strong linkage to meteorological conditions, particularly precipitation, temperature and wind action.[267] Just as today, such events must have occurred during the medieval period

267 JAEDICKE et al. 2009: 475.

although they are rarely recorded in contemporary documentary accounts. Their relative scarcity in the written record is presumably strongly correlated with the types of location where they are most likely to occur. Predominantly, these would have been upland areas with steep slopes. These type of landscape have generally only ever been thinly populated, reducing the likelihood that the occurrence of landslides would severely affect human communities or, relatedly, attract the attention of contemporary chroniclers. Where events which might be categorized as landslides are documented from the medieval period in British sources, it is notable that they exclusively occurred within settlements or affecting ecclesiastical institutions, both of which understandably attracted the attention of contemporary chroniclers. An example of one such event is the collapse of a mound at Perth in 1209, which seems to have been destabilized by sudden-onset high-magnitude flooding.[268] Another is the Cleeve Landslide, Somerset, which caused the destruction of a chapel dedicated to the Virgin Mary and is discussed more fully below. In both cases, the rationale for recording these events, which would have been relatively unremarkable occurrences had they struck in remote areas far from human habitation, is clear.

The mid-15th century landslide at Cleeve, Somerset

Heavy rains precipitated a cliff collapse on the Somerset coast at Old Cleeve in the mid-15th century. This slope failure caused the destruction of a chapel dedicated to the Virgin Mary which had been located at the site since at least 1320[269] – although, as later documents reference the chapel's antiquity,[270] it may have had significantly earlier origins. This chapel seems to have been located on low lying ground at the base of sea cliffs[271] and belonged to the Cistercian house of Cleeve Abbey. By 1398 it was receiving frequent damage from the sea and was in need of structural repairs.[272] This remained the case in 1400[273] and by 1455 the chapel had been rebuilt – as the Bishop of Bangor was to lead the dedication of the new chapel to the Virgin Mary.[274] Then, in a Royal grant to Cleeve Abbey dated 1466, it is stated that "through the continuous rain a great hill lately fell upon a chapel founded long since upon the sea-shore in their manor of Clive … and [the Abbey has] begun to build another chapel in another place within the precinct of the manor".[275] WEAVER interpreted this sequence of events

268 CORNER et al. 1998: 457.
269 HOBHOUSE 1887: 186.
270 RANDOLPH-HINGESTON 1886: 62; MAXWELL LYTE 1897: 527.
271 BLISS, TWEMLOW 1904b: 400.
272 RANDOLPH-HINGESTON 1886: 62.
273 BLISS, TWEMLOW 1904b: 400.
274 WEAVER 1906: 10.
275 MAXWELL LYTE 1897: 527.

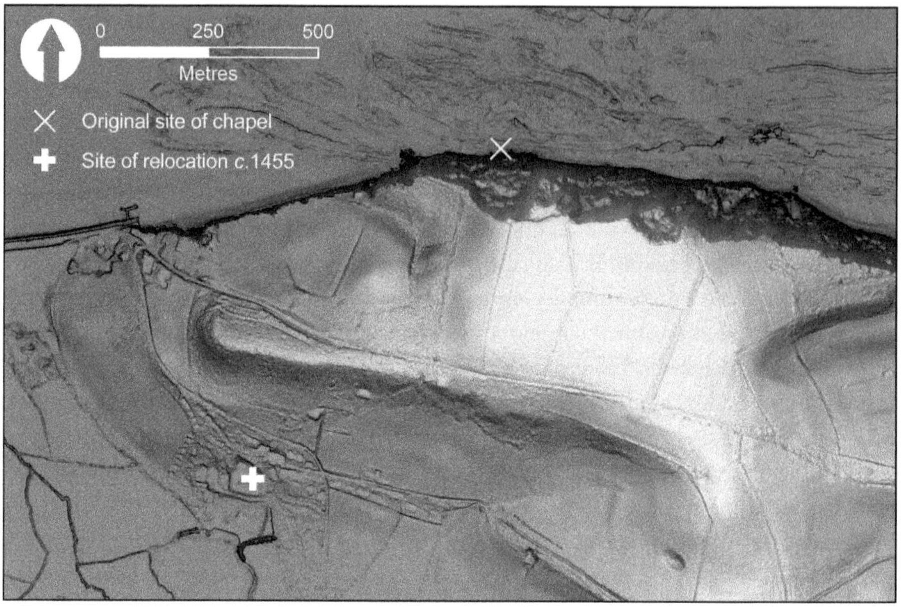

Fig. 3.16: Map of Chapel Cleeve, Somerset, indicating the re-foundation of the chapel of St Mary at a new site in response to a coastal cliff collapse at some point shortly before 1455.

as indicating three chapels dedicated to the Virgin Mary had existed at Cleeve – the first, had been founded close to the sea shore at the base of the cliff before 1320, the second, indicated by the 1455 dedication of a new chapel, on top of the cliff above the original site, and the third, around 1466 after the reported destruction in the Royal grant, at a new location[276] – the site of which is known from later evidence to have been at Chapel Cleeve *c.*900m southwest of the original site.[277] This interpretation, however, is overly complex and does not fit well with the available evidence. As the 1466 grant specifically mentions that the chapel destroyed by the cliff collapse had been "founded long since upon the sea-shore"[278] it is unlikely that this would be describing a very recently reconstructed chapel. In addition, this document does not state exactly when the cliff collapse occurred. A preferable interpretation of the documentary record, therefore, is that the cliff collapse occurred shortly before 1455 and the new dedication, for which the Bishop of Bangor was required, was in fact intended for the relocated chapel at Chapel Cleeve. It is more likely, therefore, that there were only ever two chapels

276 Weaver 1906: 9.
277 Somerset Historic Environment Record 1984.
278 Maxwell Lyte 1897: 527.

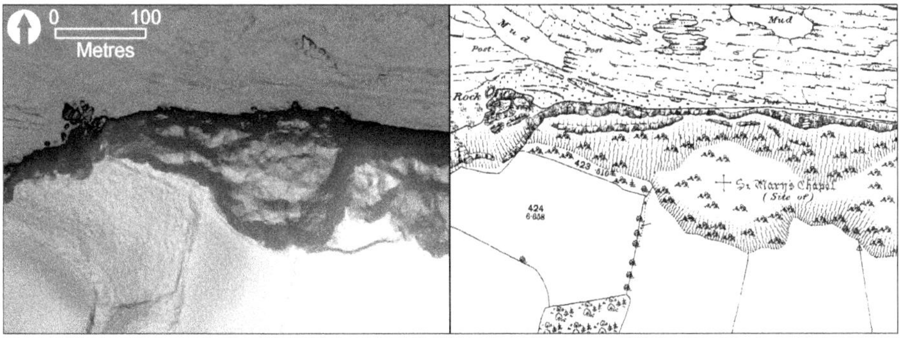

Fig. 3.17: Comparison of modern LiDAR data with 1st edition OS map of the site of the first St Mary's Chapel at Chapel Cleeve, Somerset, which was destroyed during a cliff collapse in 1466. Note that a number of rocks visible in both maps to the west must have eroded out of the hillside at some point. Wave action has presumably eroded and removed evidence closer to the site of the chapel.

dedicated to the Virgin Mary at Cleeve – the presumed sites of both are indicated on the area's first edition Ordnance Survey map (see Figs. 3.16 and 3.17).

Assuming the interpretation presented above is accurate, the precise date of the pre-1455 landslide event is unknown. The disaster is only described in the 1466 grant, which certainly gives an impression of a high magnitude slope failure as the chapel was apparently "crushed to the earth with the adjoining buildings" although, remarkably "the image of the Virgin and the altar of the chapel [were] uninjured".[279] Additionally, this document specifies that the cliff collapse was brought on by sustained rain.[280] Rainfall is the most common cause of landslides but the physical factors behind such events are complex. In addition to the intensity and duration of precipitation, the local topography, geology, soils, land use and hydrological properties are all important factors.[281] As this appears to have been a coastal cliff collapse, it is possible that, while perhaps an extreme rainfall event was the catalyst, erosion by the sea at the cliff's base over the long-term may have made such a failure inevitable. Although no archaeological evidence is known at the original site, or indeed for the replacement chapel constructed at Chapel Cleeve,[282] LiDAR data provides a useful illustration of the cliff collapse (see Fig. 3.17). Conceivably, the slope failure could have preserved evidence of the original chapel beneath the landslide deposit which archaeological excavation could recover – although, equally, coastal erosion since the occurrence of the disaster may have resulted in the disruption and destruction of any remains.

279 Maxwell Lyte 1897: 527. It may have been important, in a case such as this, to emphasise that the chapel's relics had survived the disaster as these were a significant draw to potential pilgrims.

280 Maxwell Lyte 1897: 527.

281 Iverson 2000: 1897.

282 Somerset Historic Environment Record 1984.

3.5 Summary

This chapter has explored what happened when natural hazards struck medieval communities. A combination of documentary and archaeological evidence allows these events, and how they affected contemporary populations, to be explored in detail. For those individuals and communities directly affected, these severe events frequently had stark implications but, in many cases, there remains much that is difficult or impossible to reconstruct from the available evidence. While knowledge of the occurrence of a disaster is often quite limited, these events were only the start of a chain of events that, in some cases, continued for years after the original disaster. As a result, the decisions taken in the aftermath of a disaster, or disasters, generated a further body of evidence which is the subject of the chapters which follow. Chapter 4, therefore, considers the practical measures taken to ameliorate the conditions created by disasters as well as the steps taken to protect against the possible recurrence of disaster. Then, Chapter 5 explores the spiritual and religious measures adopted by medieval populations in order to secure protection from disasters and to guard against the risk posed by the natural world more generally. Finally, Chapter 6 considers how communal memory of these events was maintained and passed on long after the time of their occurrence and to what extent this affected medieval preparedness and responses in the face of future disasters.

4 In the aftermath: practical responses against hazards

As we have seen in Chapter 3, during the emergency phase, disasters had highly destructive impacts on affected medieval populations. The most serious disasters commonly caused human casualties, drowned livestock, diminished agricultural outputs and damaged buildings and infrastructure. As such, in the immediate aftermath of these catastrophes, medieval societies had to come to terms with altered realities and deal with the changed circumstances as best they could. Over the longer term, as well as restoring the damage and trying to compensate for the losses inflicted by disasters, they had the option to take action to try and prevent hazards from causing similar disastrous impacts in the future. This chapter, therefore, considers the practical steps taken by medieval populations, in the immediate aftermath and beyond, while also exploring the extent to which medieval people attempted to protect themselves against the recurrence of future hazards.

A number of considerations affect whether or not societies take practical steps to reduce exposure to risks. Most importantly, potential hazards must pose a credible threat – the likelihood of occurrence must exceed the costs of intervention, whatever form that may take. If disasters are perceived as inevitable – events which cannot be avoided – the impetus to take proactive preventative measures is greatly diminished. In the medieval period, the widely held view that the causation of natural hazards ultimately stemmed from God, either through His action or inaction (see Chapter 2.2), certainly caused some to question whether practical steps to protect against disasters, thereby interfering with events orchestrated from on high, could be effective. Such fatalism can be seen in the case of the 14th century Florentine politician Coluccio Salutati[1] who condemned attempts to flee from outbreaks of plague as pointless exercises – for it was impossible to escape God's will.[2] As a corollary to such views, raising a flood defence would not protect against a flood which God intended to affect a particular community – for He would only heighten the floodwaters accordingly. Running counter to this argument, however, is the idea that human reactions, and the ability to protect against natural hazards, also ultimately stem from the divine. In this world-view therefore, while hazards themselves are heaven sent, so too are the tools to defend against them: "human knowledge, ability and action".[3] Accordingly, it could be said that to fail to take practical precautions was to neglect these divine gifts. While such theoretical justification probably seemed important to some theologians and scholars, in most cases medieval people likely acted primarily from a rational standpoint to minimise

1 B. 1331 d. 1406.

2 Schenk 2017: 143.

3 Schenk 2017: 142.

https://doi.org/10.1515/9783110719628-004

their chances of being negatively affected by the hazards they knew through their own lived experience – with little consideration for any wider theological ramifications. Although, as this chapter explores, practical measures were commonly put in place in the immediate aftermath of disasters, long-term adherence to such measures – the continued maintenance of flood defences, for example – was only likely to endure where memory of their utility – and, therefore, memory of the disaster that had sparked their adoption – was actively maintained.[4]

Medieval populations' ability to respond practically against recurrent threats from hazards was limited by a number of related factors. Firstly, the limited understanding and awareness of the underlying causes and physical characteristics of hazards, secondly, the level of scientific and technological advancement of the age and, finally, broader practical and economic considerations. Illustrations of all three of these limitations are evident throughout contemporary documentary sources. In Udine, Italy, in 1511, for example, reliance on the Aristotelian explanation for the occurrence of earthquakes provoked legislation to re-open disused wells across the city to allow the escape of subterranean winds which were believed to be the cause of the earth tremors – a practice which was also suggested in response to other Italian earthquakes.[5] Although, in this case, mis-identification of the root-cause of the hazard provoked a harmless policy, it nonetheless, did nothing to protect the population or reduce the chance of an earthquake occurring and diverted attention and labour that could have more profitably been applied elsewhere. Beyond understanding the mechanics of hazards, resilience was also hampered by an inability to accurately model how natural phenomena affected medieval architecture. In the absence of modern computational methods, which allow stresses and lodes within a structure to be accurately modelled, architecture in seismically active zones could only reactively take cues from what could visibly be seen to withstand past earthquakes. This process of trial and error is illustrated by the importation of Frankish Gothic architecture to Cyprus and the gradual reversion to traditional Byzantine building conventions[6] which occurred over time when it became clear that the Gothic style was unsuited to a seismically active zone.[7] Something similar can be seen in the iterative construction of other structures that were frequently damaged by hazards. In the case of bridges, for example, those designs and methods of construction that were frequently damaged, or swept away during times of flood could be contrasted against those which persisted – leading to the development of stronger and more resilient crossings. Cost was also a significant factor as major

4 This is the subject of Chapter 6.

5 Forlin, Gerrard 2017: 99.

6 These included a reduction in the number and size of window and door openings, a reduction in the height of the building overall and individual structural elements, buttresses which were stronger and more massive and a centralized more symmetrical plan – all features which increase resilience against earthquakes.

7 O'Neill 2020.

repairs were often prohibitively expensive, a fact illustrated by evidence for cost-cutting visible in campaigns of repair such as the work carried out to the spire of Norwich Cathedral following a lightning strike in 1463.[8] In spite of these limitations, as this chapter explores, a surprisingly large suite of practical interventions were available, and put into practice, during the medieval period to provide protection from nature's extremes.

4.1 The immediate aftermath

In many cases, the need to take practical action arose swiftly – in the hours and days that followed the occurrence of a hazard. This stage can be seen in both the historical and material evidence relating to the disasters considered in Chapter 3. In the immediate aftermath of the storms of 1287/88, for example, at the manor of Ebony, Kent, a boat was used to reach livestock who had been stranded on a portion of land that had become cut off by floodwaters.[9] It is easy to imagine the rapid process that must have occurred in the hours immediately following the storm during which the local community counted the cost of what had been lost and struggled to recover and rescue as much as possible – including vulnerable livestock. Nearby at New Romney, the archaeological evidence from the Southlands School site suggests a phase of site consolidation took place in the immediate aftermath represented by a deliberately deposited layer of sand, gravel and silt – perhaps to stabilise the shoreline and cover over the remains of a structure that had been badly damaged by the storm.[10] At a site off Church Road, fragments of clay flooring and walling material along with ash and other carbonised material was cleared into a large pit which the excavators interpret as the clearance and disposal of structural debris from a timber building destroyed by a storm event. Following the clearance of the structural remains, after an indeterminate period of time, a new structure was built at the site, occupying some of the previous building's footprint but with a different layout.[11] This demonstrates that, although many sites across the town were marked by a period of demolition or abandonment in the aftermath of the storms,[12] structures that had been damaged or destroyed were also repaired and rebuilt. Clearly, although the storms introduced many short-term difficulties, the inhabitants of New Romney were able to struggle through these problems and restore what had been damaged. Across Wisbech and King's Lynn, a comparable

---

8 The repairs to the spire were executed in brick faced with stone rather than pure masonry, a more expensive option, and appear to have been carefully thought out to minimise the amount of scaffolding required. Feilden 1996: 730–731.

9 Gross, Butcher 1995: 113.

10 Meddens, Wooldridge 2002: 23, 78.

11 Wilson, Linklater 2002: 10–11.

12 Discussed below in Section 4.7

phase in the aftermath of medieval flood events is apparent in which timbers from flood-damaged structures were retrieved and reused,[13] perhaps motivated by the lack of locally available timber.[14] As with the evidence from New Romney, such practices reflect the steps taken to clear the debris created by a flood event while maintaining an eye for what could be salvaged and reused in the future.

Reuse of building materials, where possible, was a highly pragmatic response as demand for raw materials as well as labour often saw short-term surges following the occurrence of a disaster. In the immediate aftermath of the January 1362 windstorm, for example, the widespread structural damage across the area of effect generated a sudden spike in demand for materials and workers to carry out repairs. This can be seen at King's Langley, Hertfordshire, where 123,500 tiles were purchased from one Roger "Tiller" for 5s 6d per thousand.[15] The Crown was clearly concerned that this rapid boom in demand would motivate tilers, roofers and other labourers to inflate their prices in order to take advantage of the heightened need for their services. To prevent this, immediately following the storm, Edward III issued a proclamation stating that "tiles and other roofing shall be sold at the price at which they used to be before Christmas last and no dearer" and "tilers, their grooms or assistants and all other their servants, craftsmen and workmen shall take no higher wage for their daily labour".[16] This proclamation seems to have been re-issued to the populace across the country – as is documented in St Albans where the Abbot ensured the townspeople were aware of the regulation,[17] as well as in Peterborough, where the Abbot wrote back to the Crown to confirm that the order was being followed within the town.[18] In London too, a writ was delivered to the mayor, the recorder and the sheriffs in late March 1362, demanding that a proclamation be made to forbid any price increases on tiles or tiling in light of the recent tempest.[19]

That not everyone was aware of this new ordinance or, more likely, that they chose to ignore it, is demonstrated by the fact that one Suffolk-based thatcher named William Champeneys was able to charge up to four times the usual rate for his services in the year of the storm.[20] Although thatchers were not specifically mentioned, the wage for tilers had been fixed to 3d per day by the Statute of Labourers in 1351[21] in response to the shortage of labour and ability of workers to demand high wages as a result of

---

13 Clarke, Carter 1977: 65; Hinman, Popescu 2012: 25, 36, 73.

14 Fletcher 2010: 73.

15 Page 1914: 265.

16 Maxwell Lyte 1909a: 238.

17 Riley 1869: 46-47.

18 Stamp 1937: 177.

19 Sharpe 1905: 138.

20 His wages went from a usual minimum of 3d to a maximum of 12d in the year of the storm. Penn, Dyer 1990: 370.

21 Raithby 1810: 312.

the high mortality caused by the Black Death. This meant that anyone earning more than this sum was doing so illegally. The continued presence of thatchers and tilers in the *Proceedings Before Justices of the Peace* from Suffolk in the storm's aftermath demonstrate that, in an area known to have been badly affected, roofing work was in high demand in these years allowing some contractors to inflate their prices. Examples include John Thatcher[22] who was paid 30 shillings, in addition to complimentary food, for work, over an unspecified period, in Little Livermere, Suffolk.[23] Others seem to have only exceeded the wage by a small amount, including the tilers John Barkesdale and Thomas Tiler, who charged 4d, in addition to food, between 1361 and 1363[24] and the thatcher Symon Wygenhale who was found to be at fault for receiving 3d in addition to breakfast.[25]

The high demand for the services of roofers, tilers and thatchers as well as other tradespeople in the storm's aftermath must have been compounded by the unfortunate coincidence of an outbreak of plague in 1361,[26] its first recurrence in England since the Black Death. Not to mention the fear and alarm that the resurgence of disease must have sparked, it would have also reduced the number of available labourers to carry out repairs, and perhaps further enhanced the ability of those who were still working to inflate the prices they charged. For those who could not afford to pay the increased market rates, or were simply unable to contract available workmen, damage was not restored for a number of years, a fact commented on by a contemporary chronicler who wrote that "houses and buildings which were … destroyed by this wind remained ruined and unrepaired because of the lack of workmen".[27]

4.2 The victims of disaster

A further strand of evidence which may relate to activities carried out in the aftermath of disasters are burials. As the historical sources considered throughout Chapter 3 widely document, disasters frequently resulted in significant numbers of casualties among both human and animal populations.[28] Where the disaster did not also make these bodies inaccessible, as may have sometimes occurred during severe floods, it would have been necessary to retrieve them for burial. For medieval Christians, burial was imperative for the human dead for religious reasons and also offered an expedient

22 Iohannes Thecchere.
23 Putnam 1938: 376.
24 Putnam 1938: 372.
25 Putnam 1938: 367.
26 Galbraith 1927: 50; Martin 1995: 184–185.
27 *Mansionesque et edificia per dictum ventum sic diruta pro defectu operariorum irreperata deformiter remanserunt.* Scott-Stokes, Given-Wilson 2008: 118–119.
28 This is particularly the case in relation to the storms and floods of 1287/88.

way to clear away the carcasses of livestock which helped to prevent the spread of disease among surviving animals.

The unusual nature of fatalities caused by disasters may occasionally leave a physical signature that can be identified by archaeologists. KACKI, for example, has explored to what extent it is possible to relate excavated burials with documented crises.[29] Where levels of preservation are suitably high, scientific techniques may permit the identification of concrete evidence relating to diseases such as plague.[30] Where the extraction of such evidence is not possible, or in the case of hazards other than disease, skeletal analysis as well as the context of the burial hold the potential to distinguish routine burials from abnormal cases likely to be linked to a disaster. Perhaps most obviously, the treatment of the dead in the aftermath of a catastrophe may differ from 'normal' burial rites. For example, in many contexts identified as plague burials, multiple individuals were interred together, differing markedly from the standard, individual inhumations which characterise medieval burial practices.[31] This alone, however, is insufficient evidence for the occurrence of a disaster as multiple burials can be explained in other, non-catastrophic ways – for example as family groups buried together. Further indicative evidence, therefore, may be found through the analysis of the remains of the interred population. Compared to 'normal' patterns of mortality, which disproportionally affect the very old and the very young, disasters are likely to be more egalitarian killers. Although both young children and elderly adults are also perhaps less able to flee from disasters, making them especially vulnerable,[32] rapid-onset hazards such as floods, storms and earthquakes are significantly more indiscriminate causes of death than old age or childhood disease. In the case of plague, excavated evidence suggests that those who succumbed to the disease were healthier than would be expected from a non-disaster related burial context, with less evidence for pre-existing health conditions. Relative to a 'normal' burial population, therefore, a clear demographic profile can be extracted from excavated plague victims characterised by a significantly greater proportion of individuals between the ages of 5–14, late-adolescents and young-adults along with a relative scarcity of fatalities younger than 5 years old–suggesting that excavated plague burials "mirror the overall demographic composition of the populations from which they are derived".[33]

One could reasonably expect mass graves arising from rapid-onset hazards to follow a comparable mortality signature. To make such an identification, a statistically

29 KACKI 2020.

30 This is accomplished through either the extraction of aDNA relating to the pathogen or, in the case of *Yersinia pestis*, palaeoimmunological techniques may allow the identification of the F1 envelope glycoprotein antigen unique to that particular pathogen.

31 KACKI 2020: 260–262. This was notably also the case in mass graves associated with war dead. See for example ANNIS et al. 2018: 25–31.

32 LEE, VINK 2015.

33 KACKI 2020: 267.

meaningful number of individuals would be required to differentiate between 'normal' and disaster-related burials. Although no mass graves arising directly from rapid-onset natural hazards have been identified from the medieval period, an example is known which may relate to the secondary impacts of such a hazard. At Spitalfields Market, London, a large group of mid-13th century burials exhibit a typical profile for famine victims thought to relate to a historically documented episode of high mortality[34]. Its close coincidence with the 1257 or 1258 eruption of Samalas Volcano, Lombok Island, Indonesia, one of the highest magnitude eruptions of the Holocene which may have substantially disrupted crop growth patterns around the world, provides a possible driver for the onset of famine conditions. CAMPBELL, however, cautions that the difficulties experienced in the late 1250s largely stemmed from a harvest failure which pre-empted, and was ongoing at the time of, the Samalas volcanic eruption rather than one that was caused by the atmospheric impact of the volcano.[35] Although the eruption may have exacerbated the crisis, therefore, its roots likely originated in longer term climatic trends.

While it is unclear that rapid-onset hazards sparked mass graves in the same way as plague or famine, severe weather may have occasionally provided an impetus for the mass burial of livestock. One interpretation of a pit containing the remains of at least seven cattle discovered at Shapwick, Somerset (see Fig. 4.1), for example, is that, following an extreme episode of cold, which decimated the herd, the victims were disposed of collectively. GIDNEY argues that such a scenario can only be envisioned when livestock lacked sufficient fodder as, in ruminants (*e.g.* cattle and sheep), the main source of body heat comes from bacterial fermentation of food in the rumen.[36] In times of dearth, therefore, when sufficient fodder was unavailable, livestock would have struggled to maintain core body temperatures making them particularly vulnerable to sudden cold snaps. Support for this interpretation comes from both the historical record – the Anglo-Saxon Chronicle describes several cold spells which severely impacted cattle[37] – as well as modern analogues – the winter of 1947, for example, brought blizzards which caused high losses among cattle herds in some areas.[38] Mass mortality caused by a blizzard or sudden cold weather certainly remains a possible if, in this case, unverifiable possibility to explain the Shapwick pit but a competing interpretation must also be the outbreak of disease amongst the herd. This could have had a similar effect and the fact that the pit in which the cattle were buried is an abandoned lime-kiln lends some support to this view – since lime is often thought to accelerate the decomposition process which may have been deemed desirable for the safe disposal of such an unusually large number of cattle. This is more likely to have been the case if

34 CONNELL et al. 2012: 230.
35 CAMPBELL 2016a.
36 GIDNEY 2020: 332.
37 GILES 1894: 418, 485–486.
38 GIDNEY 2020: 332–333.

Fig. 4.1: Pit containing the remains of at least seven cattle excavated at Shapwick, Somerset. Photograph by Christopher Gerrard. Reproduced with permission.

the cattle fell victim to a disease, rather than dying *en masse* in a cold spell, as concerns for the disease spreading from their carcasses to surviving members of the herd make more sense in this context.[39]

4.3 Agricultural management

As some of the case studies considered in Chapter 3 make clear, disasters frequently disrupted the agricultural cycle even, in some cases, causing the total destruction of the harvest. This was despite the fact that, as discussed in Chapter 2, elements of the medieval agricultural regime were arranged to minimise the impact of disasters. Where these failed, plunging communities into crisis, there appear to have been discrete strategies adopted to ameliorate difficult situations brought on by the sudden and unexpected impact of natural hazards. For example, a switch from cereal cultivation to pastoralism, although a general trend of the late Middle Ages, was particularly applicable to coastal wetlands due to their high fertility – which provided abundant pasture.[40] As well as tying into wider economic trends, such a switch may have afforded communities in locations vulnerable to flooding a greater degree of resilience as, where crops are unlikely to survive a severe inundation, livestock, given sufficient warning, may be herded to safety. Thus at Appledore and Ebony, Kent, the sheep were, as much as possible, kept away from areas liable to flooding, during the winter months when

39 Gidney 2020: 332.
40 Rippon 2001: 27.

they were most vulnerable, and a portion of the Ebony flock was routinely moved to Appledore for the winter where there was more high ground to accommodate them.[41]

The introduction of legumes too appears to have offered tangible benefits in areas which had suffered recent inundation. In addition to providing a nutritious, and easily preserved, source of sustenance for humans and livestock, beans are more tolerant of saline soils than cereals and play a role in improving soil fertility by replenishing nitrates – a fact which was appreciated by medieval farmers.[42] Beans appear to have been disproportionately introduced to wetland areas in the late Middle Ages[43] and were, most likely, a deliberate agricultural strategy in these flood-prone areas. Throughout the 14th and 15th centuries, legumes formed an important component of the agricultural output in Fenland[44] while at Ebony the introduction of beans in 1288/1289 may have been a pragmatic reaction to the storms of 1287/88.[45] Other crops saw a marked decline around the occurrence of the storms and floods – wheat and oat yields in 1288/89 were almost half what they had been two harvests before while, although initially experiencing an upturn in 1287/1288, rye also saw a decline in 1288/1289. Where traditional crops may have struggled in areas that had recently been exposed to marine flooding, therefore, the introduction of beans might have allowed farmers to turn a viable crop while gradually restoring the ability of the land to support less saline-tolerant crops. In addition to the positive role that legumes themselves played in improving soil fertility, other soil improvement strategies were a key area of investment on the lands of Canterbury Cathedral Priory at this time[46] and, in addition to the deliberate cultivation of beans, the historical evidence suggests that lime was applied to the soil, which was expensive and reduced soil acidity, while expenditure on weeding and manuring increased. Clearly, landowners with sufficient resources, such as Canterbury Cathedral Priory, were willing to make the high investment necessary to ameliorate the effects of the storms and enhance the profitability of their estates.

An additional agricultural response to crises is suggested by Gidney particularly in reference to events which sparked high mortality among livestock herds although perhaps with wider significance.[47] Where herds were decimated by natural hazards the only way for landowners, and their tenants, to quickly replace these losses was to purchase replacement animals – as breeding, the traditional method of maintaining or increasing herd numbers, would be too slow to account for such drastic reductions

---

41 Gross, Butcher 1995: 112.
42 Gross, Butcher 1995: 109; Rippon 2001: 30–31.
43 Rippon 2001: 30–31.
44 Rippon 2001: 29–30.
45 Campbell 2007.
46 Gross, Butcher 1995: 109–110.
47 Gidney 2020.

in livestock numbers.[48] This occurred on the lands of the Bishop of Ely following the drowning of 373 sheep due to a marine inundation in December 1398 at Wisbech High Fen, Cambridgeshire.[49] Such events, therefore, placed great strain on financial resources which, in turn, may have influenced contemporary agricultural practices. Indeed, immediately after these significant loses at Wisbech High Fen, the provision of pasture and, correspondingly, the management of arable cultivation on the Bishop of Ely's neighbouring estates was re-organised.[50] Comparably, GIDNEY suggests that the livestock panzootic of 1315–1322 provided the impetus for the switch from the traditional two field system, in which the fields alternated between sown crops and fallowing on an annual basis, to a more flexible three field system.[51] For example, up until 1333, in the two field system operated at Podimore, Somerset, the fallow seems to have also provided pasture for livestock.[52] The reduced grazing requirements of a diminished herd, however, might have freed up land permitting the subdivision of the manor into three fields which could be farmed more efficiently.[53]

Relatedly, FOX interpreted the irregular cultivation of areas usually devoted to pasture, such as manorial waste, in post-medieval Devon and Cornwall, as a means of maximising returns when grain prices were high. Such a system of 'outfield' cultivation is well documented from the medieval period[54] and confirmed through fieldwalking.[55] In coastal wetlands, meanwhile, un-embanked areas of marsh were sometimes exploited in a similar way by marshland communities.[56] Such opportunistic cultivation of these 'outfields' may have permitted populations hit by a major mortality event to raise the capital required to restock their herds.[57] Since there would be far fewer animals to take advantage of the pasture that these areas usually provided, and the labourers who usually tended the animals would be free to cultivate extra areas beyond the core areas of cultivation or 'infields', this seems to be a logical recourse that made the most of a parishes available resources. In the aftermath of a damaging natural disaster, therefore, a 'bonus' yield from such areas could have allowed communities a chance to recover from severe shocks to the established order – such as the loss of large numbers of livestock.

---

48 Note also that, although breeding was certainly cheaper than buying new stock it was more likely to lead to the spread of disease amongst herds. If mass mortality amongst livestock had occurred due to an outbreak of disease, this would have meant that purchasing new, healthy animals would have been doubly important to prevent the disease from spreading further. STONE 2005: 151.

49 STONE 2005: 123, 151.

50 STONE 2005: 143–144; note 36.

51 FOX 1986: 530–532.

52 FOX 1986: 532 – 533.

53 GIDNEY 2020: 335–336.

54 RIPPON 2002a: 54.

55 JONES 2004.

56 RIPPON 2002a: 64–65.

57 GIDNEY 2020: 333–334.

4.4 Structural responses

The structural repairs required in the aftermath of disasters are another type of practical response against natural hazards which the surviving material and documentary records allows us to observe. By their nature these types of interventions were usually reactive, *ad hoc*, solutions carried out in the immediate aftermath of a calamity. Where hazards led to repeated damage and destruction, however, the process of repair and reconstruction could foster a degree of experimentation, with attempts to create more resilient structures. Two classes of structure in which this is particularly visible are bridges, which, due to their riverine locations, frequently fell victim to severe floods – which either caused significant damage or total destruction – and ecclesiastical structures. Although the latter category of structures were less likely to be situated in hazardous locations, when they were affected by floods and storms the damage was often recorded, assisting the interpretation of any structural adaptations introduced in the aftermath of such events. While, no doubt, adaptations took place in many other architectural contexts, more so than most other types of structure, medieval bridges and ecclesiastical structures remain standing (to a greater or lesser extent) meaning that these types of structure can be particularly revealing in this line of enquiry.

4.4.1 Bridges

The repeated need to repair and replace bridges was cause for considerable investment. Naturally, this gave rise to consideration as to the best method of construction as can be seen in well documented examples.[58] Incidents which befell bridges as a result of floods or other hazards are often documented and the archaeological evidence can provide confirmation of episodes of repair or bridge renewal across a span. During the medieval period, bridges were most commonly constructed of either timber, stone, brick or a combination. All materials had their advantages and drawbacks – timber bridges, although relatively cheap and easy to construct, were relatively impermanent, flammable and required frequent maintenance.[59] Stone and brick bridges, on the other hand, were more durable but expensive and difficult to construct, requiring firm foundations which could cope with problems of scour and increased flow in times of flood.[60] Over the course of the medieval period, bridge construction evolved somewhat – reflected by a general shift from piers supported on piles driven into the river bed towards solid foundations constructed using coffer-dams.[61] In addition, features such as cutwaters,

58 Boyer 1981.
59 Harrison 2004: 86.
60 Harrison 2004: 87–91.
61 Boyer 1976: 151–153.

to prevent scouring around the piers, and flood arches, to enable the structure to cope with increased flows in times of high water, became more sophisticated.

One possible example of a bridge which suffered structural damage as a result of flooding, as well as the subsequent phase of repair, comes from the medieval bridge at Heigham, Norfolk. The evidence suggests that this bridge may have been a victim of the 1287/88 floods. John of Oxnead, a monk of the Benedictine Abbey of St Benet at Holme, Norfolk, describes in detail the effects of the floods caused by the 26th December 1287 storm to his own abbey as well as in nearby Hickling[62] – while, according to other chroniclers, the neighbouring towns of Waxham, Horsey and Martham were also affected by a storm (or storms) in December 1287.[63] Given that the flooding directly affected Oxnead's abbey, and the surrounding area, his account is clearly one of local events and he was probably, therefore, an eyewitness to the events he describes.[64] One of the actions of the Abbey in the aftermath of the flood is indicated by a petition, roughly dated by handwriting to *c.*1275–*c.*1300, which notified the King that the Abbey of St Benet at Holme had imposed a toll on the nearby crossing at Bastwick, where the bridge had been "broken by a storm at sea", which was negatively affecting the local inhabitants and especially the poor. The petitioner added that this toll rightfully belonged to the King, not the Abbey, and sought to re-appropriate the funds raised to repair the bridge.[65] If this petition does relate to the storm recorded by John of Oxnead, which – given the location and close-dating of the petition to the time of the storm – seems a plausible interpretation, then it appears that, in the aftermath, the Abbey attempted to raise funds through imposing a toll on the crossing.[66]

The most likely candidate for the bridge in question is Heigham Bridge, located between the parishes of Repps with Bastwick and Potter Heigham, which is only 4.5 km from the Abbey. The bridge's location, between the Abbey of St Benet at Holme and the nearby towns which were also affected in the December 1287 floods, is consistent with the damage described in the petition – as the low lying river valley in which the bridge is situated would have been particularly vulnerable during a severe marine inundation (see Fig. 4.2). Structurally, the bridge is formed by two narrow chamfered Gothic arches, one at either side, with the gap between spanned by a wider rounded arch in between. An assessment by the Norfolk Archaeological Unit in 1985 concluded that the side arches date to the 13th century while the central arch belongs to a later

62 Ellis 1859: iii; 270.

63 Luard 1859: 168.

64 Although note that his date for the event, 26th December, does not agree with the date given by Bartholomæi de Cotton, 17th December, who seems to describe the same event and agrees with a source from the Netherlands.

65 TNA SC 8/69/3407.

66 A possible explanation for the Abbey's actions without first gaining the King's assent is that Edward I was away in Gascony between 1286 and 1289 which may have made it easier for the Abbey to overlook proper procedure during this period. Burt 2013: 240.

phase[67] (see Fig. 4.2). While unconfirmed, therefore, the material evidence is consistent with the documentary record – suggesting that the original bridge was composed of the two Gothic arches at either side with a, now no longer extant, central portion that was damaged by the 1287/88 storms. It is unclear if this central section was completely destroyed by the storms, meaning that the 'crossing' on which the toll was imposed must have been a ferry or other alternative means of traversing the river, or whether this section remained standing albeit in a significantly damaged state, perhaps threatening its structural integrity and making it more difficult for heavy traffic such as carts and livestock to pass. At a later date, the damage was repaired through the insertion of the wide rounded arch visible in the standing structure today.

A comparable case comes from the succession of bridges at Hemington, Leicestershire, on the River Trent. Here the remains of three medieval bridges dating between *c.*1111–*c.*1310 were destroyed or abandoned as a result of both historically documented high-magnitude flood events and channel migration,[68] an unusual occurrence in post-Iron Age British lowland rivers.[69] This series of bridges (see Fig. 4.3) preserves possible evidence of experimentation in bridge design with the first bridge, built in the late 11th century, supported by timber-constructed caisson plinths, filled with sandstone rubble, which supported a timber superstructure above. Following a flood in the early 12th century, attempts were made to repair this structure using simple oak piles. Channel migration in the decades that followed meant that this crossing was only short-lived, however, and during the late 12th century, a new bridge formed of braced-wooden piles was constructed upstream. This was replaced during the 13th century by a bridge with masonry supports on the northern side, masonry piers on sandstone rubble bases and a timber superstructure. Unfortunately, none of these bridges was able to withstand the forces of flood and river migration over the long-term.

Ballingdon Bridge, Sudbury, Suffolk, offers a further example of bridge evolution attested through both documentary and archaeological evidence. The earliest known structure was a masonry bridge which is known only through structural fragments unearthed from the riverbed.[70] That this structure was likely damaged beyond the point of repair is suggested by its replacement by a brick-built bridge in the late 15th century. The shift in building material may have reflected, as well as the changing fashions of the time, an attempt to build a stronger crossing than the one that had stood before. Alas, this structure too was damaged by a historically documented flood which struck on 4th May 1521, with associated structural repairs evident in the archaeological record, and another flood destroyed the bridge entirely in September 1594.[71] In the cases of the crossings at both Sudbury and Hemington, therefore, the continued replacement

67 NHER 8525. http://www.heritage.norfolk.gov.uk/record-details?MNF8525 [Accessed 04 June 2016].
68 Ripper, Cooper 2009: 38, 222.
69 Brown et al. 2001: 77.
70 Gill 2007: 21–22.
71 Gill 2007: 21–22.

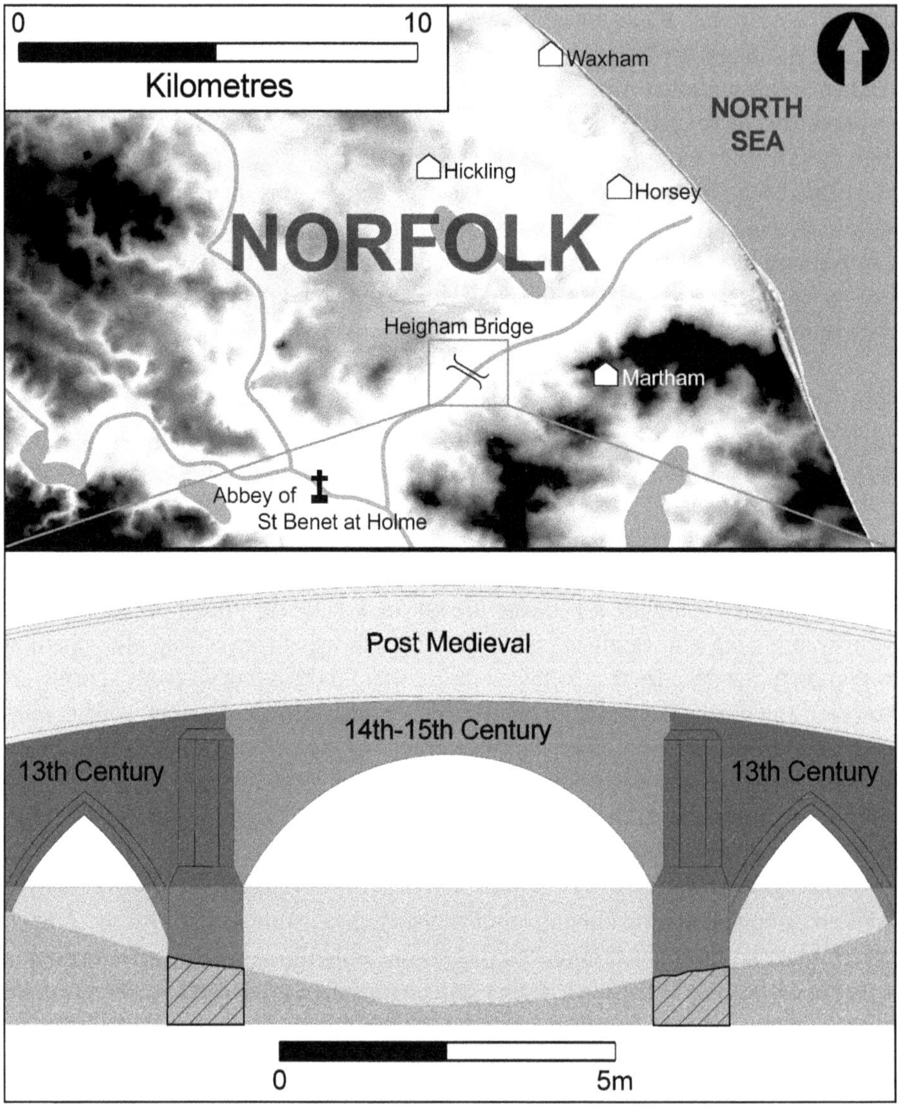

Fig. 4.2: Map of the location of Heigham Bridge, Norfolk, with a schematic diagram. Above: The location of Heigham Bridge in relation to other locations which contemporary chroniclers report were affected by the floods of December 1287. Created by the author. Below: An interpretation of the structural phasing of Heigham Bridge. Redrawn by the author after GUY 2011.

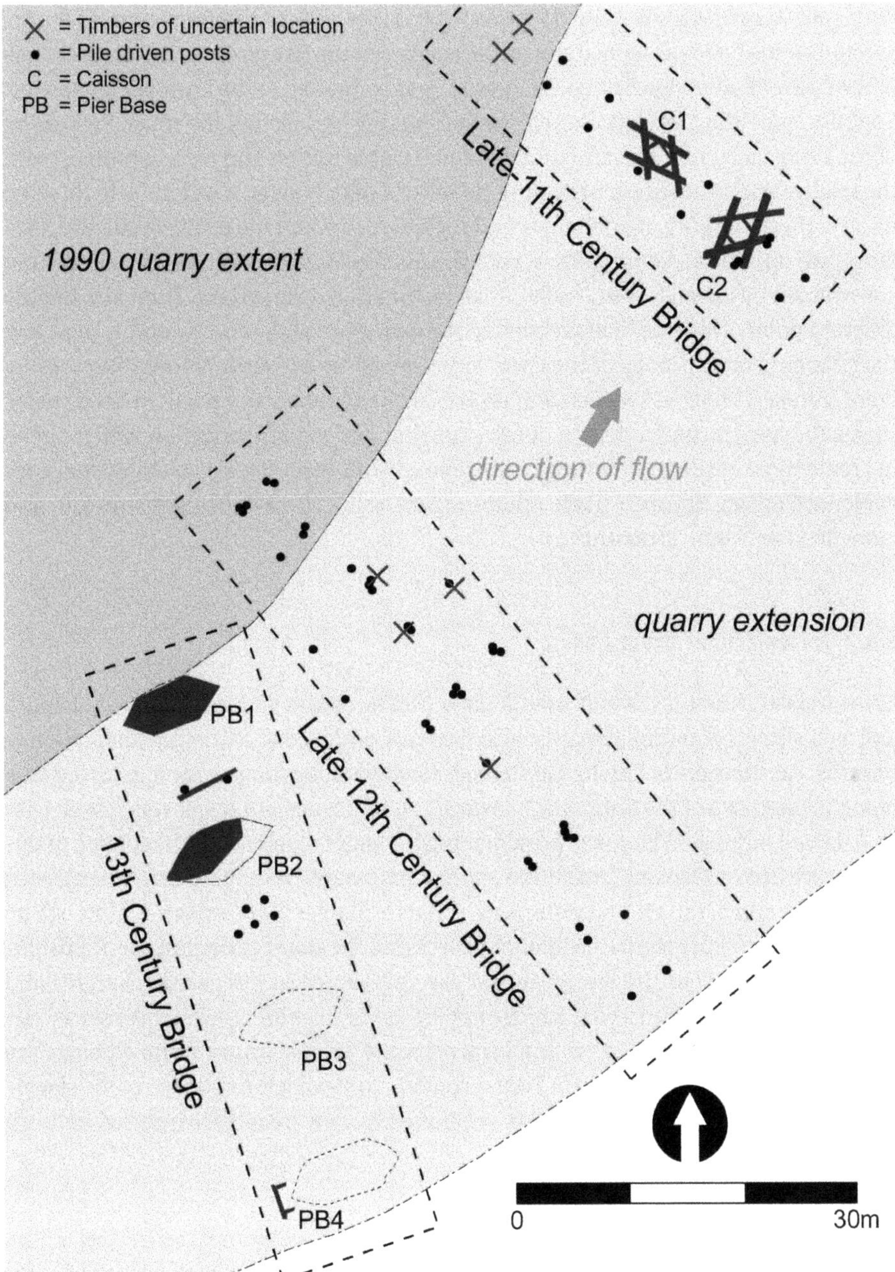

Fig. 4.3: The succession of medieval bridges on the River Trent at Hemington Fields, Leicestershire, discovered during expansion of a gravel extraction quarry. Redrawn by the author after COOPER, RIPPER 1994: 155; Fig. 2.

and renewal of bridges despite the repeated damage and destruction incurred during severe riverine floods demonstrates an acceptance of the risk posed at certain locations. In the case of Ballingdon Bridge, its location was connected to the layout of the town of Sudbury, providing an important route for entering and exiting the town. Relocation of the bridge may not, therefore, have been a viable option and this, combined with the relatively infrequent occurrence of floods of a high enough magnitude to threaten its structural integrity, made repair and reconstruction on the existing site the most attractive option. At Hemington, where the precise location of the crossing was not constrained by urban layout, each bridge incarnation spanned the Trent at a slightly different point. This was likely related to the migration of the course of the Trent and the changing morphology of the river banks, meaning the most favourable crossing point shifted with time. The frequent replacement of bridges as a result of flood events seems to have spurred a degree of development and experimentation which aimed to create more resilient crossings. The general shift from timber to stone crossings evidenced across Europe[72] likely relates to this desire to make bridges stronger, and more flood resistant, structures.

4.4.2 Ecclesiastical structures

Ecclesiastical structures, which are similarly well furnished with documented histories and well studied standing remains, also preserve evidence of structural adaptation to hazards. An example is Chichester Cathedral where, following a storm in 1210 which brought down two of the Cathedral's towers, the south-western tower was rebuilt with an enlarged buttress.[73] This was presumably an attempt to reduce the likelihood of this new tower from collapsing in a future storm. Comparable responses can be suggested in the aftermath of the 1362 windstorm explored in Chapter 3. For example, although no known contemporary sources reference the effect of the storm at Hitchin, Hertfordshire, PARKER suggests that the storm caused severe damage to the parish church[74] and interpreted the large buttresses which support the 13th century tower, inserted shortly after the time of the storm, as a structural response against future storm damage (see Fig. 4.4). More certainly, at Austin Friars, London, following the collapse of the steeple in the 1362 storm,[75] a secondary arch, supported by new piers, was inserted beneath

72 DALY 2006: 42–43.

73 TATTON-BROWN 1996: 49.

74 It is unclear exactly why Parker held this opinion. It is possible that an orally transmitted memory of storm damage had survived or, alternatively, some historical evidence that the storm affected the church at Hitchin may exist which is unknown to the author. Note that the history of the nearby church of All Saints', St Paul's Walden, Hertfordshire, given on the church's website (http://stpaulswaldenchurch.org/history/) [Accessed 03 March 2017] also states that Hitchin was damaged in 1362, although it has not been possible to trace the source of this information. PARKER 1904: 38.

75 GALBRAITH 1927: 50.

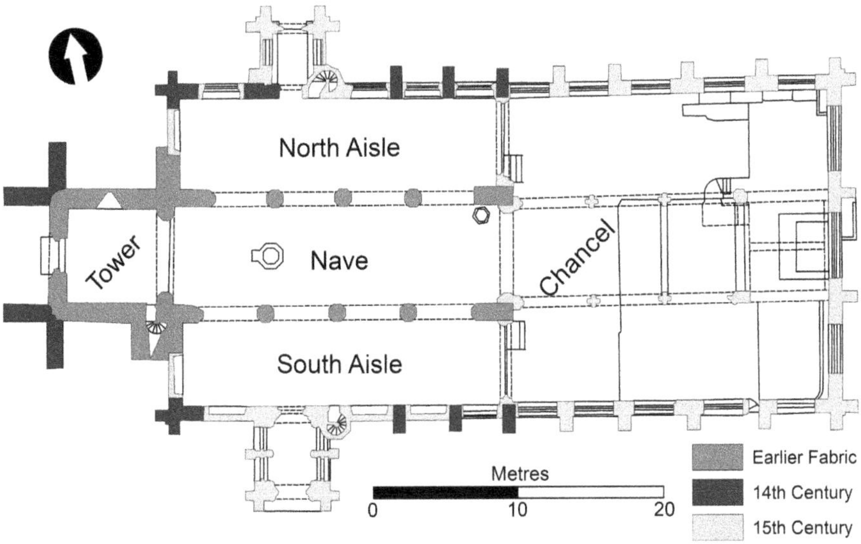

Fig. 4.4: Plan of St Mary's Church, Hitchin, Hertfordshire. Note the large buttresses supporting the west face of the tower which were added in the 14th century. PARKER suggested these were a structural response to the storm of 1362. Redrawn by the author after PARKER 1904: 37.

the original arch which had supported the steeple above.[76] This course of action was presumably taken both to make the structure safe as quickly as possible in the storm's aftermath as well as to avoid further costs by reusing as much surviving structural fabric as possible. It is also possible that this structural intervention aimed to make the new spire more stable and resistant in the face of any future high magnitude storms. Compared to the engineering problems associated with bridges, reinforcing buttresses and arches was relatively simple and was conducted as and when the need arose. The same type of *ad hoc* interventions were also called on when structures experienced problems with subsidence or other structural weaknesses as was the case at Wells Cathedral after the heightening of the tower, between 1315–1322, introduced subsidence and instability, threatening its collapse.[77]

As with the repairs at Austin Friars, London, the Abbey of St Albans also sought to make the most of the existing structural fabric from the buildings which had been damaged in 1362. The Abbey's gatehouse and an almonry were probably among the structures damaged or destroyed by the storm[78] and in the aftermath of the event a new

76 I'ANSON 1866: 71.
77 DE BLASI 2008: 204–205.
78 RILEY 1869: 387; NIBLETT, THOMPSON 2005: 254.

gatehouse was constructed by Abbot Thomas de la Mare,[79] presumably on the site of these earlier structures. The new gatehouse (see Fig. 4.5) incorporates elements thought to be reused from these earlier structures[80] which were damaged by the storm.[81] These elements include ribs in the vaulting of one of the ground floor chambers as well as decorative corbels (see Fig. 4.6) now in place on the second floor. While, stylistically, the triple-roll ribs suggest a 13th century date, the corbels find a close parallel in the corbels from *c.*1400 which support the roof of the church of the Holy Trinity, Penn, Hertfordshire.[82] Although a number of the St Albans corbels do display signs of damage, as might be expected from sculptural fragments reclaimed from another structure, it is impossible to confirm at what time this occurred. Taken together, the evidence suggests the corbels probably post-date the storm of 1362[83] and, therefore, most likely date to the period of reconstruction in the storm's aftermath. Although the corbels were probably created for the post-storm gatehouse, the ribs attest to the pragmatic re-use of available building material from structures that had been damaged by the storm. An eye to the future may further be suggested by the documentary account of the construction of the new gatehouse which emphasises that this building was "roofed with a very strong covering of lead"[84] perhaps indicating that the old roof had fared especially badly in the storm and, despite its enormous weight, lead was thought to provide a more resilient roof covering.

Resilient adaptations in roof re-design may be most evident, on the heels of the occurrence of a natural hazard, at Salisbury Cathedral, Wiltshire. Here, the storm of 1362 damaged the uppermost *c.*9m of the spire as well as the free-standing belfry. As the belfry was demolished in the late 18th century it is impossible to analyse the impact of the storm in this structure but the spire is extant and the repair pattern observable. Perhaps most interestingly, dendrochronological analysis of the timbers in the spire's internal scaffold return felling dates between 1344–1376, at least a generation after the initial construction of the spire. The scaffold has therefore been interpreted as an insertion necessitated by storm damage in 1362,[85] which is corroborated by the documentary record.[86] Although this scaffold may have facilitated the repair of the spire, a possible interpretation is that it was intended as a structural reinforcement – providing greater rigidity and better enabling the spire to withstand future storm winds. This

79 Riley 1869: 387.

80 RCHME 1982: 31.

81 I am grateful to Nigel Wood-Smith for providing information on the structural history of the gatehouse of St Albans Abbey and for kindly showing me around the building, now part of St Albans School.

82 Pevsner, Williamson 1994: 595.

83 Pevsner 1977: 196.

84 *Et fortissimum tectum ipsius cum plumbo cooperuit.* Riley 1869: 387.

85 Miles et al. 2004: 20–22.

86 Bliss, Twemlow 1902: 462–463.

Fig. 4.5: The gatehouse of St Albans Abbey constructed after the storm of 1362, replacing buildings damaged by the St Maur's Day storm and making use of structural fragments from these earlier structures. Photograph by the author.

Fig. 4.6: Structural evidence thought to pre-date the 1362 storm at St Albans. Left: Rib vaults in one of the ground floor chambers. Right: Decorated corbels now located inside chambers on the second floor of the gatehouse of St Albans Abbey. Photographs by the author.

possibility is given credence by both the location of the scaffold, internal rather than external, as well as the choice of material, oak rather than lighter a lighter wood, such as alder or pine, more commonly used in temporary scaffolding.[87]

Flooding affecting ecclesiastical structures provoked a different set of structural adaptations. At Clare Priory Church, Suffolk, for instance, the wall footings were cut into made-ground which had been imported or redeposited during the construction process to raise the site by up to 1m – presumably to provide protection from floods from the nearby River Stour. The redeposited soil contained charcoal, broken tile and mortar, suggesting that it may have been demolition waste from an earlier structure – pragmatically spread across the site to reduce the risk that the new structure would suffer inundation.[88] Similarly, at the Cistercian foundation of Bordesley Abbey, Worcestershire, the internal floors of the south transept were raised by 20–26cm in the aftermath of a flood event in order to prevent the same thing happening in the future.[89] In some cases, an alternative strategy may have been to allow flood-waters into the lower levels of a building by design. This seems to have been the preferred strategy in the Fenland villages of Tydd St Giles, Cambridgeshire, West Walton, Norfolk, Terrington St John, Norfolk, and Long Sutton, Lincolnshire, which all have (or originally had) free-standing bell-towers[90] with open arches that support the tower above. One interpretation of this design is that it allowed the region's frequent floods to simply wash through resulting in only minimal damage.[91] In addition to raising ground, and floor, levels and structural features which were designed to cope with frequent inundation, as the next section explores, there were many other practical ways in which medieval communities adapted to the risks associated with dwelling in flood-prone areas.

4.5 Flood defences

In many parts of Britain, the risk of flooding is never far away. Whether it be coastal flooding, through storm surges, riverine flooding, as a result of a rapid downpour or sudden snowmelt, or groundwater flooding as a result of sustained rainfall – a flood can occur quickly and without warning. In order to reduce risk to land which sits at a level threatened by flooding, therefore, artificial intervention is required. As seen in the examples above, this may involve the artificial raising of the local ground level or internal floors beyond the reaches of flood-waters. Due to the substantial labour and material costs involved, however, this strategy is only practicable over a relatively limited area. Though the work involved is still far from trivial, the erection of dikes and

87 Miles et al. 2004: 22.

88 Gill 2013: 20.

89 Hirst et al. 1983: 54-55.

90 A relatively rare feature in English parish church design.

91 Parker, Pye 1976: 107.

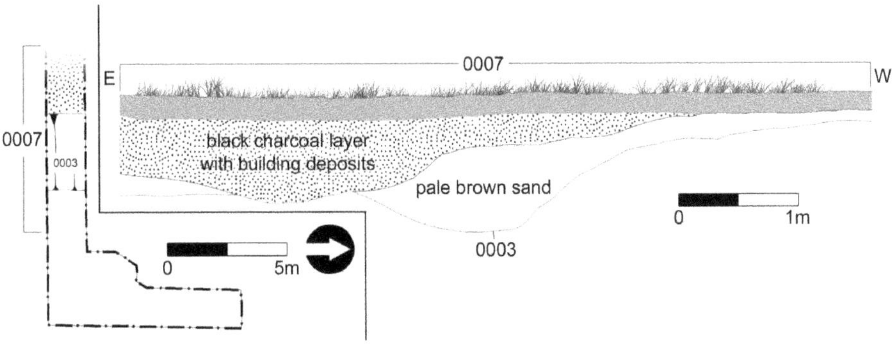

Fig. 4.7: Ditch 0003 at Flempton, Suffolk. Interpreted by the excavators as a flood defence ditch to protect against flooding from the River Lark. Redrawn by the author after GILL 1997: Figure 2.

excavation of drainage ditches allows larger areas to be protected from flooding. This section assesses the archaeological and historical evidence for the application of such defences during the medieval period.

In flood prone areas, human settlement – from individual land holdings right the way up to villages and towns – relied on some form of flood defence. In general, these took the form of raised embankments, drainage ditches or a combination of the two. These features are commonly encountered through archaeological excavation and/or field survey. Medieval property boundaries often take the form of ditches or, less commonly, embankments, which, as well as demarcating property ownership, likely often provided a degree of flood protection.[92] At Swavesey, on the edge of the Cambridgeshire Fens, for example, property boundaries within the medieval town were delineated by, or oriented in relation to, drainage ditches – although a number of these ditches contained molluscs which indicate that, at some point, they became flooded by a large body of water.[93] Similarly, at Flempton, Suffolk, a large ditch – ditch 0003 – located on the edge of the medieval green, detected during a routine archaeological assessment, seems to have provided medieval settlement at the site with flood protection against the seasonal flooding of the River Lark (see Fig. 4.7).[94] At Bordesley Abbey, Worcestershire, meanwhile, in addition to the raising of internal floor levels, analysis of the upstanding earthworks reveals a bank between the abbey precinct and the River Arrow which would have provided protection against flooding.[95]

Doubtless, embanking – the process of erecting flood defence barriers, took place on the largest scale in low-lying wetland environments where these banks, alongside

92 COLEMAN 2004: 290–291.
93 SAYER 2009: 139–141.
94 GILL 1997.
95 ASTON 1972: 135.

associated drainage ditches, allowed the fertile soils which characterised these dynamic landscapes to be exploited for agricultural purposes. Indeed, such defences were a 'pre-condition' for the settlement of these areas above a certain threshold[96] and were usually constructed from earth – both an economical and readily available resource. The required earth was usually obtained by digging a ditch parallel to the intended line of the bank – most commonly on the landward side of the defence. This had the added benefit of creating a drainage ditch to allow excess water to drain away from enclosed agricultural land.[97] Earthen banks were often reinforced with wood or stone either to provide a supporting internal core, which was then covered over with earth, or a strong facing to provide greater strength against strong waves or the increased power of currents in times of flood.[98] An example of one such flood defence is the 'Pokediche' in Marshland, Norfolk, which bridged the land between the Nene and the Ouse, protecting the low-lying land of the parishes to the north from freshwater inundation – and seawater ingression when severe marine floods brought seawater upriver.[99] Its path is marked on an anonymous map of *c*.1582 and is still labelled on the 1st edition OS map as 'Old Podike Bank'. According to DUGDALE, this feature was constructed in *c*.1223 although some evidence suggests it may have been in existence at an earlier date.[100] Since the 'Pokediche' provided protection for the whole area to the north, including all of the Marshland parishes, each parish was responsible for its upkeep and contributed accordingly towards the cost of its maintenance.[101]

In addition to raised embankments, it was also necessary to drain freshwater from within areas of embanked land. As well as a network of drainage channels, therefore, sluices or flood gates[102] were required in order to control the level and flow of water. These were pipes which ran through the flood banks and sea walls, equipped with doors so that they could be closed when flood risk was high.[103] Thus, at Bicker, Lincolnshire, the flood gate was to be closed "in time[s] of mighty floods from the Sea".[104] Such features are mentioned in the historical sources discussed in Chapter 3.2.2 – "a certain Floud-gate of Waltone, called Nobeche gote",[105] for example, is described in the sea wall which bounded the parish of Walton to the north and west. 'Nobeche gote' is unmarked on the early maps of the area but a field abutting the sea wall in the parish

---

96 SOENS 2011: 333.
97 ALLEN 1997: 5, 9.
98 ALLEN 1997: 5.
99 SILVESTER 1988: 32.
100 DUGDALE 1662: 245; HALLAM 1988b: 157.
101 DUGDALE 1662: 258–259. Maintenance was frequently required both as a result of damage caused by flooding and human interference, as discussed in Chapter 3.2.2.
102 In the Fens these were also known as 'gotes', 'goles' or 'clows'. HALLAM 1988a: 501–502.
103 HALLAM 1988a: 502–503.
104 DUGDALE 1662: 200.
105 DUGDALE 1662: 258.

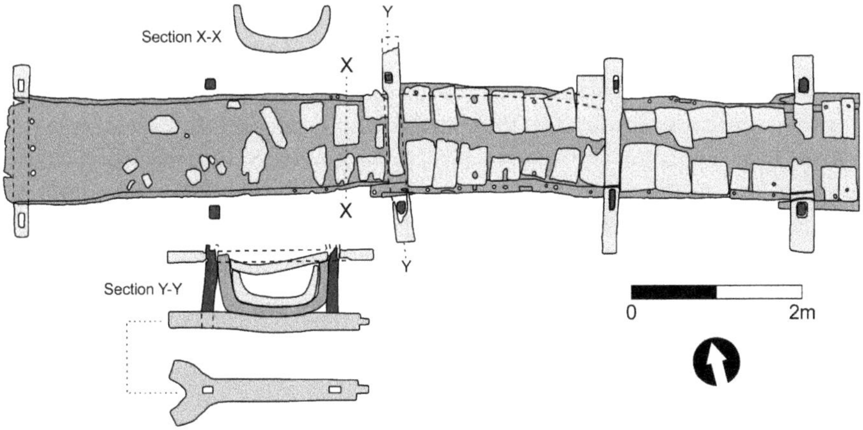

Fig. 4.8: A 13th century culvert excavated in 1977 within the medieval sea wall at Newton, Cambridgeshire. Redrawn by the author after TAYLOR 1977: 64, Fig. 1.

of Walton shown on the *c.*1610 map of Marshland[106] is named 'Nobitch', giving a likely approximate location for this flood-gate. A parallel for such a structure may be the culvert formed of three hollowed out tree trunks, tree-ring dated to AD 1250±40, which was excavated *c.*5km away within the medieval sea bank at Newton, Cambridgeshire (see Fig. 4.8). As with the flood gates described by the documentary sources, originally this culvert was probably furnished with a door allowing both water to be drained from behind the flood defences and excluded during flood events.[107]

By their very nature, flood defences (embankments, drainage ditches and flood gates) require frequent maintenance to remain effective. In complex flood defence systems covering large areas, the failure of defences protecting a neighbour's land can result in knock-on consequences – even the flooding of land owned by others. As a result of such situations, ensuring that all parties responsible for the upkeep of specific flood defences were contributing to their maintenance fairly and regularly was a continuing concern during the medieval period. To this end, often at the request of local landowners in areas where flooding posed a recurrent risk, Royal Commissions *de walliis et fossatis* were established with powers to compel landowners to pay for repairs, press charges against those who failed to comply with regulations and even to

106 British Library Cotton Augustus I.i, f. 79. http://www.bl.uk [Accessed 05 February 2018].
107 TAYLOR 1977: 65.

arrange for and oversea the execution of necessary works.[108] One such commission was established in the aftermath of the storms of 1287/88 to investigate the extent of flooding and ensure defences were properly repaired and maintained in Romeny Marsh, Kent.[109]

The complex system of flood defence infrastructure, which emerged over time in reclaimed wetlands, encouraged the concomitant development of social systems to organise their regulation and maintenance. Analogous to the arid regions of southern Europe, especially those influenced by Islamic traditions, where land came with associated rights to irrigation,[110] in many reclaimed wetlands the rights to specific parcels of land came with the duty to maintain or contribute towards the flood defence infrastructure. In Marshland, for example, in an early 13th century land grant to "Adam the priest of Walton" came the attached condition that "he is also to do[111] half a perch of sea bank and half a perch of drain".[112] Where a dike or drain needed construction or repair, the community who benefited from the protection these defences afforded came together to carry out the necessary work (see Fig. 4.9). This obligation was known as 'menework' and provided the labour required to maintain the flood defence infrastructure. This system is clearly described in the records of the Commissions for Sewers in Wiggenhall in the early 14th century:

> If anyone's dike is defective, a day is to be arranged provided that no danger is expected ... Anyone who has not come to the summons for repairing 'grundegole'[113] shall give 6d. Anyone who has not come to the summons for simple menework shall give 4d. and nevertheless do his menework on the morrow.[114]

Such regulations were enforced in Marshland by 'dike reeves', officials elected by each town,[115] whose role included the financial administration, organisation and overseeing of flood defence and drainage works in addition to liaising with Royal officials.[116] Each parish looked after its own defences with some of the major banks and ditches, such as the main sea-wall and the Pokedike, maintained in concert by all the Marshland towns. The cost of maintaining the flood defences was divided among the parishioners proportionally to individual land holdings while penalties were imposed for failure

---

108 Galloway 2009: 178.
109 Anon 1726: 155–160, Maxwell Lyte 1893: 309.
110 Gerrard 2011: 19.
111 *Faciendo.*
112 Bullock 1939: 51.
113 This refers to a breach in a dike, something that was more urgent to repair than the usual maintenance or 'menework' and therefore carried a greater financial penalty for failure to respond to the summons.
114 Owen 1981: 51.
115 Wiggenhall elected two annually in the 1320s while Tilney and Islington had 3. Owen 1981: 42; Dugdale 1662: 251.
116 Owen 1981: 42–44.

Fig. 4.9: A family supports a dike against the forces of the sea illustrating the communal nature of dike maintenance. Illustration from the Oldenburger Sachsenspiegel, produced in 1336. LBO CIM I 410, 58r. Redrawn by the author.

to make the repairs according to one's responsibilities – and these escalated upon failure to carry out the work.[117] Along with the relatively complex physical system of flood defence infrastructure, therefore, an equally sophisticated, and somewhat bureaucratic, system existed in Marshland to regulate and enforce dike maintenance and collective flood-security.

Of course, in some cases maintenance was not carried out properly or flood defences were simply too weak or too low to withstand flood waters. In such an event, a breach with potentially devastating consequences for the land behind occurred. Such failures severely disrupted the defences themselves. As an example, an archaeological evaluation of a sea wall thought to have been built in the 16th century on the Trimley Marshes, Suffolk, revealed that the internal structure of the flood defence contained reddish-brown clays reinforced with wooden poles, chalk rubble and brick, which were interpreted (based on analysis of the types of brick and the presence of saw marks on the timbers) as 20th century reconstructions in the wake of the 1953 storm surge.[118] Those defences which have suffered recurrent damage as a result of floods and storms, therefore, are likely to have been repaired and renewed at frequent intervals. As a result,

117 Dugdale 1662: 254.
118 Everett 2000.

unaltered defences from the medieval period are likely to be relatively rare survivals in the modern landscape. Additionally, in many areas, especially reclaimed landscapes, the medieval sea walls, dikes, ditches and banks are very often now located behind more recent defences as subsequent reclamation has proceeded seaward, as at Broomhill, East Sussex.[119] Today, therefore, flood defences dating to the medieval period have very often lost their local importance as flood protection infrastructure. As this most often occurred during the pre-modern period, when the earlier flood defences were not recognised as historical monuments of any value, they have often ceased to be maintained or have been subsumed into modern roads, tracks or to provide reliably dry building plots for later structures.[120]

Such issues mean that detecting archaeological evidence for the failure of medieval flood defences is a difficult task though there are a number of ways in which such events have left a distinctive physical signature. Reclaimed land which becomes inundated following a breach may preserve identifiable evidence of former agricultural use as creeks often form along existing drainage channels forming distinctive and unnatural creek networks.[121] Repairs following a breach also often leave a distinctive pattern which can be discerned through analysis of historic maps. Most repairs follow one of three possible routes: either a reconstruction of the original path of the defence or a redesign, encircling the breach externally or internally (see Fig. 4.10). A succession of the former type of encircling repairs can be seen at Broomhill, East Sussex, which were likely constructed in response to continued floods in the 1450s and 1460s.[122] The latter type of repair – encircling the breach internally – is a less common response as it typically requires a longer, more costly and manpower intensive, alteration and results in the permanent surrender of some of the ground affected by the breach. Nevertheless, a medieval example can be found at Horseshoe Corner, on the Thames near Dagenham, which is probably a repair necessitated by a flood in 1409.[123] A similar, although post-medieval, example can be observed in Marshland, Norfolk, from an area enclosed in 1775 and, therefore, breached at some-point after this date. In this case, however, repairs were made with an external encircling bank but, in common with the breach near Dagenham, the dike failure came to be known as 'Horseshoe Hole'.[124] An earlier breach in the area is marked on William Hayward's map in the sea wall between Terrington St Clement and Walpole Cross Keys and marked simply as 'The Breach' providing a *terminus ante quem* of 1591.[125] That this particular area

119 GARDINER, HARTWELL 2006: 151–152.
120 ALLEN 1997: 6.
121 GARDINER 2002: 106.
122 GARDINER, HARTWELL 2006: 155–156.
123 GALLOWAY 2012: 82.
124 This is the toponym given on the 1st edition Ordnance Survey map of the area.
125 The *c.*1610 map held by the British Library is thought to be a later copy of a no longer extant original dating from 1591.

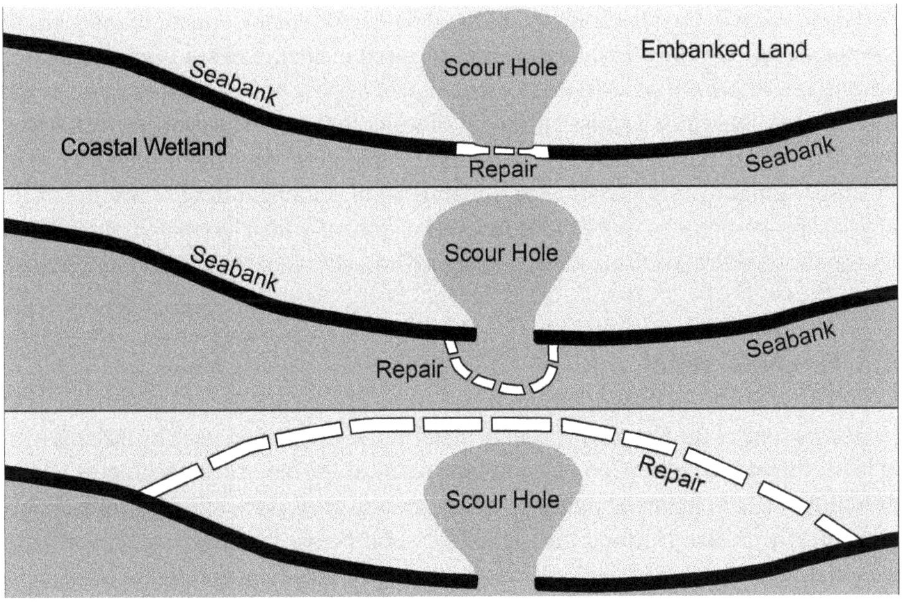

Fig. 4.10: Schematic diagram showing different ways of repairing a breach in a flood defence. Redrawn by the author after ALLEN 1997: 15.

was liable to damage is confirmed by the fact that by the mid 18th century it had been abandoned to the sea, and the route of the sea-wall had been re-aligned.[126] This re-adjustment has erased evidence for the former route of the defences depicted in Hayward's map – making it difficult to assess the relative chronology of the breach and the sea wall.

When breaches occurred, or the incidence of flooding increased, medieval communities adapted to meet the challenge. A variety of evidence attests to the building, and rebuilding, of flood defences – including their strengthening and redesign – by medieval communities in response to episodes of flooding. At Thornton Abbey, North Lincolnshire, for example, a bank interpreted as a flood defence was twice breached by flooding from the nearby Skitter Beck, leaving behind layers of alluvium, but, each time, the bank was subsequently heightened – presumably with the aim of preventing future flooding.[127] Similarly, at Broomhill, East Sussex, in the aftermath of the 1287/88 floods, the sea walls, the breaching of which contributed to catastrophic flooding in Walland Marsh, were redesigned.[128] In this case, however, in addition to a desire to prevent such a disaster in the future, the significantly altered landscape no doubt affected the

126 SILVESTER 1988: 41.
127 WILLMOTT, TOWNEND 2012: 19.
128 GARDINER, HARTWELL 2006: 151–152.

decisions taken in the rebuilding of the flood defence system. Contemporaneously, in the nearby Brede Valley, East Sussex, the increased incidence of sea floods in the 13th century seems to have sparked the construction of a linear embankment known as the *Damme*, through which a sluice allowed freshwater to flow.[129] The evolving challenges imposed by flooding on medieval communities, therefore, were met with both highly practical solutions – which display a high level of understanding of water and its physical properties – as well as complex social systems which attempted, though not always successfully, to ensure the flood defence infrastructure remained fit for purpose.

4.6 Financial relief

Money was one of the foremost concerns in the minds of those affected by hazards – or, at least, this is the conclusion gleaned from many of the historical documents which primarily relate to financial matters. As has already been seen, agricultural management largely revolved around the profitability of different farming practices and the pay-off that investment in particular areas, through embanking and soil improvement strategies, could bring. An unexpected hazard such as a major flood which disrupted the agricultural system, therefore, could quickly plunge communities into difficult situations. Following the 1338 flood in Marshland the affected communities appear to have been particularly vulnerable from a financial standpoint. The historical sources emphasise the general impoverishment of the people in the wake of the flood[130] and, in order to remedy the situation, a petition was sent to the King by 15th January 1338, just 3 days after the floods had struck.[131] In response to this petition, on the same day, the king ordered a commission *de walliis et fossatis* to investigate the situation in the Fens.[132] Since 1337, a tax of a fifteenth of the value of all movable goods had been imposed throughout the realm[133] for the following three years to pay for the ongoing Scottish Wars.[134] The inhabitants of Marshland successfully petitioned the King to order their movable goods to be reassessed on account of the flood of 1338 in order to reduce their contribution.[135] Several elements of this petition suggest an attempt to exaggerate the plight of the community in order to maximise the concessions they might receive from the Crown. The lands of the parish of Wiggenhall, for example, were described as being

129 Gardiner 1995: 131.
130 Maxwell Lyte 1915: 62.
131 Maxwell Lyte 1915: 61–62.
132 Maxwell Lyte 1895a: 581.
133 The tax was levied as a fifteenth for rural dwellers but only a tenth for those who lived in urban areas.
134 Maxwell Lyte 1915: 50.
135 Maxwell Lyte 1915: 61–62.

split in two by the sea.[136] As Wiggenhall was made up of land on both sides of the Ouse (see Fig. 3.6), this may have been an exaggeration of the normal state of affairs to make the problems with flooding appear even more precarious than the reality. Equally, the loss of areas of marsh, as the parishioners of Terrington complained had occurred at Rushemershe and Newemershe,[137] may be another case of exaggeration as such areas were usually outside the embanked area and were largely undeveloped – perhaps used for pasture. As a result, the local community would have been well aware of the high flood risk of these areas and it is, therefore, unlikely that the inundation of these areas amounted to the significant loss that it appears in the parish's petition to the Crown. Although the reassessment of the area's tax burden seemingly ameliorated the situation for the communities of Marshland, the tax was subsequently mistakenly collected based on the pre-flood assessment – rather than the reassessment carried out with Royal consent in the aftermath of the disaster[138] – although this was eventually rebated.[139]

In petitioning the King for a tax concession in 1338 the people of Marshland were far from unique. To cherry pick some comparable examples, the people of Saint-Sever, Landes, within the English possession of Gascony, were granted a tax exemption lasting for two years due to the damage caused by floods to the town's defensive walls in the late 1420s.[140] Comparably, when the buildings of the Augustinian nunnery of Lacock Abbey, Wiltshire, were severely damaged by a fire ignited by a lightning strike in 1447, the Abbess was able to negotiate an exemption from "all tenths and other subsidies, quotas, aids and contributions whatsoever for a term of 40 years".[141] In a similar vein, in the aftermath of the 1362 windstorm, the Black Prince (the heir to the throne) was willing to forgo a percentage of rent from his Cornish tenants who had been ill-affected by the storm[142] while, compared to earlier periods, the exchequer was particularly amenable to requests for the remission of debts in light of storm damage.[143] Such remissions of rent and tax obligations facilitated recovery from short-term shocks, while for landlords and authorities, loosing out on a portion of rent or tax may have made good economic sense over the longer-term by preventing their tenants from suffering so acutely that they were plunged into total poverty – rendering them unable to make any future payments.

In some ways hazards brought relief with them. In the aftermath of the 1362 windstorm, for example, the trees blown down by the storm provided not only a ready source

136 Dugdale 1662: 258.
137 Dugdale 1662: 259.
138 Maxwell Lyte 1903: 394–395.
139 Dugdale 1662: 260.
140 TNA: C61/125: 26. http://www.gasconrolls.org [Accessed 12 February 2018].
141 Maxwell Lyte 1909b: 86.
142 Dawes 1931: 189.
143 Ormrod 1996: 169.

of raw materials to make repairs but also a potential source of income. This can be seen widely across the royal parks belonging to the Black Prince. Only one month after the storm, the Prince instructed his steward at Cottingham, East Yorkshire, to sell the trees felled by the wind at Cottingham and Kirbymoorside, North Yorkshire, for profit. At the royal park at Byfleet, Surrey, meanwhile, the Black Prince gave orders that the wind-felled timber was to be sorted into wood suitable for timber, wood that could be used as park paling and firewood[144] and the wood felled in Cornwall was to be used to make repairs across the Prince's estates – though some was also to be given to the Prince's tenants to allow them to make repairs.[145] At the parks of Berkhamsted and Gaddesden, Hertfordshire, the timber was to be sold "as profitably as possible"[146] and this also evidently took place at Clarendon, Wiltshire.[147] The unprecedented number of trees felled at Berkhamsted even led the Prince to renegotiate the long-held custom that his parker, the administrator who oversaw the management of the park, was entitled to the profits from wind-felled trees which, after the 1362 storm – so great was the quantity of trees felled – was changed to a fixed annual sum of 100 shillings.[148] Of course, access to the wind felled timber was unequal across society – although landowners, such as the Black Prince, were free to make a profit from the fallen timber on their estates, the rights to trees on land held by tenants typically belonged to the landowner. That the storm exacerbated poverty among the lower classes is suggested by a petition to Parliament in 1362 which, in light of the storm, sought to secure tenants' rights to sell such wind-felled timber "and make a profit",[149] although how Parliament responded to this petition is unknown.

Although landowners sometimes assisted their tenants, often it seems to have been the tenants responsibility to make any repairs necessitated by storm damage. Details of a court case from London conducted four years after the storm's occurrence demonstrate that, at least in the case of the burler Henry Maynburgh,[150] tenants were expected to pay for repairs to rented properties which, for some, must have been a significant financial burden. The difficulties of the less well off proffered opportunities for those in a position to take advantage of the situation. This is hinted at by a description of the activities of the Archbishop of Canterbury, Simon Islip, in the aftermath of the storm of 1362. That year, Islip acquired land in Oxford in order to found Canterbury Cathedral

144 Dawes 1933: 416.
145 Dawes 1931: 189.
146 Dawes 1933: 431.
147 TNA: E 101/698/4.
148 Dawes 1933: 431, 464. This change guaranteed the parker an income but it was rather less than that derived from wind-felled trees in an average year.
149 Ormrod et al. 2017: 147–148.
150 The court case, conducted in 1366 concerned damage to the roofing of the property rented by Maynburgh in St Lawrence Lane, London. The rental agreement proscribed that the tenant was financially responsible for making the repairs, which, although Maynburgh attempted to do, led to a disagreement with the landlord. Thomas 1929: 61–62.

Priory's college at the University, which was named Canterbury Hall. The source which describes this purchase strongly implies that the land was available, and perhaps cheaper, as a result of the damage caused by the storm.[151] It therefore seems to be the case that the Archbishop seized on the opportunity that the storm had provided by purchasing land below the market value, taking advantage of the short-term financial needs of the previous owner(s). This illustrates that for those with sufficient financial capital to hand, such an event could offer attractive opportunities.

Ecclesiastical authorities were particularly well placed to raise additional revenue as a result of the damage wrought by hazards. Importantly, for example, the Church was often a beneficiary of gifts from the aristocracy and wealthy middle classes in the wake of disastrous events. Following the storm of 1362, for example, the Black Prince donated timber blown down by the storm across his estates in Lamarsh, Essex, to enable the repair of the chancel of the church at Great Henney, Essex.[152] Similarly, wood for fuel was donated to the Dominican friary of Dunstable from wind-felled timber in the park of Berkhamsted.[153] That such largesse may relate to a wider Christian tradition is suggested by the fact that gifts of timber or rights to take wood were also frequent donations to churches affected by natural hazards in medieval Iceland.[154] Gifts of this nature were not simply charitable donations, as most donors had clear expectations when they made these kinds of contributions to the Church. Amongst a populace inherently concerned with the final destination of their souls, and those of their relatives, such gifts could provide peace of mind. This is demonstrated by a local alderman's donation of £100 towards the construction of a new bell-tower for the Augustinian friary of Clare, Suffolk, in 1363.[155] In this case, the size of the donation along with the "munificent largesse in alms and other benefits" of the benefactor led to the appointment of a priest to hold prayers for him, his parents and any to which he "is obliged, both in life and in death".[156]

Where the donations of wealthy individuals were insufficient to cover the cost of repairs, other fund-raising methods were required. *In extremis*, this could include the sudden development of an attraction to stimulate a tourist industry – as occurred at Glastonbury in 1191 when the bodies of King Arthur and Queen Guinevere were miraculously discovered a few years after a destructive fire, providing a ready flow of

151 Wharton 1691: 415.
152 Dawes 1933: 432.
153 Dawes 1933: 417.
154 Although the scarcity of timber in Iceland must have given these gifts particular importance in this setting. Swedo 2012: 150–153.
155 There is no certain evidence that the Augustinian Friary of Clare was a victim of the storm of 1362 although this would be consistent with both the date of this donation and the friary's geographical location.
156 Harper-Bill 1991: 85.

both visitors and cash.[157] Another atypical example may be the Royal grant to Cleeve Abbey, which came in the aftermath of the cliff collapse that destroyed the chapel of St Mary the Virgin,[158] which gave the Abbey the right to hold a weekly market and fairs on appointed days of the year[159] in order to raise additional revenue to recover from the impact of the disaster. More dependable and universally applicable methods to generate capital in response to disasters were also available. Indulgences allowed the Church to elicit donations of money and labour from the lay populace in return, theoretically, for reductions in the amount of time an individual's soul would spend in purgatory.[160] Particular indulgences were connected to specific locations or causes and reduced the stay in purgatory by varying lengths of time, with the indulgences promising the longest reductions attracting donations of higher amounts and from further afield than those which made more modest claims. Indulgences were commonly issued in the aftermath of disasters as can be seen at the Benedictine priory at Molycourt, Norfolk, located immediately south of Marshland, which was granted an indulgence of 40 days by the Bishop of Ely in 1385 due to poverty exacerbated by storms and floods.[161] Similarly in 1394, Ankerwycke Priory, Buckinghamshire, was granted an indulgence for "relief of [the] poor nuns ... whose goods [have been] destroyed by floods"[162]. Two indulgences were issued to stimulate the donation of alms towards the repair of the chapel of Saint Mary the Virgin, at Cleeve, Somerset, (which was later destroyed by a cliff collapse)[163] to remedy the damage caused by marine floods.[164] and in 1485 a Papal indulgence was granted to raise money for repairs to flood defences in the Isle of Ely.[165]

Indulgences are particularly visible as a response to the widespread damage caused by a disaster in the aftermath of the January 1362 storm. Thus, from 1363, those who contributed towards the restoration of Norwich Cathedral saved themselves seven years and seven quadragene[166] in purgatory.[167] At Salisbury Cathedral, another victim of this storm, no indulgence was advertised but the Papacy granted the appropriation[168]

157 Rippon 2004: 128–129.
158 This is discussed in Chapter 3.4.
159 These were the feast of Saint James (1st May) and the feast of the Exaltation of the Cross (14th September) – and the three days that followed. Maxwell Lyte 1897: 527.
160 Swanson 2007: 61–63, 149–160.
161 Page 1906: 349.
162 Gibbons 1891: 399.
163 This is discussed in Chapter 3.4.
164 The first was issued in 1398 by the Bishop of Exeter while in 1400, either as a result of continued problems with flooding or the failure of the first indulgence to raise the necessary funds, the Papacy granted an additional indulgence of three years and three quadragene. Randolph-Hingeston 1886: 62; Bliss, Twemlow 1904b: 400.
165 Lee 2015: 142.
166 A period of 40 days.
167 Bliss 1896: 418.
168 Income from tithes.

for six years of the nearby church of St Thomas, which annually amounted to £10, towards the cost of repairs.[169] Lesser ecclesiastical foundations attracted correspondingly proportioned indulgences in the wake of the 1362 storm: Cloyne, County Cork, and Stone, Kent, were both issued indulgences of three years and three quadragene[170] while Whitechapel, London and Colchester, Essex, both attracted indulgences of one year and one quadragene.[171] Although indulgences might be considered a spiritual reaction to natural hazards, their ability to stimulate the donation of funds and manpower from amongst the ranks of medieval society made their issuance a sensible and pragmatic response, employed widely throughout Christendom,[172] in the face of severe destruction.

4.7 Migration, Abandonment and Decline

To what extent disasters had long-term effects, influencing the trajectories of individual settlements, wider landscapes and society at large, is an important question. Investigating how medieval settlement and activity developed in the aftermath of disaster, however, is far from a simple task. Traditionally, historians have relied on tax records to estimate economic or demographic change in regions and individual towns but such sources are beset by a number of problems – the lengthy gaps between available records and the discrepancy between how taxation was levied in urban and rural areas to name but two.[173] The archaeological record offers some alternative routes to shed light on these questions – through, for example, excavated evidence for desertion or continuity and trends in material culture – such as a decline, shift in production, or a change in style.[174] A combination of the historical and archaeological evidence readily demonstrates the diversity in possible outcomes in the aftermath of a disaster but, as will become clear, isolating the contribution made by an individual disaster from wider, pre-existing endogenous forces proves to be a challenging task that can rarely be satisfactorily resolved.

In the moment of disaster, one of the primary responses open to medieval communities was to flee – to leave the area of danger until it was safe to return. Relocation away from the usual place of settlement, in the face of environmental change is a complex topic which, in light of contemporary climate change, has generated a vast literature. Within studies examining modern populations there is disagreement over the extent to which environmental migration as a concept exists – with different research-

169 Bliss 1896: 462–463; Bliss, Twemlow 1902.
170 Bliss 1896: 414, 421–422.
171 Bliss 1896: 444, 468.
172 For example see Rohr 2003: 138.
173 Rigby 1979.
174 Jervis 2017: 6–7.

ers emphasising the environmental aspect, as opposed to economic, social, political or cultural concerns, to any given migration to varying degrees.[175] In the case of medieval examples, the archaeological and historical sources rarely furnish us with the level of detail required to make a nuanced assessment of the factors and decisions behind an apparent shift in population. In the most extreme cases, however, the evidence is stark. Sudden-onset and unexpected high magnitude hazards left medieval populations with no choice but to 'up-sticks' and relocate to less vulnerable locations. What they chose to do next however, is more nuanced. Few displaced populations would readily renounce their connection to the land they had previously called home. After a landslide at Leirfall, Stjørdal, Norway, in *c.*1200,[176] for example, even though the landslide probably totally destroyed the pre-existing farmstead, the site was subsequently re-settled, likely indicating that those displaced by the disaster were unwilling to relocate and chose to resettle the area once it was safe to do so.[177] This was perhaps also the case during the medieval period in southern Iceland, close to the Mýrdalur mountain range. Here, the historical evidence suggests that pre-14th century settlements were abandoned as a result of floods, though the presence of post-medieval farms in the area indicates a subsequent resettlement must have occurred in the aftermath.[178]

Permanent relocation was only a favoured option when factors made it impossible for resettlement to take place. Certainly, in some cases, natural hazards made it so: during the excavation of the church at Broomhill, East Sussex, the lack of evidence predating the late 13th century,[179] suggests it had been relocated following the storms of 1287/88. If correct, this would suggest that this flood disaster forced the permanent abandonment and relocation of human settlement. Nearby, at Winchelsea, an earlier series of storms in the 1250s instigated a gradual process of migration from the town's original coastal situation to a new planned town occupying a nearby hilltop.[180] As conditions deteriorated at the town in the wake of the mid-13th century storms, Edward I took an active role in the planning and redesign of the new town.[181] Although this process was certainly underway by 1280, the official transfer of land at the new site to the Winchelsea residents did not occur until July 1288[182] – a process which may well have been brought forward due to the occurrence of the storms throughout 1287

175 Lübken 2012: 11.

176 Unger, Huitfeldt-Kaas 1889: 1.

177 The toponym 'Leirfall' may reference the landslide event and stands out amongst surrounding toponyms which are Iron Age in origin. Its appearance in a 15th century land register indicates the area had been resettled by this time. See Sognnes 2011: 186; Jørgensen 1997: 201.

178 Vésteinsson 2000: 100–101. It should be noted, however, that the account which describes these events includes folkloric themes, and seems to have been composed at a significantly later date, casting doubt over its historicity.

179 Gardiner 1988: 125.

180 Martin, Martin 2004: 4–6.

181 Lilley 2014.

182 Martin, Martin 2004: 5.

and early 1288 – which probably made what remained of the old town uninhabitable. The particularly vulnerable location of Winchelsea, therefore, forced the town to be abandoned though, in this case, through the support of the Crown, a nearby location was developed to replace what had been lost. Somewhat comparably, following the Saint Elizabeth's Day Flood of 1421, which permanently submerged a number of villages in South Holland, many of the survivors appear to have relocated to nearby towns such as Dordrecht.[183]

Several additional examples of permanent abandonment are known in the face of wind-blown sand. The settlement at Forvie, Aberdeenshire, in existence from at least the 12th century, appears to have been besanded suddenly in the early 15th century. The site was never subsequently resettled and a 'best guess' for the location to which the survivors relocated comes from the foundation, shortly after the occurrence of the sand inundation, of a chapel 8km away at Leask dedicated to Saint Adamnan – the same dedication as Forvie's parish Church and a relatively rare dedication locally.[184] While the pull-factors of the new place(s) settled by the Forvie residents are unknown, it seems unequivocal that the sand inundation provided the primary impetus for their relocation. In this case a substantial portion of the town's agricultural land may have also been affected which would have made any attempted resettlement particularly challenging, ensuring that what may have been initially intended as only a temporary migration, morphed into permanent abandonment.

In a similar vein, in the mid 16th century a sizeable settlement, attested by a significant archaeological assemblage,[185] at Meols, Merseyside, appears to have been erased suddenly by a sand inundation. The archaeological evidence supports a shift in activity to the nearby location of Great Meols, which regression analysis of later tithe maps suggests was most likely an area of marginal land, re-organized after the disaster to provide a new area of settlement.[186] In this case, however, the precise causation of the migration to Great Meols can be questioned. The assemblage from the Meols site is indicative of an economic decline in the century before the final abandonment[187] and the shift to Great Meols could have begun before the inundation of the Meols site by sand. It is possible, therefore, that, while the natural hazard likely provided a short-term catalyst, there may have been underlying endogenous factors behind the relocation which pre-dated the onset of disaster. While in the most extreme cases, therefore, it certainly seems that disasters forced temporary or permanent abandonment and migration, the resulting depopulation and economic decline may have had other root causes–such as a shift in the availability of markets or shifting regional trading

183 Pollmann 2017: 122–123. Note that the disaster which caused this abandonment is depicted in Fig. 6.4.

184 Brown 2015: 142.

185 Griffiths et al. 2007: 434.

186 Griffiths et al. 2007: 414, 409–411.

187 Griffiths et al. 2007: 435.

trends–which may have been entirely unrelated or, at least, less closely correlated with the occurrence of disasters.

The relocation of monastic foundations provide further examples of site-based abandonment and migration–particularly in the face of flooding. Monasteries actively managed water supplies for practical purposes such as milling, waste disposal and the provision of drinking water. As a result, they were frequently sited in locations which were liable to flood. One example comes from the Premonstratensian house of Leiston Abbey, founded on the Suffolk coast in *c.*1182, which in 1363 was moved further inland as a result of damage caused by marine inundations. To make the most of the existing assets, the original foundation appears to have been carefully demolished so that the building materials could be reused in the structure of the new Abbey buildings. This activity necessitated the transportation of structural fragments over 3km inland. On the highest ground on the original site, a chapel was constructed to make use of the abandoned land while avoiding, as far as possible, further problems with flooding.[188] Such relocations affecting monastic foundations, especially those of the Cistercian order, in response to flooding and other environmental, or political concerns, were relatively common.[189] Most often, however, monastic foundations were re-sited as a result of routine and recurrent low-level problems rather than one-off, unprecedented disasters.

The many cases described above illustrate that disasters did force abandonment and the migration of populations but undoubtedly more common was continuity and the persistence of settlement. In such cases, however, disasters still exerted an impact. Such a picture may be traced across New Romney in the aftermath of the 1287/88 storms through the survey of archaeological 'grey literature' discussed in Chapter 3.1.3. The period directly following the occurrence of the storms was marked by a phase of abandonment and demolition visible at many excavated sites across the town.[190] This presumably corresponds to a dramatic reduction in both population and economic activity as a result of the damage wrought by the storm–which obliterated the town's harbour[191] as well as many of the structures abutting the beachfront[192] – in addition to the ongoing siltation problems that threatened the harbour in the pre-storm period. The presumed decline that followed is evidenced at the Southlands School site, by a phase of low-intensity activity[193] while buried soil horizons overlain by windblown deposits identified at St Martin's Field, dated through ceramics to the mid 13th–early 14th century, could relate to areas which fell out of use in the storm's aftermath.[194] At 16

188 Boulter 2008: 40–41.
189 Donkin 1959: 253–255.
190 E.g. Draper, Meddens 2009: 68.
191 Canterbury Archaeological Trust 2010: 8.2.
192 E.g. Wilson, Linklater 2002: 14–15; Draper, Meddens 2009: 67–68.
193 Meddens, Wooldridge 2002: 23.
194 Gollop 2008: 38.

High Street the excavators suggest that the notable lack of material datable to the time of the storm (*c*.1250–1350) relates to the impact of the disasters on the town.[195] In the north-east of the town at 'The Elms', glazed peg tiles were interpreted as evidence for the primary deposition of roofing in the demolition of a high status building dating to the 13th century[196]–perhaps indicating a period in which buildings fell out of use and into disrepair in the aftermath of the storms. This could relate to a phase of demographic and economic decline or stagnation which saw a significant proportion of the population migrate from New Romney to other more prosperous towns – leaving buildings, such as the one detected at this site, to fall into a state of disrepair.

The shift in the style of pottery in use between pre- and post- storm contexts may also hint at one of the storms' negative impacts. The local pottery industry, which up until the time of the storms appears to have supplied the majority of the town's ceramics, seems to have ceased production – with the pottery from contexts post-dating the storm instead imported from the nearby production centre at Ashford.[197] The decline of the local ceramic industry may be explained either by the destruction of pottery production facilities as a direct result of the storms, a lack of available manpower in the storms' aftermath, the fact that damage to the town's harbour removed the local potters' access to important markets – making their business unsustainable – or a combination of more than one of these factors.

The survey of the archaeological 'grey literature' relating to New Romney[198] revealed 15 sites with evidence interpreted as relating to decline in the post-storm era (see Fig. 3.2). In addition, a reclassification of those sites where the storm(s) directly caused damage or deposited discrete layers of sediment,[199] based on any evidence for what transpired after the occurrence of the storms, added an additional 10 sites with evidence for decline or abandonment in the aftermath. To these commercially funded excavations may also be added the 1987 discovery of two 13th century medieval structures on the town's southern periphery, which show evidence for abandonment in a time-frame consistent with the aftermath of the storms[200] – giving a total of 27 sites indicative of a post-storm decline in the town. These sites are plotted in Figure 4.11 where they are contrasted with those sites which indicated continued, undisrupted occupation in the town – of which there were a total of 92. Interestingly, the sites interpreted as indicating disruption or decline display some patterning with a pronounced cluster discernible in the town's southern margin. This may indicate a sector of the town which was hit particularly severely by the storms, although this is not an obvious conclusion based on the results presented in Figure 3.2. Alternatively, as a relatively

---

195 Wessex Archaeology 2012a: 18.
196 Stevenson 2006: 21.
197 Canterbury Archaeological Trust 2010: Appendix 5.
198 This is discussed in detail in Chapter 3.1.3.
199 Those in classes 1 and 2 as defined in Chapter 3.1.3.
200 Wilson 1987: 204.

marginal area, buildings in this part of the town may have been more readily abandoned in the aftermath of the disaster. Although, much of the archaeological evidence for decline is relatively circumstantial, a sizeable body of evidence suggests that the town was gripped by some level of decline in the aftermath of the 1287/88 storms and it is probable that the occurrence of the storms was at least partially responsible for this outcome.

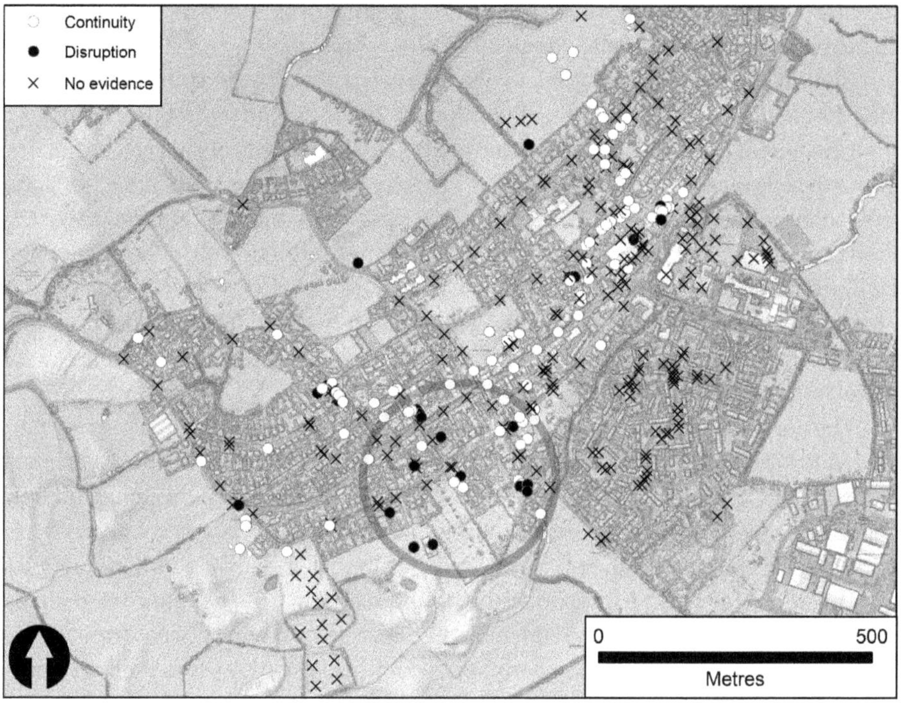

Fig. 4.11: Disruption and continuity in New Romney after the storms of 1287/88. White dots indicate sites with evidence for undisrupted occupation at the time of the storms while black dots indicate sites with evidence for abandonment, disruption or decline. The dark circle highlights a cluster of sites interpreted as indicating disruption. Created by the author.

It should be noted, however, that the apparent evidence for decline detected across New Romney may be misleading. High uncertainties in dating the archaeological contexts across the town make it possible that the occurrence of this decline overlapped with the Black Death (1348–1352). Although the storms of 1287/88 would have compounded any subsequent decline as a result of the epidemic, the acute impact of plague on demographic and economic activity resulted in a general trend of decline visible in the archaeological record more generally – at least in East Anglia.[201] Furthermore, it should be noted that the majority of sites with evidence for continuity amongst those considered in the 'grey literature' survey – those taken to indicate continuity in Figure 4.11 – did not reveal any obvious evidence for the impact of the storms, either directly or through a subsequent phase of decline. Due to the low resolution of the chronologies which can be established based on the ceramic evidence, whether these sites relate to pre- or post- storm activity cannot be reliably differentiated. In addition, the survival of evidence for a relatively short phase of disruption and decline in the aftermath of the storms is likely to be the exception rather than the rule – while later site clearance and construction can be expected to have widely destroyed or truncated contexts relating to this period. These considerations mean that, while the interpretation of many of the sites as indicating continuity of occupation throughout the occurrence of the storms and their aftermath is probably reliable, the precise form that this continuity took in the short-term is difficult to gauge. Moreover, the picture revealed by the evidence presented in Figure 4.11 may over-emphasise the extent to which continuity characterised life in the storms' aftermath – at least in the short-term, which must have been characterised by substantial disruption.

This contrasting picture of decline and continuity is further muddied when the town's longer-term economic and demographic trajectory is considered. Shifts in the wider economy of Romney Marsh over the long-term – principally the switch from arable agriculture to less labour intensive pastoral farming combined with the reduction in maritime trade as a result of problems with siltation and protracted conflicts with the French – led to a general trend of depopulation.[202] Additionally, in reference to the archaeological evidence for decline in the neighbouring ports of East Sussex, (Rye, Winchelsea, Hastings, Pevensey, Seaford and Shoreham) which were also affected by the 1287/88 floods, Jervis highlights the diversity in the long-term trajectories of each town.[203] While some experienced depopulation and economic adversity, many aspects of life continued and civic administration appears to have remained strong throughout the medieval period.[204] Exactly what component the storms of 1287/88 played in the decline apparent through the archaeological record across New Romney, therefore, is difficult to gauge – a problem which is compounded by the low-chronological resolution of much of the evidence indicative of decline, the likelihood that large quantities of related evidence have been destroyed in later periods and the alternative explanations for decline operating across different time-scales. The long-term impact of disasters in settlements, therefore, is challenging to disentangle. Pre-existing endogenous processes, as well as other near-contemporary catalysts for decline – such as the Black Death – may have been more important factors in shaping long-term trajectories of

201 Lewis 2016.
202 Rippon 2001: 31–33. See also Chapter 7.2.1.
203 Jervis 2017.
204 Jervis 2017: 21–26.

change. Although quantitative assessments of a disaster's impact on individual settlements are plagued by a wide array of issues, therefore, disasters such as the 1287/88 storms undoubtedly did play a critical role in the short-term that disrupted life and led to repair, rebuilding and re-organisation across a wide variety of spheres of activity.

4.8 Summary

Clearly, despite the theological predispositions of the age, which emphasised the inevitability of disasters as events brought about according to God's plan, medieval populations were not fatalistic and widely adopted practical responses to mitigate the impacts of hazards. This is evident across many strands of evidence including pragmatic agricultural management strategies, structural repairs and adaptations and the construction, repair and re-design of flood defences. It is clear from the level of organization evident in the historical record relating to communities that dwelt in flood-prone areas, and the systems of flood defence – which are, in some cases, attested by upstanding remains – that flooding as a hazard was well understood as a recurrent risk that could be prevented through human action. This is precisely the same interpretation reached by MORGAN in relation to responses to flooding in early modern England;[205] clearly, therefore, this was not a phenomenon of the early modern period but a system with roots that were well-established in medieval Britain.

205 MORGAN 2015: 48–49.

5 Systems of protection: religious and superstitious defences against hazards

This chapter investigates the development of beliefs associated with protection and the material and historical evidence for these practices. Where possible, a number of case studies are explored in order to more fully understand how religious and superstitious rituals and practices were believed to provide protection from natural hazards during the medieval period. From much of the evidence discussed in Chapter 4 it certainly appears that populations occupying locations which were regularly exposed to risk from natural hazards appreciated their vulnerability. Despite this fact, natural hazards do not appear to have widely been regarded as a discrete category against which special precautions were required. Instead, natural hazards seem to have been unconsciously grouped together with other 'naturally' occurring misfortunes such as disease, crop-failure, blight, pestilence and general ill-luck. To combat against these diverse problems an equally broad suite of both overtly Christian as well as superstitious behaviour, and a blurred combination of the two, was practised to gain protection, avoid risk and provide peace of mind. These apotropaic and ritualistic practices believed to provide protection, which are often difficult to understand from a modern perspective, are the subject of this chapter.

An important and understudied area of medieval ritualistic behaviour are superstitions connected to protection – of which a wide array relate to contemporary material culture. Although it is very difficult to trace the origins of these superstitious practices, many have roots in earlier pagan beliefs. Certainly, various early medieval and Viking-Age artefacts have associations with apotropaic or amuletic powers.[1] For example, according to DICKINSON, the animalistic designs on Anglo-Saxon shields added "a supernatural protective layer to the shield itself, and hence to its bearer".[2] Relatedly, an early shield boss from Thorsberg, Schleswig-Holstein, Germany, bore a runic inscription that referenced 'hail', perhaps channelling the powerful and dangerous associations of this type of weather to protect the shield bearer. Similar inscriptions invoking the power of hail are also found on later inscriptions such as a pendant from Ølst, Denmark, and a small soapstone tablet from Kinneve, Sweden.[3] Coins too held magical significance in this period with both Byzantine and northern European traditions attributing protective qualities to coinage from ancient Greece and Rome.[4] The popular Norse God Thor was also frequently invoked for his ability to control the elements, especially thunder

1 FUGLESANG 1989.

2 DICKINSON 2005: 161.

3 MACLEOD, MEES 2006: 89–90.

4 MAGUIRE 1997.

https://doi.org/10.1515/9783110719628-005

and lightning. This can be seen on an amulet from South Kvinneby, Öland, Sweden which called on the God's lightning for protection.[5]

Early medieval documentary evidence also specifically attests to superstitious beliefs connected to protection from hazards. For example, the early English text *Medicina de Quadrupedibus*, which was known across Europe from Late Antiquity to the Reformation, tells how the teeth of a badger, properly kept and carried, provide protection as "neither heavenly body, nor hail, nor strong storm, nor evil man" will cause harm to the bearer.[6] Similarly, the late Roman writer Palladius described how hail could be warded off with bloody axes lifted towards heaven, owls fixed to the ground with outstretched wings or by coating knives used for pruning plants with bear's fat.[7] This account also reveals an interesting dimension to superstition, which is that for such activities to be effective, it was necessary that they remained secret. This may relate to a social taboo around admitting belief in the efficacy of superstitious activity which explains why such behaviour is only relatively rarely included in documentary accounts and, in some cases, archaeologically attested practices appear to be completely undocumented. Clearly, during the early medieval period, an extremely wide gamut of beliefs existed related to different aspects of the natural world. Such beliefs came to be regarded as superstitious due to their relation to Pagan worship or, as Christians perceived it, the worship of daemons posing as alternate deities.[8] Early Medieval writers, such as Isidore of Seville,[9] widely dismissed these superstitions and advanced Aristotelian explanations for the occurrence of natural hazards. During the later Middle Ages although official attitudes towards superstitious magic stiffened,[10] some scholars have argued that these beliefs actually became more prevalent during the 15th and 16th centuries. Certainly, the witch hunts which began during this period were characterised by authorities extolling the abilities of witches and magicians to exert control over demonic and diabolic forces.[11] According to Bailey, however, superstitions were essentially a continuum of belief which operated from at least the early Middle Ages right through to the 18th century when they were largely swept aside by the Enlightenment.[12] In some cases though, elements of superstitious folklore appear to have persisted from the the pre-Christian era at least until the late 19th century.[13]

5 Macleod, Mees 2006: 27–29. The inscription is fully detailed on the website of the Skaldic Projet at: http://skaldic.abdn.ac.uk/db.php?id=15517&if=runic&table=mss [Accessed 01 February 2021].

6 Cockayne 1864: 327.

7 Owen 1807: 49–50.

8 Bailey 2009: 638.

9 Barney et al. 2006: 174, 181–183.

10 Bailey 2007b: 100–101.

11 Bailey 2001.

12 Bailey 2009: 658–660.

13 Heanley 1898: 186.

The boundary between superstitious and Christian material culture was permeable and open to interpretation. Many practices, which at first may be regarded as superstitious, are clearly connected to Christian liturgy and beliefs.[14] While, in writing, the orthodoxy was swift to condemn such practices, in actuality the Medieval Church often appears to have sanctioned and even encouraged the use of charms, amulets and other apotropaic material culture. As THOMAS highlights, for example, the official Church stance had no aversion to the use of holy water or church bells to drive away storms.[15] At a local level, though, it is likely that some parish priests misinterpreted or even disregarded official Church guidelines. An example can be seen in the initial veneration afforded to the Holy Sacrament, allegedly discovered near Passau, Bavaria, Germany, in the late 15th century by the local clergy which was quickly denounced when higher Church officials were made aware of the situation.[16] Such inconsistencies would have resulted in a patchy and shifting application of official Church guidelines with some parishioners receiving the impression that certain practices were legitimate while those from other parishes regarded them as fraudulent or even heretical. The development of beliefs was also far from unilateral with the laity exerting some control over the Church's attitude to particular issues. This can be seen in the examples given by ZIKA of German parishes in which priests were shunned by their parishioners for attempting to reform traditional practices, including processions of the Eucharist to bless crops against hail.[17] Unlikely artefacts may even have held a Christian, although superstitious, significance. Prehistoric arrowheads, for example, were frequently interpreted as 'elf-shot' which was believed to protect against disease, thunder and lightning. HALL, however, theorizes that a Bronze Age example, discovered at Perth, could have had its protective powers enhanced by being touched against a relic of St Sebastian – who was widely held to protect against both physical archery and elf-shot due to his martyrdom by arrows.[18] Such practices however, operated within a difficult grey-area as misunderstandings or improper interpretations of genuine Christian rites ran the risk of inviting in demons or the devil.[19] By the Reformation many of the practices of the monasteries were considered superstitious[20] and the switch to Protestantism saw the end of many officially sanctioned Church festivals and rituals connected to protection. Equally however, many practices, and crucially, beliefs did continue, sometimes in an adapted form, into the post-medieval period.

The extent to which archaeology can provide an insight into past beliefs has long been a major debate within the discipline. Hawkes famously described archaeological

---

14 BAILEY 2009: 657.
15 THOMAS 1971: 303.
16 ZIKA 1988: 28–29.
17 ZIKA 1988: 35.
18 HALL 2011: 94.
19 BAILEY 2009: 645.
20 HEALE 2007.

research on the religion of past societies as "the hardest inference of all".[21] Despite this early negativity, religion and belief have risen to become major themes in contemporary archaeological research. Although, for the medieval period, many of the problems encountered by prehistorians are mitigated by documentary evidence, providing a primary record of contemporary rituals, beliefs and theological debates, there remain many practices revealed by the archaeological record that have no obvious basis in documented beliefs.[22] Material and documentary evidence are therefore most valuable in unison as while archaeology, in some cases, can provide an insight into the practices of individuals or groups, such as the peasantry, which are usually marginalized in the surviving documentary sources, so too documentary sources testify to ephemeral activities, such as processions, which do not leave a significant material signature.

Documentary sources, therefore, combined with the careful interpretation of archaeological evidence, allow many types of artefact to be associated with beliefs about protection from particular dangers and hazards. In medieval Europe, these included personal items such as jewellery, coins, symbols of pilgrimage and a wealth of items relating to the cult of the Saints. Where beliefs related to protection associated with specific types of artefact were documented it is logical to assume that the owner, wearer or bearer expected to receive these protective benefits though inevitably questions remain as to the universality of specific beliefs and how widely these may be applicable. In Britain, a useful resource for this research is the Portable Antiquities Scheme (PAS), an online database of finds reported voluntarily by the general public throughout England and Wales. To date, the database holds almost a million records,[23] 23.5% of which are medieval in date. This mass of data, however, is not without its problems. The uptake of the scheme has been far from uniform and as a result some areas are over represented, such as the Isle of Wight, Norfolk and Lincolnshire, while others have relatively few reported finds, as is the case throughout most of Wales as well as in the north and south-west of England. This can be seen visually in Figure 5.1 which shows the density of medieval finds reported to the PAS across England and Wales. To some extent this disparity may reflect genuine disparities in the archaeological record as both Norfolk and Lincolnshire were among the most populous counties during the medieval period.[24] Through comparison with the map on the right in Figure 5.1, however, it is clear that some areas, notably the south coast, are overrepresented while the counties around London seem to be reporting less medieval material than would be expected based on estimated contemporary population levels. In addition, to the spread of population during the medieval period itself, the PAS database, in common with other national archaeological datasets,[25] is biased by the collection of data in the present.

21 Hawkes 1954: 162.

22 See for example Champion 2015a: 9–10.

23 As of 01 February 2021 there were 973,655 records. www.finds.org.uk [Accessed 01 February 2021].

24 Broadberry et al. 2015: 25–26.

25 Van der Veen et al. 2013: 154–156.

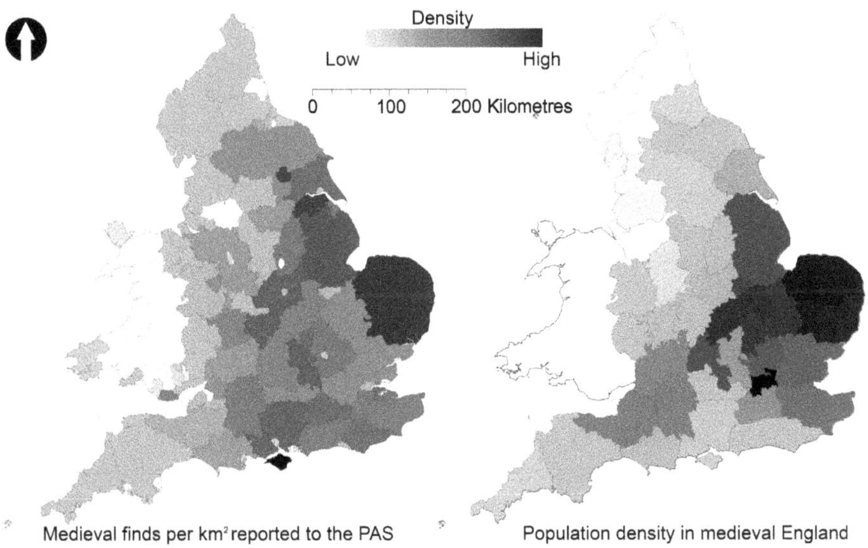

Fig. 5.1: Left: The density of medieval finds (finds/km^2) reported to PAS database by current administrative areas. Medieval finds are defined as those dating from AD 1000–1600. Right: 'Heatmap' representing average medieval population by historic English county after BROADBERRY et al. 2015: 25–26. Population data not available for Wales. Average population is the sum of BROADBERRY et al.'s figures for 1066, 1290, 1377 and 1600 divided by four. This figure was then divided by the area of the historic county. Created by the author.

The accessibility of potential metal-detecting sites, land-ownership and permissions as well as the differential reporting of finds to the scheme by detectorists in different areas all impact the number of finds recorded.[26] While such problems must be acknowledged, the PAS remains an enormously useful resource and, where appropriate, data derived from the database is discussed below in relation to different categories of artefactual evidence relating to medieval beliefs in protection from natural hazards.

5.1 Saintly relics

The saints, through their material remains, were regarded as intercessors who had the power to avert crises in the material world. This belief was reinforced by a multitude of Church teachings and Christian popular culture, including saints' lives, mystery plays, wall paintings, stained glass (see Fig.5.5) and sculpture (see Fig. 5.6) which provided examples of the marvellous acts of individual saints on the behalf of penitents. That

26 ROBBINS 2013.

assistance from the saints was often the first recourse for medieval Christians seeking deliverance in times of need is widely demonstrated throughout the historical sources. When the windstorm of 1362 blew in above St Augustine's Abbey, Canterbury, for example, the chronicler's account details how, during the storm, a chaplain named Ralph sought shelter in a chapel dedicated to Saint Pancras where he prayed before the image of the Virgin Mary. Unfortunately, his prayers were not immediately answered as the force of the storm blew one of the chapel's roof beams down into the chapel fatally injuring Ralph, although miraculously "the image [of the Virgin] re-mained untouched".[27] Similarly, at Longstanton, Cambridgeshire, the fierce conditions of the storm drove two parishioners to pray for their safety inside the church but, again, they were killed when a tree, blown by the wind, fell against the church, knocking masonry down on to the two parishioners at prayer below.[28] That the only detailed accounts of prayer for protection from the storm relate to fatal accidents indicates that, in such circumstances, prayer was so routine that it was only remarkable when it coincided with calamitous events.

Prayer for protection in ecclesiastical settings was believed to be especially efficacious due to the fact that all altars contained saintly relics.[29] These were the personal remains, including body, clothes and possessions, of a saint. The location of the body itself constituted a sacred place or shrine and was usually outfitted with an impressive reliquary. Small parts of a saint, such as hairs, small bone fragments or possessions, were usually kept in smaller, more modest, portable reliquaries.[30] Through their connection to the saint, these relics were considered to be powerful – almost magical – objects and it is clear that they were widely believed to offer an effective means of protecting against, or influencing, the natural world. In London in *c*.1313, for example, when the cross in the bell tower of the cathedral of Saint Paul's was taken down for repair, a collection of relics – presumably deliberately placed to protect against lightning and fire – was discovered attached to the cross.[31] Relics were clearly believed to offer some 'passive' protection but their 'active' use allowed more targeted protection against specific threats. Processions, for example, which often included relics carried by participants, were a common 'active' method to guard against, or mitigate the impact of, natural hazards. This practice probably emerged out of the Roman practice of *robigalia*, an annual procession to secure protection for crops, which was later combined with a 5th century Church tradition in which processions had been held in supplication against the occurrence of earthquakes and lava flows in Auvergne, France.[32] Thus, according to local custom in the parish of Fintray, Aberdeenshire, the silver head of

27 Davis 1934: 564.
28 TNA: JUST 2/18/58.
29 Snoek 1995: 185.
30 Nilson 1998: 3–5.
31 Riley 1869: xxxiii. See also Aungier 1844: 38.
32 Stilgoe 1976: 15.

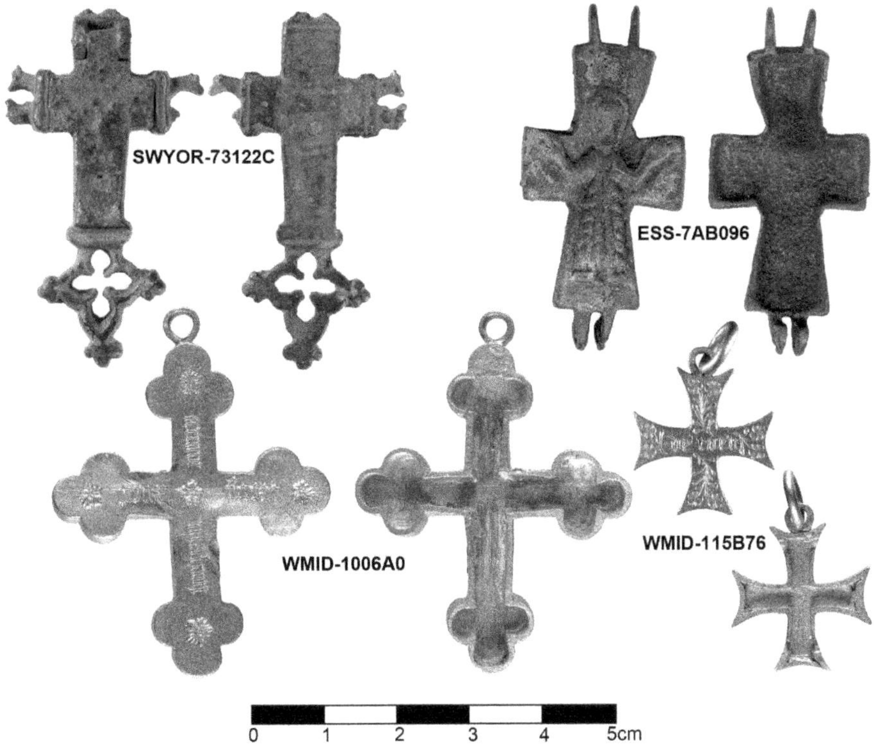

Fig. 5.2: Four reliquary crosses from the PAS database. SWYOR-73122C: a silver reliquary cross dating from 1400–1600 found at Stockton-on-the-Forest, North Yorkshire. ESS-7AB096: a late medieval copper-alloy Byzantine reliquary cross from Tendring, Essex. WMID-1006A0: a 15th century hollow gold cross from Chicheley, Buckinghamshire. WMID-115B76: a 15th century hollow gold cross, which may have contained a saintly relic, from Harlaston, Staffordshire. Photographs reproduced from the PAS website, www.finds.org.uk, under CC BY-SA 4.0 licence: https://creativecommons.org/licenses/by-sa/4.0/.

the patron saint, St Modan,[33] was processed through the parish in order to summon rain or put an end to bad weather[34] while the *Vita Sancti Columbae* describes how processions of Columba's tunic could conjure rain.[35] Likewise, during a severe dry spell, the procession of the relics of John of Beverley around the church caused a sudden rainstorm – easing the drought.[36]

33 Also spelt: 'Meddan'.
34 Leslie, Leslie 1845: 168.
35 Reeves 1857: 174–176.
36 Raine 1879: 269–271.

Alongside the Church's very public use of relics, some individuals possessed their own personal relics contained in, sometimes lavishly decorated, portable reliquaries. A small number of these are documented within the PAS database though, when this research was compiled,[37] only *c.*20 medieval reliquaries were listed – attesting to their rarity. Half of those recorded took the form of crucifixes, perhaps the most obvious Christian symbol, which itself held some connection to protection from hazards. In Jacobus de Voraigne's popular hagiographical compendium he describes the use of, in this case, processional crosses to combat storms,[38] a practice which was also recommended by the 11th century guidelines of Abbot William of Hirsau.[39] The other half of the portable reliquaries contained in the PAS database were a variety of different shapes including lozenges,[40] circles,[41] rectangles[42] and one in the shape of a miniature arched doorway.[43] In some cases it is difficult to be sure whether an object contained a relic or not as only part of the object survives. Where this is the case, the possibility can usually be inferred through comparison with analogous artefacts. For example, CAM-2136A3, a sub-circular disc bearing the sacred monogram 'IHC', appears to be the top plate of a reliquary very similar in design to YORYM-6CE0B1 which, in turn, is closely reminiscent to an example from Gleaston, Cumbria.[44] In addition, it should be noted that the PAS database also records over 100 mounts which may have originally been attached to reliquaries in the form of caskets as fittings and ornamentation. As this identification cannot be confirmed, however, these objects were not included in further analysis.

By the 15th and 16th centuries many of the functions of relics, including as foci for processions in order to accrue protection, had been assumed by the sacred host,[45] though relics did continue to attract lay devotion. Due to the low number of examples in the PAS database, the geographical spread of the artefacts' find spot locations does not demonstrate a noticeable pattern, other than a notable cluster near York (see Fig. 5.3). In terms of chronology, as with a number of other categories of artefact discussed below, portable reliquaries were most popular during the 15th century (see Fig. 5.4) though it should be noted that, due to the low number of portable reliquaries recorded

---

37 23/01/2017.

38 "When storms come up, the cross is brought out of the church and held up against the tempest … and the bells are rung that the demons who are in the air may flee in fright and desist from harassing us". Ryan 1993: 288.

39 Herrgot 1726: 524.

40 YORYM-09903A.

41 YORYM-6CE0B1.

42 PAS-5BA841.

43 NMGW-9E8024.

44 Enticott 1996: 10.

45 Zika 1988: 33–34.

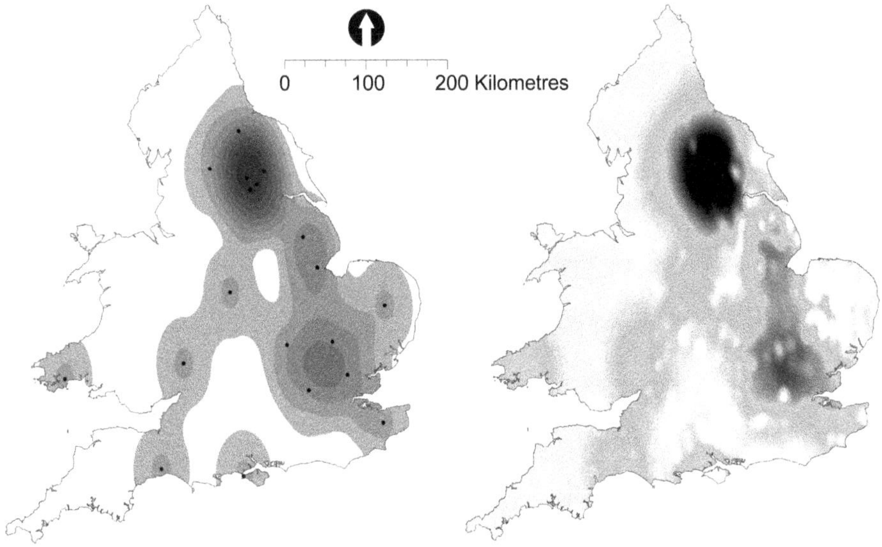

Fig. 5.3: Left: The geographical distribution of reliquaries recorded in the PAS database as of 23/01/2017. Right: A comparison of the distribution shown to the left with the wider distribution of all medieval period finds in the PAS database. Darker areas correspond to higher densities relative to the PAS database as a whole while lighter areas denote areas which are under-represented. Created by the author.

through the PAS as well as the, sometimes considerable, uncertainties in the dating of these artefacts, these results might not be indicative of any wider trends.

5.2 Images of the Saints

As described above, saints were believed to offer protection from the hazards of the world. Just as their relics were powerful, so too were their images. This is perhaps best exemplified by St Christopher – whose image was commonly displayed in churches due to the belief that by looking on his visage the onlooker would gain protection from sudden death or misadventure for the rest of that day. Leaving little doubt, the wall painting of St Christopher at Woodeaton, Oxfordshire, is accompanied by a painted caption, which states: "He who sees this image shall not die an ill death this day".[46] Sudden deaths, such as those caused by lightning, earthquakes or floods, caused particular concern for medieval Christians as death without proper preparation, with repentance and the administration of last rites, put the soul in danger of failing to

46 Tristram 1955: 115.

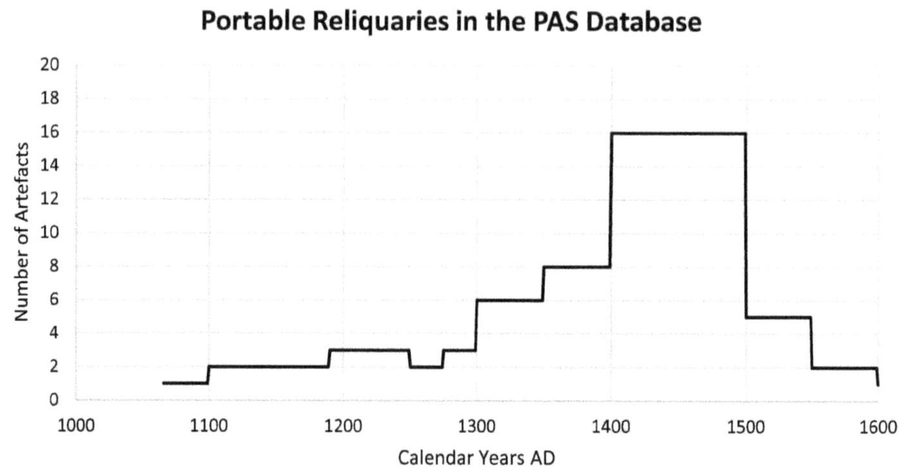

Fig. 5.4: The chronological distribution of reliquaries recorded in the PAS database as of 23/01/2017. Created by the author.

successfully pass into heaven. Therefore, it was especially important to guard against a sudden death and, in addition to beholding the image of St Christopher, there were a number of ways in which protection could be gained. Somewhat comparable to the practice associated with the image of St Christopher, John Myrc's 15th century *Instructions for Parish Priests* states that sight of the raised host during the celebration of mass conferred protection from sudden death upon all witnesses for the duration of the day.[47] A number of textual amulets, discussed below, also claimed to offer protection from death without preparation. In addition to St Christopher, the cults of St Barbara and the Three Kings (or Magi) of Cologne were particularly believed to protect against sudden death. Through an analysis of the material relating to these three cults in the PAS database it is possible to draw some conclusions as to the distribution of material, both geographically and chronologically, similarities and differences between the three cults and some more detailed aspects to the popularity of their cults in medieval Britain.

5.2.1 St Christopher

As detailed above, the cult of St Christopher is an obvious starting point. The Saint's most famous deed is the widely retold story of how he helped a child cross a river. During the crossing St Christopher experienced difficulties due to the strong current and the weight of the child, who seemed to have become unbelievably heavy. With

47 Peacock 1868: 10.

difficulty, St Christopher was able to cross safely, whereupon it is revealed that the future saint had, in fact, borne Christ and the weight of the world during the crossing.[48] As a result he was particularly regarded as a protector of travellers and misadventure but the riverine setting of the story, combined with an early textual tradition that his later burial had protected the local town from flooding, also meant that he could be relied on to protect against floods.[49] At Norton Priory, for example, after a spate of flooding in 1331, a statue of the saint may have attracted particular significance for its promise to protect against future floods.[50] Furthermore, as described above, images of the saint were held to protect the viewer from sudden death for the duration of the day. For this reason, images of the Saint were commonly placed in prominent places such as the north wall of a church opposite the entranceway so that passers-by could see him without even entering the church.[51] This was the case at the church of All Saints', North Street, York, where a stained glass window depicting St Christopher (Fig. 5.5) was originally set into the north window opposite the door.[52] Unfortunately, precise beliefs associated with individual images are very rarely recorded so where helpful captions, such as the one at Woodeaton, are lacking it must be assumed that the same beliefs applied as those documented at other locations, or more generally in regard to the Saint's cult. This probably holds true for popular international cults such as St Christopher's but beliefs are more likely to have been geographically variable where local and less popular cults are concerned. Certainly, it is a strong possibility that where his image occurs on personal artefacts, the intention was that his image would protect the wearer – and especially against floods and sudden death.

5.2.2 St Barbara

St Barbara was another saint whose cult offered protection from sudden death though in her case she was also particularly associated with lightning. The basic story of her life concerns how her heathen father, Dioscorus, imprisoned her in a tower, initially to protect her from unsuitable male wooers. During her captivity, Barbara became a Christian but when her father discovered her conversion he had her persecuted and eventually executed for refusing to renounce her faith. After the execution her father was struck by a bolt of lightning which burnt up his body leaving no trace of his existence.[53] Due to this final detail, Saint Barbara's cult became widely associated

48 Ryan 1993: 398.
49 Pridgeon 2010: 9.
50 Greene 1989: 67.
51 Tristram 1955: 115.
52 Although it has since been moved to the east-end of the church. Pedersen 2000: 38.
53 Wolf 2000: 2–3.

Fig. 5.5: Left: The stained glass window depicting Saint Christopher at the church of All Saints', North Street, York. Right: A painted panel depicting Saint Barbara at the church of Walpole St Peter, Norfolk. Photographs by the author.

with protection from lightning, fire and sudden death.[54] Although it is not specified that her images offered similar instantaneous protection to those of St Christopher, her iconography is similarly prevalent in later medieval material culture with images of Saint Barbara frequently found on pilgrim badges, rings and as decoration in churches (see Fig. 5.5). One possibility is that these images themselves were believed to bestow protection upon the bearer. If this was not the case then presumably possession of the object was believed to increase the chance of the bearer receiving intercession and protection from the saint.

5.2.3 The Magi (The Three Kings of Cologne)

The Three Kings of Cologne, or the magi, were the three kings who visited Jesus at birth, bestowing gifts of gold, frankincense and myrrh. Their names, recorded in the

54 Lockwood 1953: 24.

medieval texts which document their lives, were Balthazaar, Caspar (or Jasper), and Melchior. Although commonly depicted in religious art from the early Christian period, interest in their cult appears to have been ignited in 1154 when their relics are reported to have been discovered in Milan and ten years later these were translated to Cologne.[55] The 'Adoration of the Magi', the episode in which the three kings visited the newborn baby Jesus, became a popular scene in religious sculpture and painting which would have been familiar to lay populations throughout medieval Christendom (see Fig. 5.6). Various traditions relate their cult to protection from sudden death, perhaps as a result of the way in which their deaths are described in the most popular medieval account of their lives, the: *Historia Trium Regum*, which dates to 1375. An old English translation of this text describes Melchior's death as follows:

> So in the feast of the circumcision Melchior, that was king of Arabia and of Nubia, said a mass solemnly in the church; and [when] he had said his mass, before all the people he laid him[self] down and without any disease he yielded up his spirit to the father of heaven, and so died."[56]

The others are described as dying in similar circumstances, Balthazaar 5 days later, and Jaspar 6 days afterwards.[57] Perhaps as a result, in addition to their association to the birth of Christ, charms and amulets bearing their names, rather than their image, were regarded as efficacious against an unprepared death.[58] In addition, their cult was also believed to protect against epilepsy and general misadventure.[59] As well as several examples known through excavation, the names of the Magi are found on a number of rings, brooches and other items of jewellery recorded in the PAS database.

5.2.4 Summary

These three cults, St Christopher, St Barbara and the Magi, can be investigated through the material culture in the PAS database which bears pictographic or textual references to the saints. This reveals some important differences in the audience and veneration of these three cults. St Christopher was a popular saint with *c.*30 items in the PAS database bearing his image. The distribution of these objects suggest a particular focus in towns such as Norwich, Gloucester and Chester with lesser clusters around York and in Hampshire (Fig. 5.9). The different types of artefact associated with Christopher,

55 Harris 1959: 23–24.

56 Spelling modernized by the author. Original text: "so in þe feest of þe circumcisioun Melchior, þat was kyng of arabie and of Nubye, seyde a masse solemplich in þe chirche; and [whan] he had seyde his masse, tofore alle þe pepil he leyde hym downe and withoute any dissese he ȝelde vp his spirit to þe fadir of heuene, and so deyde. Horstmann 1886: 118.

57 Horstmann 1886: 118–120.

58 Gilchrist 2008: 126.

59 Hildburgh 1908: 83.

Fig. 5.6: Alabaster relief depicting the Adoration of the Magi from Holy Trinity Church, Long Melford, Suffolk. Photograph by the author.

predominantly straps and buckles as well as jewellery (Fig. 5.10), suggest his image appealed to a broad audience who must have been keen to take advantage of the general protection he offered against misadventure and sudden death to those who beheld his image.

St Barbara's cult was similarly popular but appears to have had a more generally southern orientation, with a particular focus around London (Fig. 5.9). Pilgrim badges make up the majority of the artefacts with mounts and jewellery the next two most popular object categories (Fig. 5.10). This suggests her cult had a more continental focus, with a large proportion of the material culture perhaps brought back by pilgrims returning from popular European pilgrimage centres. This is supported by the fact that pilgrim badges depicting St Barbara are particularly common in France and the Low Countries – although exactly where many of the pilgrim badges were cast and sold is unknown.[60] This agrees with the historical evidence which attests to the strength of Saint Barbara's cult in France and Belgium.[61] By contrast there is relatively sparse evidence for her veneration in medieval England, with only one church dedication, at Ashton-under-Hill, Tewkesbury, known with certainty. On the other hand French claims that Barbara's cult was "tremendously popular in late medieval England"[62]

60 As recorded in the database of pilgrim badges at www.kunera.nl [Accessed 01 February 2021].

61 Wolf 2000: 29.

62 French 2008: 140.

while LOCKWOOD highlights the popularity of her cult in Norfolk and East Anglia.[63] With three artefacts associated with her cult discovered in East Anglia, the PAS evidence may corroborate LOCKWOOD's theory though it should be noted that the majority of objects were discovered in the central south of England. Certainly, the PAS evidence demonstrates that her cult attracted greater veneration than might be inferred from Church dedications alone.

Of the three, the Magi seem to have been the least popular, with only 15 linked items but their cult was widespread (Fig. 5.9). The vast majority of objects were items of personal jewellery (Fig. 5.10) – finger rings, brooches and pendants – many of them silver or gold, suggesting the Magi were a more exclusive cult that appealed to a wealthier sector of society. In a number of towns, such as Dublin and Aberdeen, the Magi were the subject of urban plays organised by the local goldsmiths' guilds – which may suggest their cult was particularly popular among the makers of these valuable items.[64] Although pilgrim badges depicting the Magi are absent from the PAS database, a low number originating from Cologne, Germany, have been discovered in London, King's Lynn, Canterbury and Huntingdon[65] indicating that this cult likely attracted low numbers of English pilgrims. In terms of chronology, the three cults seem to have peaked in popularity during the 15th century, though St Christopher was most popular at the beginning of the century (see Fig. 5.7), while St Barbara and the Magi appear to have peaked towards the end (see Fig. 5.8).

These types of material culture are usually dated based on developments in iconography. In the case of St Christopher, for example, the position of the Christ-child; at the waist in the earliest depictions and carried over the head in later images, as well as the Saint's characteristics, becoming more muscular and heavy-set with time, are drawn on in order to provide a rough date.[66] Interestingly, when the chronology of St Christopher images gained from the PAS database is compared against that of datable English and Welsh wall paintings depicting the saint (see Fig. 5.7), obtained from PRIDGEON's database,[67] it becomes clear that the two chronologies exhibit clear divergences. Notably, a large spike in wall paintings during the late 15th century is not reflected in the material evidence from the PAS. This indicates that the trend in wall painting at this time was out-with the popularity of St Christopher's cult more generally. In the case of the Magi, the chronology of the material culture from the PAS database reflects wider trends in the popularity of the cult in Britain.[68] St Barbara's cult too, at

63 LOCKWOOD 1953: 23.
64 WILLIAMSON 2009: 165.
65 See note 60.
66 PRIDGEON 2013.
67 PRIDGEON 2010: 307–338.
68 WILLIAMSON 2009: 174–179.

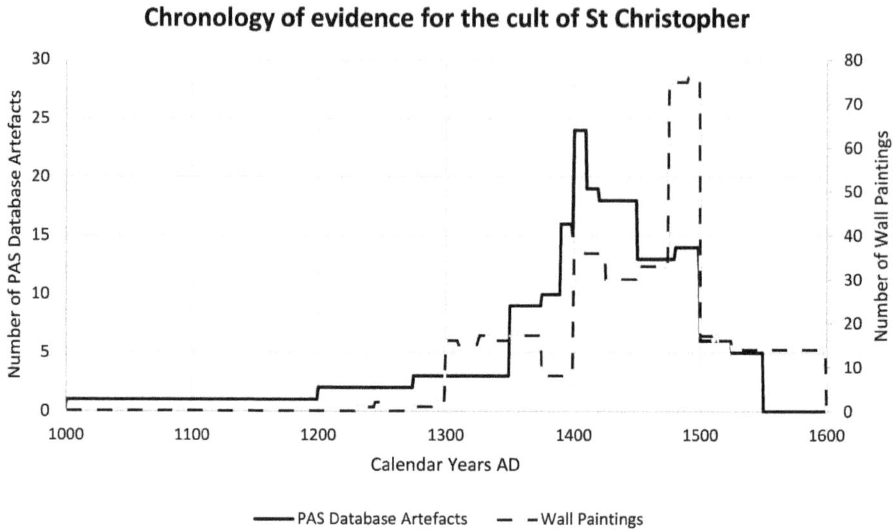

Fig. 5.7: The chronological distribution of artefacts in the PAS database related to the cult of St Christopher, as of 03/10/2016, against the chronology of wall paintings depicting the saint in England and Wales. Created by the author. Wall painting data is derived from the database compiled by Pridgeon 2010: 307–338.

least in a western European context, reached its apogee contemporaneously with the spike in material culture reported to the PAS.[69]

5.3 Pilgrim badges

These saintly insignia were widely available at the shrines of saints and other pilgrimage sites across medieval Europe. As well as indicating a pilgrim's status to society, pilgrim badges were also believed to hold apotropaic powers due to their connection to the saint, of whom many bore an image. The power of the image alone was sometimes enhanced through touching the badge against relics of the saint or perhaps through submersion in holy water. As a result, these pilgrimage souvenirs became mini-relics which bestowed the protection of the saint upon the wearer, bearer or owner. Accordingly, pilgrim badges were valued by medieval people as objects with spiritual and perhaps protective significance. This is borne out in a number of ways – some, for example, have been found in burial contexts,[70] often in positions which suggest amuletic significance. In

69 Wolf 2000: 41, 45.
70 Gilchrist 2011: 164.

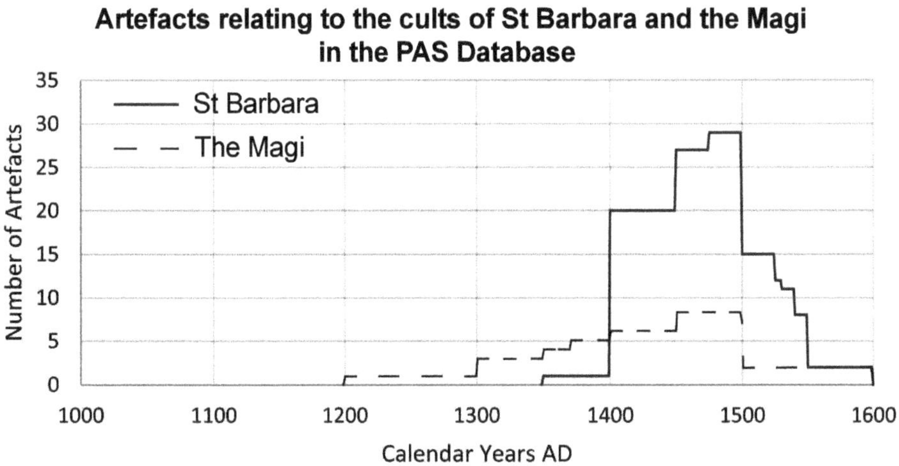

Fig. 5.8: The chronological distribution of artefacts in the PAS database related to the cult of Saint Barbara as of 03/10/2016. Created by the author.

an interesting case from East Haven, Angus, a 13th–14th century lead pilgrim badge was modified by cutting off parts of the badge to form a crucifix, suggesting the badge itself was considered an item of holy significance.[71] Furthermore, from the mid 15th century, many pilgrim badges were collected and curated, as demonstrated by the fact that they were frequently attached to the pages of prayer books or, in cases where the actual badges were not available, artistic representations are found.[72] In these cases, the retention of badges, and their inclusion alongside prayers and holy scripture, suggest that these were treasured items.

Pilgrim badges have often been recovered from 'wet' locations, such as rivers. While GARCIA advances the possibility that this action signified a successfully completed pilgrimage[73] the traditional interpretation attaches ritual significance to the deposition of these objects in watery locations. This may represent a continuum of pre-Christian practices which are well-attested through archaeological evidence in western Europe throughout the Bronze and Iron Ages.[74] Based on a re-analysis of the archaeological, art historical and documentary evidence, however, LEE argues that there is little evidence for a ritual component to the majority of pilgrim badge depositions in watery contexts – which were, instead, simply discarded objects that no longer held any value

71 Note that it is not possible to date when the badge was modified as the item was an unstratified find discovered through metal detecting. SHIELS 2005: 21.

72 FOSTER 2011: 39–40.

73 GARCIA 2005: 6–7.

74 YATES, BRADLEY 2010a;b.

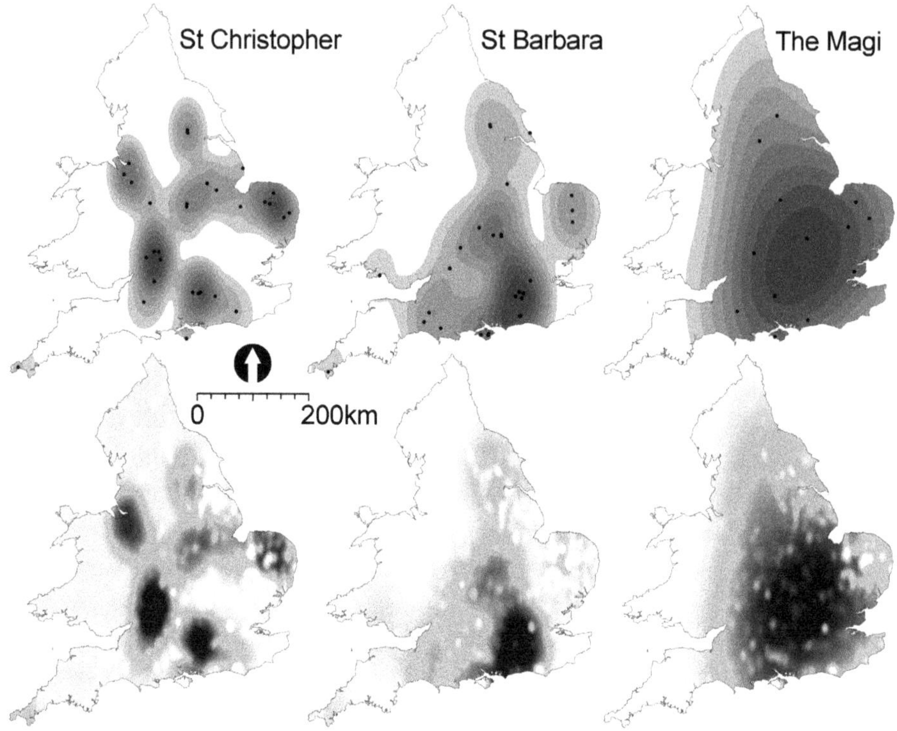

Fig. 5.9: Above: The geographic distribution of artefacts in the PAS database related to the cults of St Christopher, St Barbara and the Magi as of 03/10/2016. Below: A comparison of the distributions shown above with the wider distribution of all medieval period finds in the PAS database. Darker areas correspond to higher densities relative to the PAS database as a whole while lighter areas denote areas which are under-represented. Created by the author.

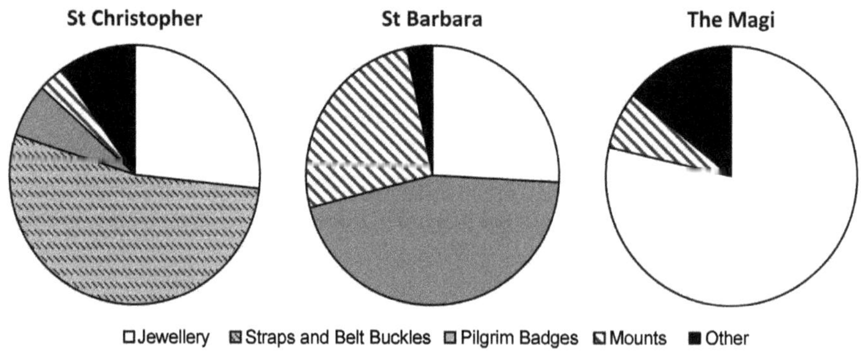

Fig. 5.10: The different types of artefact belonging to each the cults of St Christopher, St Barbara and the Magi in the PAS database as of 03/10/2016. Created by the author.

to their medieval owners.[75] This, however, contradicts the accepted wisdom that pilgrim badges were items of great value believed to hold apotropaic powers as a result of their relationship to saints' shrines[76] – which seems to be confirmed by the other contexts in which pilgrim badges have been found. These valuable items, therefore, would probably not have been disposed of in watery contexts unless this act held some kind of ritual significance.

The sudden appearance of many new forms of pilgrim badge in the aftermath of the Black Death has been interpreted as a form of ritual protection particularly motivated by fear of the disease.[77] Analysis of the chronological distribution of pilgrim badges reported through the PAS database certainly demonstrates a peak in the overall number during the 15th century (see Figure 5.11). This roughly coincides with the demographic decline instigated by the Black Death, which recent estimates gauge to have precipitated a drop in population of *c.*48%[78] matched by a concomitant decline in material culture.[79] This illustrates that, despite a dramatic drop in population, more people were going on pilgrimage and/or when on pilgrimage people were acquiring more pilgrim badges. The reasons for this are likely to be multi-faceted – a possible factor, for example, was that the rise in living standards meant people had more disposable income with which to purchase such things. It is far from unbelievable, however, to suggest that the uncertainty created by the recent trauma of the Black Death created a milieu in which superstitious beliefs and fears increased. In such an environment it is easy to understand why people would have been more likely to undertake a pilgrimage and, while away, see the value in acquiring material culture offering protection from 'natural' dangers such as those that had made such a strong impact within living memory.

5.4 *Ampullae*

Another artefact class related to pilgrimage are *ampullae*. These were vessels, usually cast in lead or lead alloy, which contained water, oil, dust or other substances from the shrine of a saint. In England, although the practice had earlier origins, *ampullae* became common following the murder at Canterbury, in 1170, and canonisation, in 1173, of St Thomas Becket. The saint's 'blood', theoretically the blood spilt at the murder,[80] mixed and diluted with water, a process depicted in stained glass in Canterbury Cathedral's

75 Lee 2014.
76 Spencer 1990: 10–11.
77 Gimbel 2012.
78 Broadberry et al. 2015: 21.
79 Lewis 2016.
80 Finucane 1995: 90.

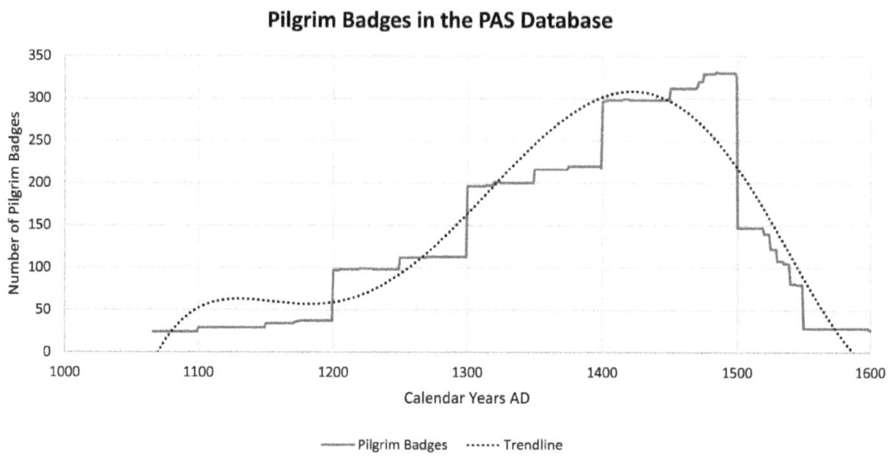

Fig. 5.11: The number of pilgrim badges reported to the PAS database, as of 12/12/2016, against time. Created by the author.

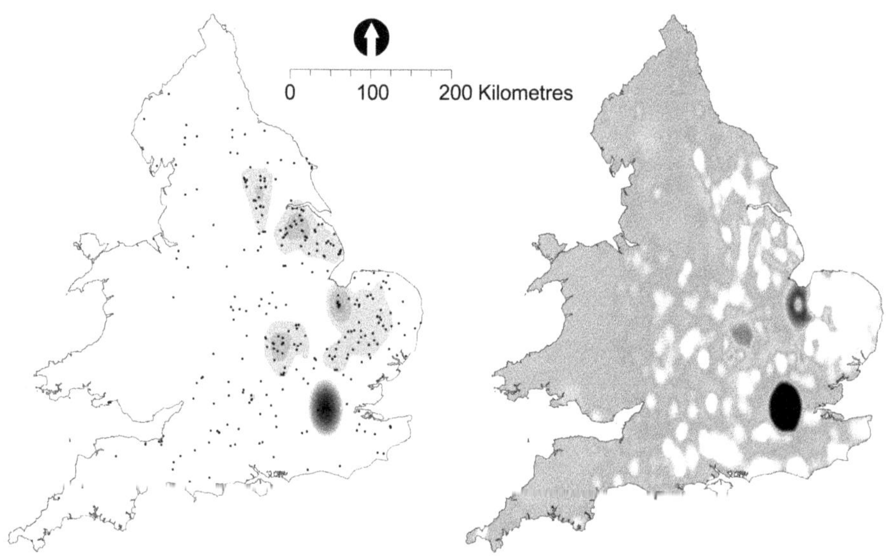

Fig. 5.12: Left: The geographical distribution of pilgrim badges reported to the PAS as of 12/12/2016. Right: A comparison of the distribution shown to the left with the wider distribution of all medieval period finds in the PAS database. Darker areas correspond to higher densities relative to the PAS database as a whole while lighter areas denote areas which are under-represented. Created by the author.

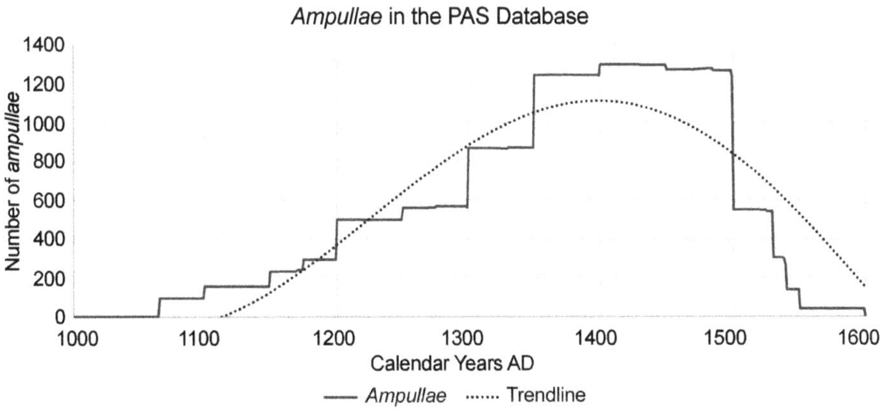

Fig. 5.13: The number of *ampullae* reported to the PAS database, as of 16/12/2016, against time. Created by the author.

Trinity Chapel,[81] was sold in *ampullae* and held responsible for miracles of healing[82] leading to high demand from the thousands of pilgrims who flocked to Becket's shrine following his canonization. As mass produced, inexpensive and widely available items, which offered to provide healing and protection from a multitude of dangers and ailments, the uptake of *ampullae* was high. This can be seen in the PAS database which, at the time of writing, records over 2000 *ampullae*, the majority of which were recovered through metal detecting. Most *ampullae* are lentoid in cross-section with a rounded base and flared opening, usually with suspension loops on either side of the body.

The form of the containers has been the subject of some debate. The traditional view has been that the majority were intended to resemble scallop shells as a reference to the widely recognised symbol of the pilgrimage to Santiago de Compostela.[83] This was seemingly confirmed by the decoration of radiating lines commonly found on the rounded base which is usually interpreted as mimicking the ridges found on scallop shells. Campbell, however, argues that only a minority actually depict scallop shells with most *ampullae* instead modelled on contemporary purses or pouches.[84] With this form, lead-cast *ampullae* aped the actual cloth or leather purses and pouches which contained glass vials of sanctified dust, water or oil. These were presumably similarly available at pilgrimage sites but were likely rarer and more expensive than their lead-cast counterparts. Furthermore, due to the comparatively poor durability of glass, cloth and leather, these artefacts have not survived in the archaeological

81 Koopmans 2016.
82 Webster 2016: 6–8.
83 Anderson 2010: 184.
84 Campbell 2015: 129.

record in comparable numbers to lead-cast *ampullae*.[85] This re-interpretation does not alter the fundamental way in which *ampullae* are viewed but it does reorient their symbolism. As only the minority were decorated with the scallop shell symbol, rather than acting as universal symbols of pilgrimage, the form of the majority of *ampullae* may have more simply acknowledged their contents, and its protective and healing power. It should also be noted that the containers were sometimes reused and refilled with other substances signalling that their curative and protective powers were not exclusively associated with the contents.[86]

Given the widespread discovery of *ampullae* in settings which are presumed to have been core agriculture areas in the medieval period, ANDERSON has convincingly suggested that the deposition of these vessels, containing substances of apotropaic significance, was deliberate with the probable intention of securing a bountiful harvest through the protection of crops from bad weather, disease, blight or other hazards.[87] The distribution of *ampullae*, as revealed by the PAS database (Fig. 5.12), are suggestive of some interesting trends with the highest concentrations found in central Norfolk and Lincolnshire – a pattern which likely stems from a variety of factors. Firstly, these areas contained important pilgrimage centres, most notably the shrine of Hugh of Lincoln, at Lincoln Cathedral, and the cults of Our Lady of Walsingham and the Holy Rood of Bromholm in Norfolk in addition to the shrine of William of Norwich in Norwich Cathedral.[88] Secondly, Norfolk and Lincolnshire were two of the most densely populated English counties during the medieval period (see Fig. 5.1). Thirdly, and perhaps most importantly, these areas are both represented in the PAS database by high levels of find reporting suggesting that, compared to other regions, more metal detecting has been carried out in these counties and a high percentage of these finds are reported to the authorities (see Fig. 5.1). As a result, although there may be a historical basis for the densities visible in the PAS data it is difficult to entirely disentangle this from the biases in finds reporting.

5.5 Charms and textual amulets

Another category of amulet from this period are those which took the form of documents. These were usually sheets, parchment or paper, of text including magical charms, seals and words which claimed, and were believed, to offer protection to the bearer. These were certainly in use up to the Reformation as demonstrated by the denunciation of their use “as a protective charm against thunder and lightning” by Martin Luther.[89] Some

85 CAMPBELL 2015: 125.
86 FINUCANE 1995: 90.
87 ANDERSON 2010: 199–200.
88 SHINNERS 1988: 133.
89 LUTHER 1912: 628; SKEMER 2006: 67.

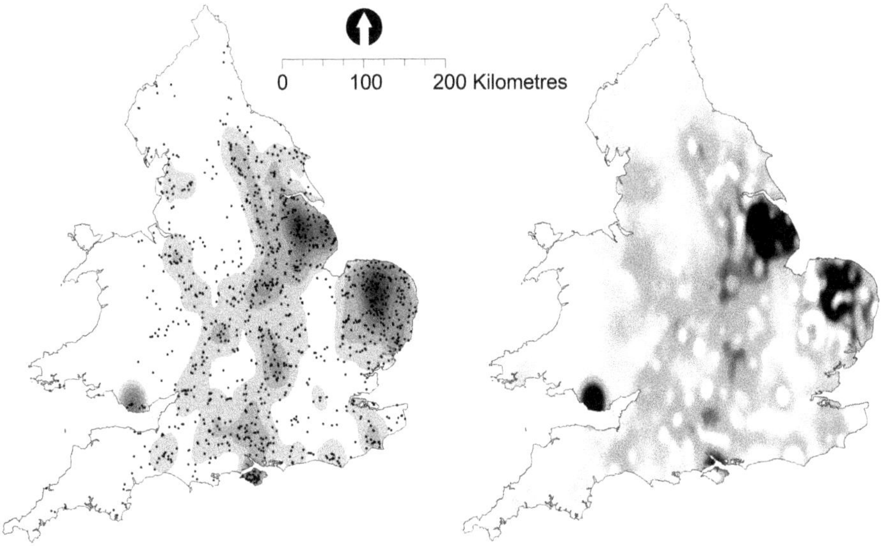

Fig. 5.14: Left: The geographical distribution of *ampullae* reported to the PAS as of 16/12/2016. Right: The distribution on the left compared with the wider distribution of all medieval period finds in the PAS database. Darker areas correspond to higher densities relative to the PAS database as a whole while lighter areas denote under-represented areas. Created by the author.

charms provided protection through recitation while others were carried about the person to provide long-term safety. One of the miracles of St Cuthbert, retold by Reginald of Durham in the 12th century, for example, tells how an engineer hired by the Abbot of Durham carried with him a sack full of textual amulets to resist evil.[90] A particularly interesting example is the Canterbury Amulet,[91] a mid-13th century manuscript, on the reverse of which are 17 seals which offer protection against particular hazards including fire, storm, flood and sudden death.[92] The high number of different protective charms which offer protection against various hazards, diseases and injuries essentially provides general protection against almost any conceivable threat. The amulet was probably folded up, allowing easy transportation so that it could be carried around but, by unfolding, the owner could view the seals and read their accompanying text for stronger protection.[93] While it is possible that the manuscript was used in this way, to provide personal cover-all protection, another interpretation is that it served as a

90 Raine 1835: 94–98.
91 Canterbury Cathedral Library, Additional MS 23.
92 Skemer 2006: 204–205.
93 Skemer 2006: 199, 206.

blueprint that could be copied in order to produce bespoke textual amulets for almost any individual or situation as and when they were required.[94]

Another type of document which may have been used as textual amulets were 'heavenly letters', which reportedly came from Christ. Their supposed provenance gave these medieval forgeries protective powers, a belief which in some regions survived into the post-medieval period.[95] One 15th century English example of the 'heavenly letter' contained a charm that promised: "who that bere this lettre vppon hym shall not deye of no euel dethe … nor water nor lightnyng nor thondryng".[96] The invocation of holy names was another way to add protective qualities to a textual amulet with some examples specifying that the list of names would protect the bearer from fire, drowning, storms and lightning.[97] Somewhat relatedly, sudden death could be avoided, by beholding the measure – or length – of Christ each day. This was a horizontal line often drawn in textual amulets, the length of which could be multiplied by a specific number to give the true height of Jesus.[98] Textual charms could also be used to augment the power of other forms of protection. For example, in a 15th century ceramic jar which was inserted into a wall in St Mary's Church, Bexley, Kent, a piece of parchment was recovered[99] which Merrifield suggests would have originally been a written charm, probably containing spells and incantations to protect the church.[100] Another possibility is that the parchment was a prayer, most likely petitioning the Virgin Mary, the patron saint of the church, for assistance.[101] As with foundation deposits, discussed below, the placement of the jar may have been motivated by the belief that such a vessel could act as a trap to protect a structure from evil spirits and the Devil. This interpretation is strengthened by the fact that the jar came from the North wall of the church which, as discussed below, held particular associations with the Devil in the medieval mind.

5.6 Papal *bullae*

Papal *bullae* are lead seals attached to documents originating from the Papal curia, either in Rome or, during the schism between 1378 and 1417, Avignon. On the obverse they are inscribed with the name of the pope at the time of issue while on the reverse an image of St Peter and St Paul is shown. This format did not change throughout the

94 Skemer 2006: 210.
95 Jones 1975: 173.
96 Chardonnens, Hebing 2010: 190.
97 Skemer 2006: 256.
98 Skemer 2006: 143.
99 Tester 1956: 260–261.
100 Merrifield 1987: 121.
101 Gilchrist 2012: 234.

medieval period, save for the papacy of Pope Paul II (1464–1471) when a new design was introduced, although the old style was re-adopted by all subsequent popes. Through their connection to the Papacy and the fact they came from Rome – one of the most important pilgrimage centres – as well as the fact that they were inscribed with images of St Peter and St Paul, some considered them to hold apotropaic significance. This may be reflected by their inclusion in burial contexts – a practice which particularly intensified during the 14th and 15th centuries. These *bullae* were likely originally attached to documents which granted indulgences, or remissions from sins, and were included with burials in order to ease the journey of the deceased through purgatory.[102]

At the time of writing, the PAS database contains records more than 500 Papal *bullae*,[103] of which a significant number were found broken in two halves or folded in a manner superficially similar to folded coins, discussed separately below. In the case of *bullae* this can likely be explained by the fact that they were originally attached to documents through a hole running through their centre from top to bottom which must have introduced a weakness to the object making it more likely to fold or break along this line. Some, however, display clear evidence of cutting as opposed to accidental folding or breaking. These cases likely reflect deliberate destruction of *bullae* to take them out of circulation and prevent their reuse on fraudulent documents.

Eight percent of the *bullae* in the PAS database analysed as part of this research, were removed from the documents to which they were originally attached and pierced with some sort of hole. While pierced holes could conceivably have been added as a way to facilitate easier attachment to documents, this is perhaps unlikely because the main value of *bullae* was as a means of authentication. Therefore, attaching them to documents in unofficial ways demonstrated that the document had not been sent out from the Papal Curia in that condition and could therefore not be trusted as an official document. This suggests that where *bullae* have pierced holes this is likely to relate to apotropaic reuse of the Papal seals – likely as amulets. Perhaps the clearest example of this is SOM-DB4323, originally a *bulla* of Pope Boniface IX (1389–1404) found in Somerset, which was carefully trimmed around the face of St Paul and a hole pierced at the bottom of the saint's face (see Fig. 5.15). Presumably this would have enabled the saint's image to be worn as a protective amulet. Another interesting example is SUSS-F5E022, a *bulla* of Innocent IV (1243–1254) from Sussex, which has a number of holes pierced through it suggesting that it may have been nailed somewhere, perhaps onto a door (see Fig. 5.15). Such an act might have been believed to bestow protection upon the structure and those who dwelt within. Standley argues that two of the *bullae* with pierced holes reported to the PAS[104] were reused as spindle whorls although in

102 Gilchrist 2008: 130–131.
103 As of 01 February 2021.
104 HESH-1517A7 and SWYOR-F52016.

Fig. 5.15: Three Papal *bullae*. Top Left: A typical Papal *bulla* from the papacy of Innocent IV (1243–1254). North Ferriby, East Riding of Yorkshire. Top Right: An amulet created from a *bulla* – St Paul's face has been trimmed and a hole pierced through the bottom. St Cuthbert Out, Somerset. Below: A *bulla* from the papacy of Innocent IV (1243–1254) which has been affixed somewhere, perhaps to gain protection. Lancing, West Sussex. Photographs from the PAS, www.finds.org.uk, reproduced under CC BY-SA 4.0 licence: https://creativecommons.org/licenses/by-sa/4.0/.

these cases she concedes that the selection of Papal *bullae* for this seemingly utilitarian purpose must have had a deeper significance.[105]

By analysing all the *bullae* that had been reported to the PAS at the time this research was conducted, it is possible to plot the distribution of this type of artefact over time (Fig. 5.16). As the pope at the time of issue is included on each *bulla* it is normally possible to date each *bulla* quite precisely – usually within a decade. In a small number of cases this is not possible because the name of the Pope is illegible or not present due to later damage. The results of this analysis demonstrates a relatively close correspondence between the numbers of *bullae* reused as amulets and the overall number of *bullae* in circulation. This correlation between *bullae* with evidence for later reuse for apotropaic purposes and the overall availability of *bullae* is further demonstrated when the *bullae* in the PAS are analysed spatially – as those with evidence for amuletic reuse show a remarkably similar distribution to the total spread of *bullae* across England and Wales. The PAS evidence suggests a notable absence of *bullae* with evidence for amuletic reuse in Norfolk and Suffolk but this may simply be a result of

105 Standley 2016: 284–285.

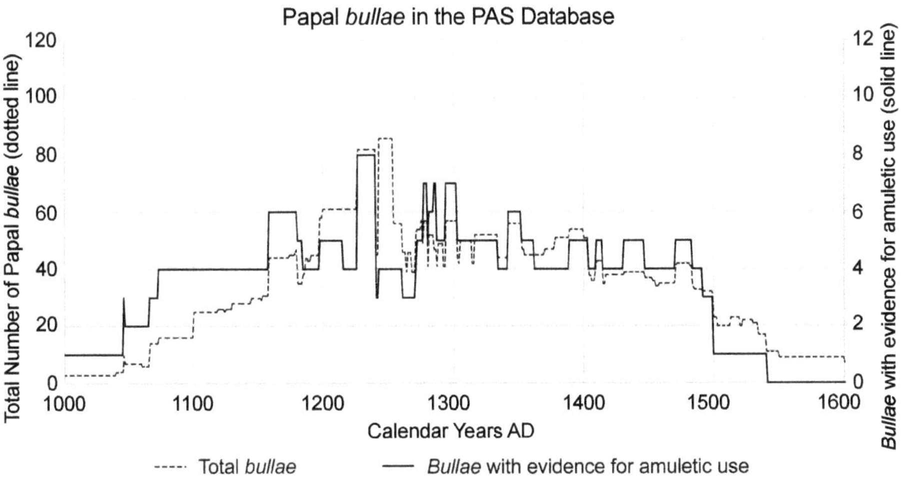

Fig. 5.16: The chronological distribution of Papal *bullae* reported to the PAS, as of 14/12/2016, against time. The total number of *bullae* is plotted in grey (left axis) while the number with evidence for reuse as amulets are plotted in black (right axis). Created by the author.

the relatively low number of artefacts which fit into this category reported to the PAS database. Certainly, *bullae* of this type have been discovered in these counties as is demonstrated, for instance, by a pierced *bulla* dating to the papacy of Innocent VI (1352–1362) found at Strumpshaw, Norfolk, in 2004.[106] An important point is that the supply of *bullae* was limited by the number of Papal documents reaching Britain which may have prevented the reuse of *bullae* as amulets from growing into a more popular practice.

Clearly, although the reuse of *bullae* for apotropaic and protective purposes can only be suggested for a relatively low percentage of the overall corpus reported to the PAS database, the practice was geographically widespread and seems to have been carried out consistently throughout the medieval period.

5.7 Church bells

During the medieval period church bells served more than simply the practical function of signalling the time of day and notifying parishioners of religious and social events. They were also imbued with ritual significance. A newly cast bell was blessed and these blessings usually included a reference to the bell's ability to protect against storms

106 Gurney 2005: 749.

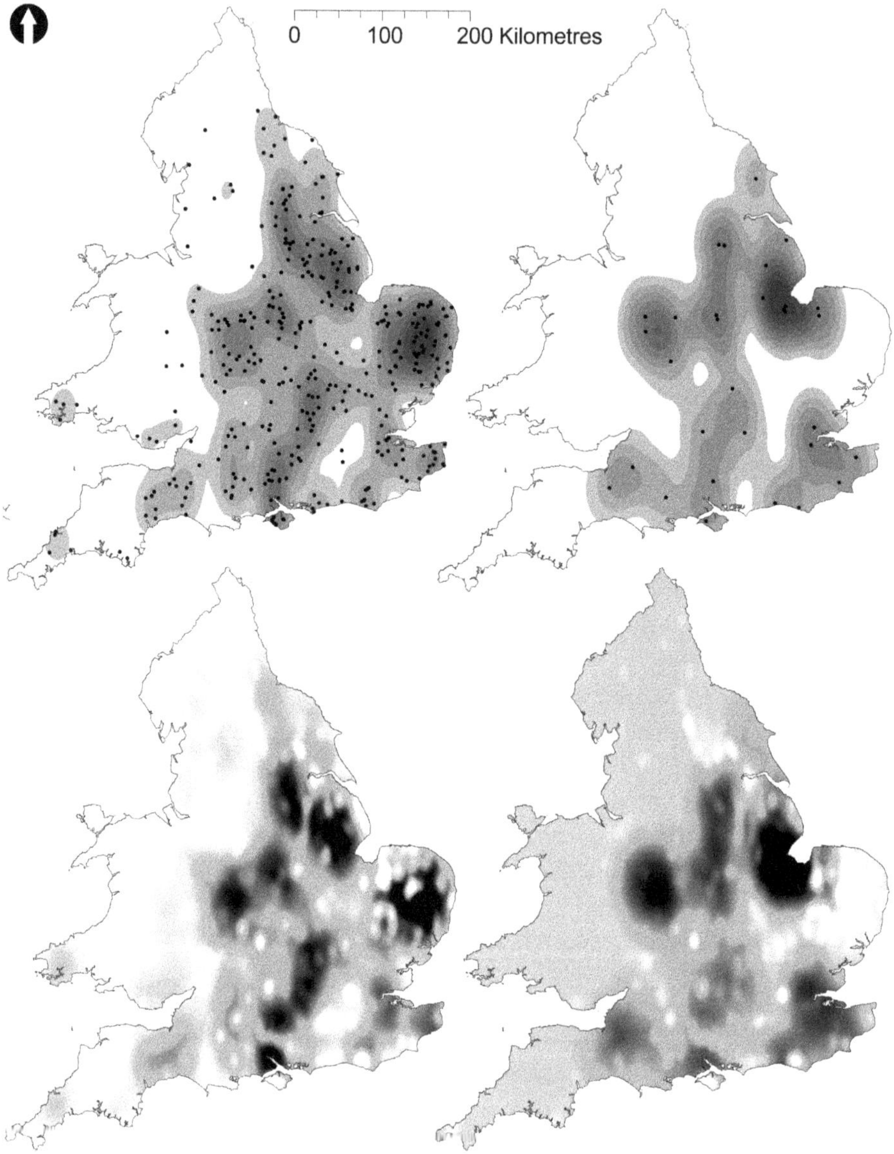

Fig. 5.17: The geographical distribution of Papal *bullae* reported to the PAS, as of 14/12/2016. Above Left: The distribution of all *bullae* reported to the PAS. Above Right: The distribution of *bullae* with evidence for reuse as amulets. Below: A comparison of the distributions shown above with the wider distribution of all medieval period finds in the PAS database. Darker areas correspond to higher densities relative to the PAS database as a whole while lighter areas denote areas which are under-represented. Created by the author.

and lightning.[107] This belief is further reflected by inscriptions found on medieval bells. For example, the inscriptions on a 13th century bell from Assisi, Italy, described the various powers of the bell including: "I break lightning"[108] and "I disperse the winds".[109] While some church bells were inscribed with such phrases, directly referencing their power to avert storms, others had more general prayers such as a church bell from Malma, Västergötland, Sweden, which simply invoked the intercession of the Virgin Mary.[110] The church bells from Burg, Schleswig-Holstein, Germany, and Garyp, Freisland, The Netherlands, bore, respectively, an inscription and a depiction referencing St Barbara.[111] Given the association with lightning in her hagiography, imbuing bells with the power of St Barbara's cult may have been thought to grant particular protection against storms. Inscriptions on English church bells also commonly invoked the assistance of a saint, often the saint of the church's dedication, while others made indirect references to the belief that church bells could drive away storms. For example, many from Devon and Somerset are inscribed with the line "With my living voice I drive away all hurtful things".[112] That these blessings were more than just words is demonstrated by the widely documented practice that bells were rung during storms to drive away the evil spirits which guided them.[113] In Britain this practice is documented in a number of sources – at Malmesbury Abbey the great bell of Saint Adelm was rung to disperse thunder and lightning[114] while surviving church warden's accounts[115] (see Table 5.1) attest to payments made to bell-ringers for their services during storms.

5.8 Foundation and special deposits

Merrifield, in his seminal text on the archaeological evidence for ritual and magic, discussed a number of types of foundation deposit commonly encountered in medieval and post-medieval structures.[116] These included skeletal or mummified animal remains (including chickens, cats and horses), shoes, coins and pots. Such cases are widely interpreted, sometimes through surviving folklore, as strategies to secure protection or good luck for the building and those who dwelt within. Cases are found widely through-

---

107 Arnold, Goodson 2012: 119.

108 *Fulgura frango*. Arnold, Goodson 2012: 120.

109 *Dissipo ventos*. Arnold, Goodson 2012: 120.

110 The inscription is given in full on the website of the Skaldic Project: https://skaldic.abdn.ac.uk/db.php?id=16751&if=runic&table=mss [Accessed 01 February 2021].

111 Wolf 2000: 52–53.

112 *Voca mea viva depello cunta nocina*. Walters 1912: 325–326.

113 Ryan 1993: 288.

114 Britten 1881: 22.

115 Cox 1913: 212.

116 Merrifield 1987: 119–136.

Tab. 5.1: English Church Warden Accounts recording bell ringing in response to storms after Cox 1913.

Year	Church	Detail	Payment
1450	St Mary-at-Hill, London	*Dat' hominibus pulsantibus in nocte Sci Petri pro tornitura*	viij d.
1457–8	Yeovil, Somerset	*In potacione dat' pulsatoribus dum tonatruat*	j d.
1464	St Mary, Sandwich, Kent	For bred and drynke for ryngers in the great thunderyng	iij d. ob.
1519	Spalding, Lincolnshire	Pd for ryngyng when the Tempest was	iij d.

out Britain, including a cistern containing an iron key buried under the foundations of a house in Oxfordshire.[117] In this case, the difference in date between the 12th–13th century cistern and 14th–15th century key suggests the deposition of a curated cistern from an earlier period, a practice evidenced through a wide range of artefact types,[118] which may have been deemed of greater significance due to their age. In relation to disasters, one of the most interesting cases is that of the medieval houses at Culemborg, the Netherlands, which, after being razed by fire in 1422, were rebuilt with jugs deposited in the foundations of the structures. The ceramic vessels may have been thought to act as a trap for evil spirits and, as such malevolent forces were held responsible for causing disasters,[119] this would have provided protection for the community against a repeat of the blaze which had recently affected their town so severely.[120] It should be noted, however, that not all ceramics buried beneath structures served a protective purpose, with documentary evidence attesting to functional reasons, including fermenting, distillation and as containers for hot ashes and embers from fires.[121]

In the case of two miniature bronze cauldrons found separately within two different medieval flood embankments in the Netherlands there is no obvious functional explanation.[122] The cauldrons can be confidently dated on stylistic grounds to the 14th century and, based on documentary evidence, the dikes within which they were buried date from the same period.[123] This suggests that they were purposefully deposited during the construction of the dikes. It is possible that, as with the jugs from Culemborg, the

117 Hinton 1968.
118 Gilchrist 2012: 241–242.
119 As discussed in Chapter 2.2.
120 Merrifield 1987: 120.
121 Moorhouse 1978: 12–13. Such a functional interpretation likely explains the complete clay vessel discovered incorporated into the floor of a medieval structure at New Romney, Kent. Linklater 2004: TP56.
122 van Vilsteren 2013.
123 van Vilsteren 2013: 173–174.

cauldrons were believed to protect the dikes from harm and make them more effective flood defences. VAN VILSTEREN theorises that such offerings were a relatively common practice which has not previously been recognised in the archaeological record.[124] The deposition of cauldrons has pre-Christian ritual significance – during the Viking Age, cauldrons were commonly deposited in lake edge settings in Scandinavia, with analogous practices in Britain and Ireland[125] – so in the case of the medieval Dutch examples, these may have harnessed traditional folkloric beliefs attached to the deposition of cauldrons in order to enhance the protection offered by the dikes within which they were deposited. Of relevance here is a well-known literary description of a foundation deposit with a similar context and motivation. In Theodor Storm's *Der Schimmelreiter*[126] Frisian farmers who are constructing a new flood embankment plan to include a still-living dog within the earth of the dike as a living sacrifice[127] – presumably to enhance the protection afforded by the flood defence.[128] While this is a fictionalised account, it is widely thought to have some basis in the beliefs and folklore of the early modern communities along the Frisian coast[129] and provides an interesting analogue for the medieval cauldron deposits which appear to have been similarly deposited within the fabric of dikes during construction.

A wide array of evidence also relates to foundation and 'special' deposits in ecclesiastical settings. One British example, recorded through documentary evidence, comes from the Abbey of St Albans where, sometime between 1326–1335, when the foundations of the new cloister were being laid, saintly relics were 'sprinkled' beneath the foundation stone.[130] An Italian parallel records how, in 1325, in addition to coins placed in the foundation of the Torre del Mangia, in Siena, stones were placed in each corner, inscribed in Hebrew, Latin and Greek,[131] to protect the tower against thunder, lightning and storm.[132] 'Special' deposits in churches are also widely known through archaeology although many of these are interpreted as acts to safely remove sacred objects from circulation[133] rather than to accrue protection for the church and local community. Special deposits in the vicinity of saintly shrines were probably intended as offerings to the respective saint for healing, protection or good fortune. The votive wax figurines and other objects discovered hidden above the tomb of Bishop Edmund Lacey (d. 1455) in Exeter Cathedral – which included body parts, strings and male and

124 VAN VILSTEREN 2013: 180.

125 LUND 2008.

126 Originally published in 1888.

127 We are also told that, in the past, children had been the preferred sacrifice for this task.

128 STORM 1908: 112.

129 E.g. VAN DE NOORT 2011: 122–123.

130 *Et sub eodem lapide conspersit fundamentum de minutis reliquiis Sanctorum.* RILEY 1867: 282.

131 The three languages used on the plaque on Jesus' crucifix, known as the *Titulus Crucis.*

132 TRAVAINI 2015: 219.

133 GILCHRIST 2012: 234–235.

female figures – likely represent such a case.[134] Lacey, although uncanonized, was held responsible for a number of posthumous miracles[135] and therefore, these objects were almost certainly left, at or near Lacey's tomb, by pilgrims in the hope of receiving his help in their temporal woes. It is unclear how they came to be hidden within the masonry surrounding the tomb – this may have simply been part of the routine clearing of articles left at the tomb or, perhaps more likely, a surreptitious act in the years of upheaval surrounding the Reformation. Presumably, though, by hiding these votive objects in a location so closely related to his tomb, the belief was that Lacey would continue to help those who had left objects at his tomb.

The practice of foundation and 'special' deposits was widespread throughout western Europe. In southern Scandinavia, for example, artefactual deposits in the foundations of houses and churches have been interpreted as offerings to protect against external forces.[136] The spatial distribution of depositions in churches, clustered to the north and around the altar, has been interpreted as evidence that these artefacts were placed to gain protection against the devil.[137] Interestingly, in comparison to earlier and later periods, medieval depositions were overwhelmingly found in liminal areas – parts of the building forming the boundary between the internal and external worlds. This likely relates to contemporary theology which emphasised the need to protect against the devil.[138] Continuity for this practice can be seen in both earlier and later periods. For example, special deposits in Anglo-Saxon England were frequently made in "transitional places (i.e. entrances and boundaries)".[139] Equally, post-medieval burn marks, the vast majority of which have been conclusively demonstrated to have been deliberate creations,[140] are frequently found on structural timbers, similarly found primarily in liminal locations throughout domestic structures such as roof timbers as well as on the north side of churches.[141] A further medieval example of this practice comes from the inscribed slates unearthed in a doorway at Nevern Castle, Pembrokeshire,[142] the graffiti on which is discussed below.

Evidently, the location in which a special deposit was made was of equal importance to the objects included in the deposition. In addition to liminal locations within

134 Radford 1949: 164. Note that although the strings have been interpreted as a method to hang the other objects around the tomb, the fact that some were "twisted thread dipped in wax not unlike candle-wicks" accords with the widely documented practice that a person, or an injured body part such as a limb, or an object such as a ship were measured using a string, which was then made into a candle and donated at the shrine of a saint. Finucane 1995: 95–96.

135 Radford 1949: 165.

136 Falk 2006: 204.

137 Falk 2006: 203.

138 Falk 2006: 202–204.

139 Hamerow 2006: 27.

140 Dean, Hill 2014.

141 Lloyd et al. 2001: 66, 69.

142 Caple 2012: 440–441.

structures, there may be some relationship between ritualised deposits and specific environmental settings.[143] For example, although Viking Age hoards have traditionally been interpreted functionally[144] – as valuable objects hidden in times of unrest with the intention of retrieval at a later date – the high number which were never recovered as well as the preponderance discovered in environments such as lakes, bogs and wetlands – where it would be exceedingly difficult to mark the location of, and subsequently retrieve, precious objects – means functional interpretations are unlikely.[145] Additionally, the high number of Viking Age ritual deposits from 'watery' contexts,[146] especially the shores of lakes,[147] advances the possibility that, in some cases, such acts were motivated by a desire to accrue protection against hydrological hazards.[148] A connection may also be drawn between the Norse custom of ritualised hoard deposition and sites which were abandoned in environments affected by wind-blown sand.[149] In these cases, it is possible that the experience of the hazard provoked an act of ritual deposition – either for future protection or to calm the unseen forces behind the mobilisation of sand.

5.9 Folded coins

Folded coins make up a low yet significant number of all numismatic finds from the medieval period (almost 1%)[150] and are routinely encountered on a variety of archaeological sites. The majority (76% of those in the PAS database) are clearly the result of a deliberate act and are folded in half, often along the arms of a cross where these are present. A much lesser number are folded in complex arrangements. These include coins folded into thirds and quarters, coins which contain other coins and coins folded in association with other artefacts. Equally rare are coins with more ephemeral folds. These more ephemeral folds could conceivably have occurred by accident or as a result of post-depositional factors yet an intentional act remains possible. Excluding this latter category, the folding of coins was clearly a well-established practice during the medieval period which, as finds from the PAS database demonstrate, was spread widely throughout England (see Fig. 5.18).

A number of explanations have been advanced to explain the folding of coins in the Middle Ages. These included functional interpretations – that the fold was a means of

143 E.g. Yates, Bradley 2010a.
144 Graham-Campbell, Batey 1998: 243, 246.
145 Lund 2008: 55–56.
146 Lund 2008: 57–58.
147 In common with some of the classes of object considered in this Chapter such as pilgrim badges.
148 As has been theorised for lake side ritual deposition in a prehistoric context. Menotti et al. 2014.
149 Brown 2015: 139–141.
150 Kelleher 2011: 1499.

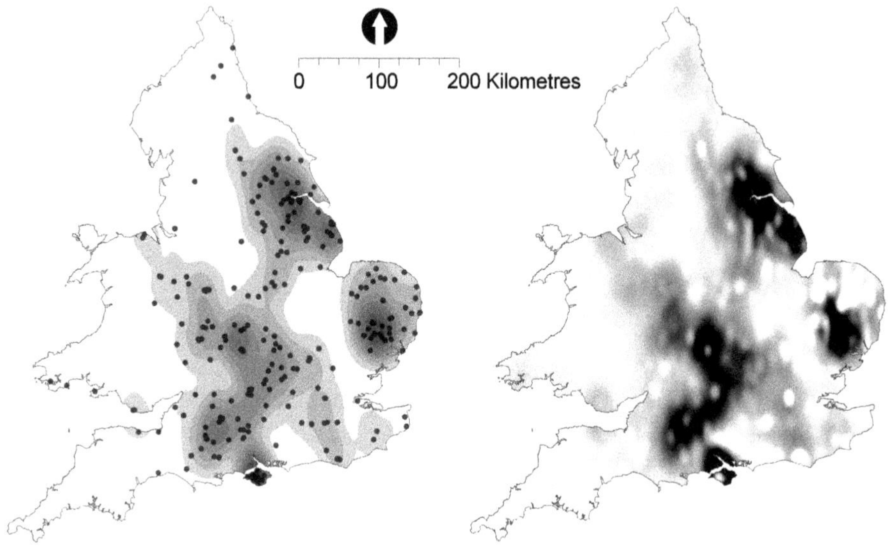

Fig. 5.18: Left: Distribution map of the folded coins in the PAS database as of 01 April 2016. Right: A comparison of the distribution shown to the left with the wider distribution of all medieval period finds in the PAS database. Darker areas correspond to higher densities relative to the PAS database as a whole while lighter areas denote areas which are under-represented. Created by the author.

testing the metal content of the coin[151] – although this can only explain a small minority of folded coins.[152] On the other hand, a large volume of evidence exists for folded coins as objects of apotropaic and ritual significance. A wealth of documentary evidence exists recording the folding of coins as a vow to a particular saint in which the coin folder promised to make a pilgrimage to the saint's shrine. In return, the supplicant would ask for, and receive, the help and protection of the saint in the material world.[153] This practice was particularly followed in times of need, for example: to extinguish fire, cure sickness or for protection in storms at sea.[154] The miracles of the canonized king Henry VI, provide a wide array of examples of the practice. One such tells how a certain Ralph Gabbott was restored to health after a long illness lasting three years by bending a coin in the name of "the blessed King Henry" while another relates how a victim of the plague was cured after bending a penny.[155] The bending of coins was described as 'the English custom' by the commissioners investigating the canonization of Thomas Cantilupe in 1307 which may suggest the practice was more common in

151 North 1994: 42.
152 Kelleher 2013: 252.
153 Duffy 1992: 183.
154 Finucane 1995: 94; Bartlett 2013: 355.
155 Knox, Leslie 1923: 192, 195.

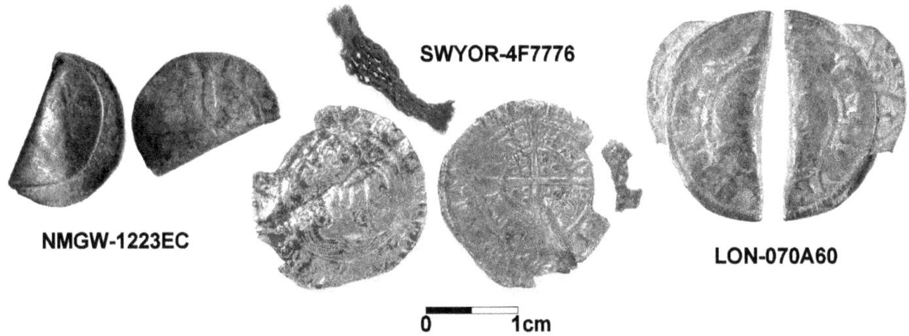

Fig. 5.19: Three folded coins. (Left to Right) NMGW-1223EC: Penny (1272–1327) found at Llanasa, Flintshire. A typical example of a coin deliberately folded in half. SWYOR-4F7776: Half-groat (1485-1509) found at Fulford, North Yorkshire. Folded (now unfolded) along with a strand of fabric suggesting use as jewellery. LON-070A60: Penny (1272–1301) found at Totteridge, London. Folded with a half-penny (1279–1377) inside. Photographs reproduced from the PAS, www.finds.org.uk, under CC BY-SA 4.0 licence: https://creativecommons.org/licenses/by-sa/4.0/.

Britain than in continental Europe.[156] A number of folded coins from the PAS database appear to have been folded in complex ways indicative of ritual behaviour consistent with the beliefs recorded in the documentary sources. For example, SWYOR-4F7776, a half-groat of Henry VII, was found folded over with a short strand of textile within the fold suggesting the coin had been converted into an amulet or piece of jewellery while SWYOR-6FA7C6 was folded within a sheet of lead. Five other finds[157] record coins folded within other coins. Furthermore, a number of folded coins are known from burial deposits – such as NARC-03B111, a silver groat of Edward IV which came from the mouth of a skull – as well as the foundation deposits of structures.[158] These practices are clearly inconsistent with a functional explanation and can therefore only be explained as ritual practice.

Folded coins are well represented in the PAS database. A search for coins of medieval date which contained the word 'fold' in the object description returned 337 results as of 1st April 2016. Each of these was assessed to confirm whether the fold was deliberate or due to post depositional factors. The folding of 23 of the coins was classed as non-deliberate and these coins were excluded from subsequent analysis. In the remaining cases, it was not always possible to decide with certainty so they were sorted into four categories based on the assessment of the fold (see Fig. 5.20). These categories were:

156 Finucane 1995: 95.
157 NMS-7BF5A3, BH-E654B1, KENT-6AB814, LIN-2BCE24 and LON-070A60.
158 Hall 2012: 79–80.

- Unknown - Where the PAS record is not accompanied by photographs and the object description is not detailed enough to allow attribution to one of the other categories.
- Slightly Folded - A minor fold, often only affecting one side of the coin rather than the entire object. These are the most likely to be non-deliberate although the fold is significant enough not to exclude the possibility that they were folded on purpose.
- Halved - A clear, deliberate fold in half. On cross pennies this is often along one of the arms of the cross.
- Other - This category includes all other non-standard folding, (for example into thirds, quarters, or in connection with other artefacts.

In order to assess how common the practice of coin folding was throughout the period, all of the finds from the PAS database were then graphed against time (see Fig. 5.21). This data was then compared with ALLEN's estimates of the volume of the English currency for the period to provide a benchmark with which to compare fluctuations in folded coin numbers.[159] The two datasets match relatively closely although there are disparities. Notably, the period *c.*1180–*c.*1310 saw a relative increase in the practice of coin folding compared to the general supply of coinage. A spike in the late 14th century also suggests a resurgence in the practice out of proportion to the number of coins in circulation, though it must be conceded that the estimates for the volume of English currency during these years lack data points. An important caveat in this analysis is that, although many of the coins can be dated with great accuracy, sometimes to particular years, this date is the date of production providing only a *terminus post quem* for when the act of folding took place. As a result, Fig. 5.21 indicates the likelihood of coins struck in different periods to be folded rather than periods when the practice of folding was most prevalent.

This can still be useful however. The folding of a coin was not the same as a coin being taken out of circulation due to general wear and tear. A coin could just as easily be folded immediately after being struck as 40 years later so, theoretically, folded coins should randomly represent coins at different stages of their lives – those that had been struck recently and those that had been in circulation for many years. One account, telling of a vow made to Henry VI, bears out the random nature of folding. The supplicant, in his haste to pledge himself to the Saint, was forced to bend a gold coin as he could not find any silver coins in his purse although he knew them to be there.[160] Due to this random element, there is no way to estimate accurately at what stage in the life of each coin it was folded. Some medieval coins remained in circulation for extended periods, as demonstrated by the contents of coin hoards – at least one

159 ALLEN 2001.

160 KNOX, LESLIE 1923: 129.

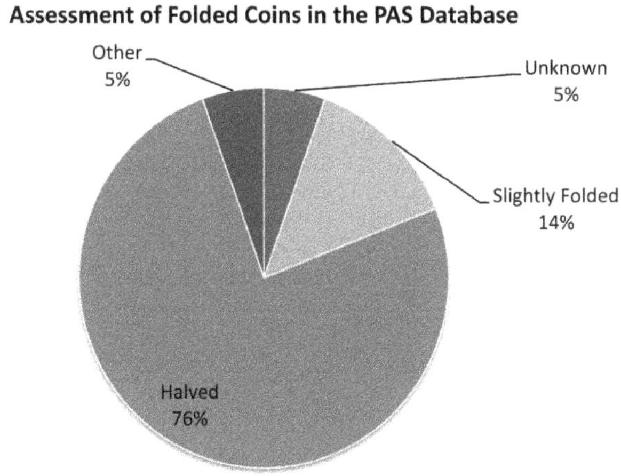

Fig. 5.20: Analysis of the type of fold found on folded coins in the PAS database as of 01 April 2016. Created by the author.

example is known of a coin remaining in use for 110 years.[161] As a result, the arbitrary application of a conservative estimate, in this case 20 years, seems the best solution. In addition, to excluding coins which may not have been folded deliberately, all those in the 'Unknown' and 'Slightly Folded' categories were removed from the analysis. With the allowance of 20 years to take account for circulation, the peaks in the numbers of folded coins more closely match those in ALLEN's estimate of the volume of the English currency (See Fig. 5.22).[162] Although the argument is somewhat circular, this makes sense because coins would be more likely to be folded when more coins were in circulation.

The extent to which the find-spot locations of folded coins are meaningful is muddled by the biases of the PAS database. Taking a smaller unit of analysis, such as the counties of Yorkshire, the spread of an admittedly small sample does appear to bear some relationship to rivers – with the vast majority of coins found along river courses. Of course in a British setting, rivers are never far away and settlements inherently cluster along their courses, focussing population and movement in river valleys. As a number of the categories discussed above have already highlighted, rivers and water bodies are widely associated with ritual deposits, both in the medieval period and earlier. Prehistoric metalwork deposited in rivers and water bodies have frequently been interpreted as ritually motivated acts[163] and the deposition of coins in springs and water bodies seems to have become a widespread phenomenon during the Roman

161 SPUFFORD 1963: 128.
162 ALLEN 2001: 607.
163 YATES, BRADLEY 2010a;b.

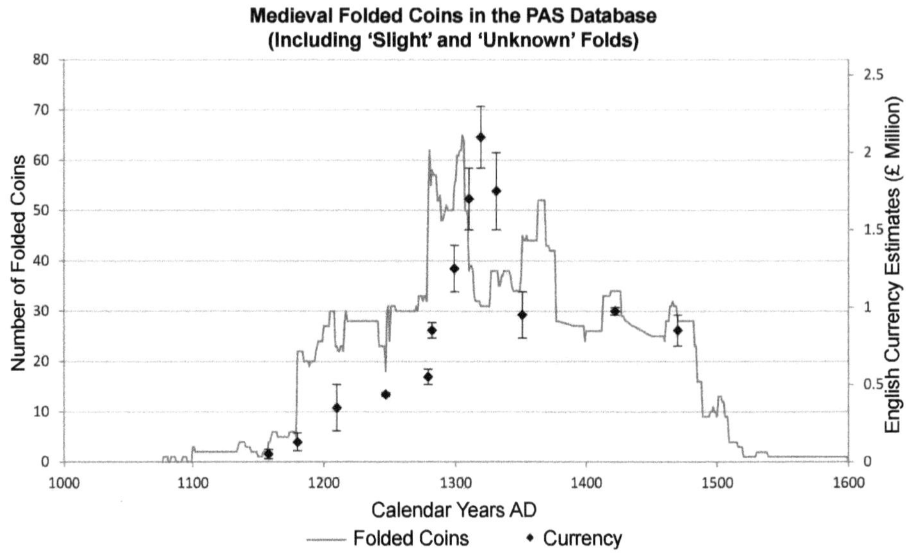

Fig. 5.21: Folded coins in the PAS database as of 01 April 2016, including those classed as 'Slightly Folded' and 'Unknown', plotted along-side estimates of the volume of the English currency after ALLEN 2001: 607. Created by the author.

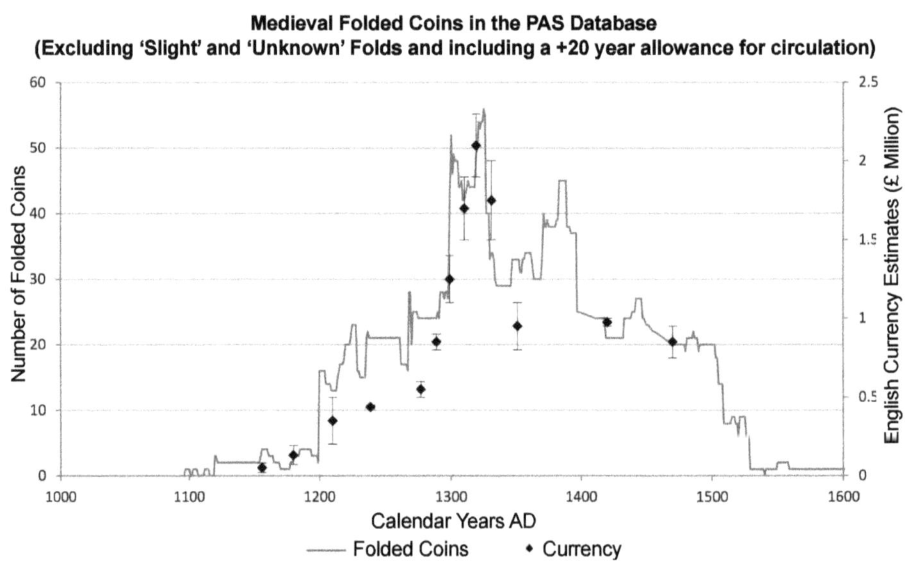

Fig. 5.22: Folded coins in the PAS database as of 01 April 2016, excluding those classed as 'Slightly Folded' and 'Unknown', plotted along-side estimates of the volume of the English currency after ALLEN 2001: 607. To take account for the circulation of coinage after production, 20 years has been added to the dates of the coins. Created by the author.

period.[164] The possibility exists, therefore, that some form of continuity, in folklore and ritual practice, persisted from prehistory into the medieval period. In light of this, the interpretation of many of the folded coins from the PAS database as deliberate depositions in areas close to rivers gains some credence though further research in this area would likely reveal a more nuanced picture.

In summary, the corpus of folded coins in the PAS database demonstrate that the ritual practice of coin folding remained popular throughout the medieval period, broadly following the trend of available coinage. The largest departure from this trend occurred in the late 14th century, perhaps reflecting a greater need for spiritual protection in the uncertain times that followed the Black Death – though this cannot be proven with certainty. There may be a relationship between the deposition of folded coins and river courses – perhaps implying that not only the act of folding itself held apotropaic significance but the location in which this occurred was also important. While the graphs shown in Figures 5.21 and 5.22 seem to indicate the practice faded out with the Reformation, this is primarily because only coins in the PAS database catalogued as 'Medieval' were included. Further work including 'Post-Medieval' coins would doubtless show that the practice continued into later centuries as many later examples are known – a number of folded 18th and 19th century coins, for example, have been recovered from the River Wear at Durham.[165]

5.10 Graffiti

Graffiti commonly encountered in medieval standing buildings offers another area in which superstitious beliefs were expressed. These included both Christian symbols such as the holy monogram and the name of God as well as more obscure patterns such as pentangles and circular compass drawn designs – all of which were probably linked to the accrual of some form of spiritual protection.[166] This is sometimes expressed pictorially – as in the depiction of a daemon at St Mary's, Troston, Suffolk, which is overlain by a pentangle (see Fig. 2.2). This appears to be a visual expression of the power wielded by the symbol of the pentangle to 'cancel out' the threat of daemons, which were widely believed to cause natural hazards.[167] While designs drawn using compasses, also known as daisy wheels, have been interpreted as purely functional designs, perhaps relating to the teaching of Vitruvian geometry, it is now generally accepted that they held ritual, and probably protective, significance.[168] The thinking

164 Sauer 2012.
165 G. Bankhead pers. comm.
166 Champion 2014.
167 Champion 2014: 249–250. As discussed in Chapter 2.
168 Meeson 2005: 46.

behind these designs may have been that their endless lines entrapped daemons and the devil who was unable to escape from their continuous loops.

Perhaps echoing the practice of depositions and burn marks described above, some examples of graffiti including protective symbols are found on churches' north facing walls such as at Baddesley Clinton, Warwickshire, where the most extensive graffiti was on the north wall of the entrance lobby.[169] Such a distribution of ritualised graffiti may relate to the fact that the northern door of a church was frequently known as the Devil's door and was thought to allow the devil an exit route from the church after being forced out of a new-born child during the baptism ceremony.[170] This association between the northern door of a church and the Devil certainly explains why these areas would have been particular foci for protective graffiti. However, doorways in general seem to have been regarded as requiring special protection. In Norway, runic and Latin inscriptions are frequently found on, or surrounding, medieval Church doorways, perhaps due to the perceived need for protection at liminal areas of the structure.[171] Comparably, slates with a wide variety of inscribed apotropaic symbols were unearthed beneath the south entrance at Nevern Castle, Pembrokeshire.[172] While the exact meaning of many of the symbols is unclear, some may be interpreted in relation to protection from nature or evil forces – for example, lightning bolts and pentangles. The fact that, of the many slates unearthed in the course of excavation, no others bore comparable inscriptions – as well as the deliberate positioning of these slates in the entranceway by the medieval construction workers – convincingly demonstrates that these particular slates held apotropaic significance and were probably believed to afford some extra protection to the castle's entrance.[173] In these cases, perhaps where structures did not have a dedicated 'Devil's door', it was thought necessary to bestow special protection on the other means of entrance and exit.

As with the other types of ritual practice described in this chapter, the precise motivation behind the inscription of graffiti is usually extremely difficult to appraise. Protection was usually sought against a broad suite of dangers rather than any one specific hazard. To what extent, if any, protection from natural hazards was a component in the cognitive process behind the creation of these marks is usually not known. At St Nicholas Church, New Romney, Kent, there is circumstantial evidence which ties graffiti within the church to the series of North Sea storm surge floods which devastated the town in 1287 and 1288.[174] The church itself was badly damaged by this series of storms – especially the chancel, which was rebuilt after the event[175] while the Norman

169 Meeson 2005: 41.
170 Champion 2015a: 41.
171 Jones 2010.
172 Caple 2012.
173 Caple 2012: 440–441.
174 Discussed fully in Chapter 3.
175 Tatton-Brown 1987: 344.

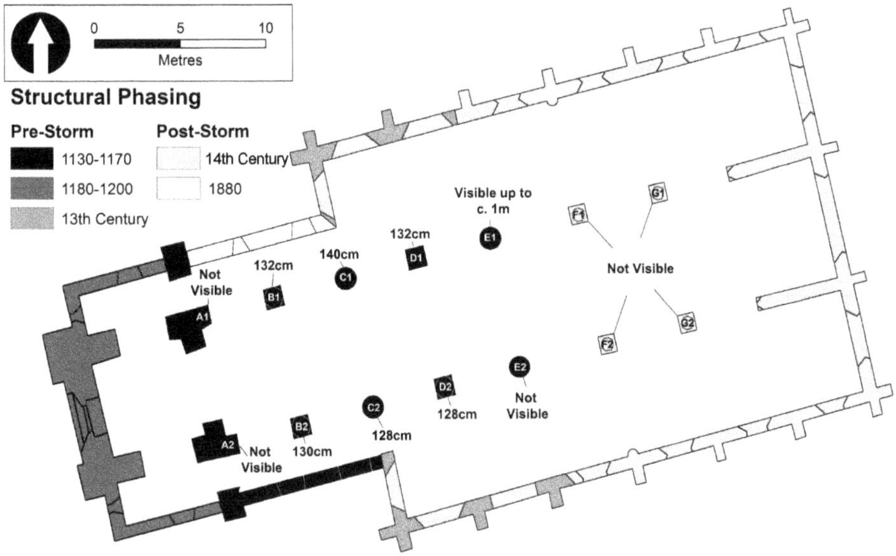

Fig. 5.23: The structural phasing of St Nicholas Church, New Romney. Redrawn by the author after TATTON-BROWN 1987: 345.

pillars in the nave were stained by the floodwaters to a height of *c.*135cm. A number of these pillars bear visible graffiti inscriptions including pillar B2, as marked on Fig. 5.23, which is inscribed with a partially-complete compass drawn circle and a difficult to interpret symbol formed of two superimposed crosses; perhaps an unfinished heraldic badge.

A particular concentration of graffiti, however, is found on pillar D2 which is inscribed with a number of inscriptions which may relate to the flood event. The symbols engraved on this pillar (Fig. 5.24) include a cross, two six-pointed stars, a compass drawn design or 'hexfoil', a set of two interlocking circles, a full circle and an unfinished circular arc. The cross could be explained as a consecration cross although its simplicity and small scale tend to suggest otherwise. Another possibility is that it was intended to ward off demonic forces, a well attested attribute of the sign of the cross. Similarly, as discussed above, the compass drawn design was a symbol of ritual protection[176] thought to entrap the Devil or demons preventing them from causing harm. The interlocking circles and other circular marks may be interpreted as simplistic and unfinished 'hexfoils' similarly intended to provide ritual protection.

Stars too were regarded as symbols of protection during the medieval period.[177] The six-pointed sign of Solomon, or hexagram, and five-pointed pentangle were syn-

176 CHAMPION 2015a: 39–44.
177 CHAMPION 2015a: 51.

Fig. 5.24: The graffiti on pillar D2 at St Nicholas Church, New Romney, with close-ups of the most important symbols. Photographs by the author. Symbols artificially highlighted in central image.

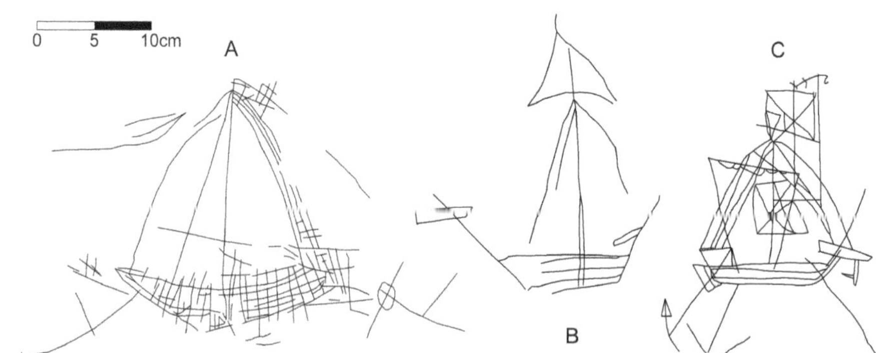

Fig. 5.25: Three examples of medieval ship graffiti from A: The south wall of the church at Broomhill, East Sussex. B and C: Two graffiti ships from Winchlesea St Thomas, East Sussex. Redrawn by the author after GARDINER 1988: 123, Fig. 10.9. and DHOOP 2016: 160, Fig. 64, ST5 and ST1.

onymously employed as protective signs, a fact attested by the late 14th century work *Sir Gawain and the Green Knight.*[178] In Scandinavia, pentangles were inscribed "above doorways and on wall posts" to protect corn in threshing barns.[179] Medieval dress accessories bearing the six pointed star are common occurrences, with the symbol usually interpreted as apotropaic as in the case of the 16th century pewter brooch from Christchurch, Dorest,[180] or a mount excavated in London.[181]

The symbols on pillar D2 are exclusively inscribed below the high water mark stained on the pillar in 1287 or 1288 and only on the side facing southeast – out to sea. Notably, both pillars with appreciable graffiti, B2 and D2 are the only two octagonal pillars on the south side of the nave, the others being cylindrical, with the flat faces of these pillars probably preferentially selected over the curved surfaces of their neighbours. Of course, the location of the graffiti may not be of great significance – it may simply relate to where the parishioners of New Romney found themselves standing during Church services – and other interpretations of the motivation behind its inscription are possible.

For any coastal community such as New Romney, in which fishing and ocean-going trade represented important contributions to the local economy, protection from the dangers of the sea would have been important and such inscriptions could relate to a more general desire for protection rather than an explicit response to the storms of the 13th century. The concentration of graffiti on D2 may also relate to the fact that, during the medieval period, a shrine to St George was situated against the south wall opposite the pillar. The cult of St George was relatively popular throughout medieval Kent and, as with all saintly shrines, the intercessory power of the saint would have been believed to offer extra protection[182] which would have enhanced the efficacy of these ritual symbols. As always with medieval graffiti, dating is extremely difficult and in this case it is not possible to say when these symbols were carved into the pillars. One possible interpretation, however, is that the memory of the flood led the parishioners to seek protection from a recurrence of such a disaster through the inscription of protective marks inside the parish church to ward off the evil spirits that might bring a future tempest to their town.

Parallels for this graffiti can be found in nearby churches. At Winchelsea St Thomas, East Sussex, for example, a corpus of 11 graffiti depictions of ships on pillars opposite

178 The description in Sir Gawain and the Green Knight is slightly confusing as the symbol is described as "the sign of Solomon" a "pentangle" and "the endless knot" GARDNER 2011: 119. On balance this description seems to best fit a five-pointed star but CHAMPION believes that six and five pointed stars were believed to offer comparable levels of protection. CHAMPION 2015a: 51.

179 MERRIFIELD 1987: 125.

180 SPENCER 1983: 81–82.

181 EGAN, PRITCHARD 2002: 203.

182 GOOD 2009: 97, 161.

the site of a chapel of St Nicholas have been recorded.[183] These graffiti ships are loosely dated by the construction of the church and the style of the ships to *c.*1290–*c.*1530.[184] Another graffiti ship was uncovered on the south wall by the doorway during the excavation of the church at Broomhill, East Sussex, (see Fig. 5.25) which was probaly rebuilt after the loss of an earlier church in the 1287/88 storms.[185] In common with the graffiti in New Romney, these graffiti depictions of ships may relate to the accrual of protection from the sea which – given the occurrence of the 1287/88 storms and other storms of the 13th and 14th centuries – would have had particular poignancy for the medieval populations of both Winchelsea[186] and Broomhill – just as they would have across a wider area for the inhabitants of Romney Marsh and neighbouring coastal areas.

5.11 Summary

From the different strands of material evidence discussed above it is clear that, throughout the medieval period, an extremely diverse web of methods were believed to provide protection from nature's extremes. For many of these it is extremely difficult to quantify the extent to which they were relied upon: a high proportion of saintly relics were lost at the Reformation, images of the Saints in churches have been lost or obscured through campaigns of iconoclasm and later renovations, medieval church bells are rare survivals and graffiti is invariably almost impossible to date beyond broad periods. For some categories of evidence, the PAS database allows some measure of quantification, although there are inherent biases within the data. Some trends suggested by the different categories of evidence are the preponderance for deposits and graffiti to be focussed around liminal areas or boundaries such as doorways and windows. Particularly in churches, the north wall, which was associated with the Devil, was also believe to require special protection. With more portable material culture thought to offer protection, such as pilgrim badges, *ampullae*, folded coins and artefacts associated with the saints' cults, there does appear to have been an increase in the circulation of these objects during the 14th and 15th centuries. Other than the three saintly cults explored above, a number of others markedly grew in popularity shortly after the mid-14th century – such as the cult of the holy name of Jesus.[187] The major changes of this period, most notably the Black Death – the aftermath and memory of which, in concert with the punctuated return of plague during the centuries which followed and the continued climatic upheaval caused by the transition to the Little Ice Age – may

183 Dhoop 2016: 158–164; see Fig. 5.25.

184 Dhoop 2016: 164.

185 Gardiner 1988: 122–125, Fig. 10.9.

186 Dhoop 2016: 223.

187 Blake et al. 2003: 177.

have created a milieu in which society as a whole accepted superstitious explanations and subscribed to apotropaic protective practices more readily. Certainly, BAILEY identifies a sea change in patterns of belief related to magic and witchcraft after *c.*1350[188] and witch hunts across Europe demonstrably increased in frequency and vehemence during this time.[189] The chronological patterns observable in some of the categories of artefact included in the PAS database, therefore, appear to reflect these wider shifts in lay devotion which affected how people sought to protect themselves against the many worldly dangers that they feared – including the events what we today describe as natural disasters.

188 BAILEY 2001: 986–987.
189 BEHRINGER 1999.

6 Memories of disaster

The preceding chapters have explored how natural hazards, and the disasters they engendered, were perceived, their short-medium term impacts and the steps which medieval society took to secure protection against them. A further important aspect to the 'experience' of disaster during the Middle Ages is the role of memory in preserving and transmitting knowledge of past events to future generations and, in turn, increasing preparedness and resilience in the face of future hazards. To what extent the occurrence of disasters feeds into a positive feedback loop, or 'adaptive cycle',[1] of this kind; in which each event provides new knowledge and experience from which successive generations can positively enhance collective social resilience, is a question of current relevance in studies of disasters, as well as change in complex systems more generally.[2] Some have questioned whether any such loop exists at all – likening the impact of natural disasters on the wider arc of history to the behaviour of aggressive bees: "they sting and then they die".[3] This argument posits that the acute short-term impact(s) of disasters, although painful, are inevitably rapidly overcome and forgotten. While this is perhaps true over the *longue durée*, especially in the case of low frequency events with a high return period,[4] on the time-scale of decades and centuries, focussing on events with a return period under 100 years, this approach seems to ignore some of the reverberations that disasters can trigger across society.

When considering hazards in long-term social memory, it is important to recall that the short-term impacts of hazards – the destruction of a town's harbour facilities by storm surge flooding or widespread flooding throughout a city centre, for example – frequently force human decisions which have long-term implications. In the aftermath of a disaster, the process of taking-stock, repair and re=organisation can often lead to adjustments, modifications or re-orientations in settlement patterns, the design and construction of infrastructure or established economic trends. In addition to such secondary impacts, certain types of hazard generate highly protracted direct physical impacts. Droughts and famines, for example, have direct impacts which can last for considerable periods of time. In the case of famine, malnutrition can leave a lasting negative effect – unborn children and infants who grew up under famine conditions in the 20th century were marked out in adulthood as, on average, shorter in stature, less economically successful and, in some cases, less healthy than those who grew

1 Forlin, Gerrard 2017.

2 Gunderson 2010; Angeler et al. 2015.

3 Uekötter 2004. "Sie stechen, und dann sterben sie".

4 Though even in the case of absolutely catastrophic events, memories do not always persist. The 8200 BP Storegga Slide tsunami, for example, which is estimated to have affected up to 3,000 individuals, during the relatively low population densities of the Mesolithic, has left no identifiable trace in communal memory – it has been effectively forgotten and is now known only through its 're-discovery' by modern scientific enquiry. Weninger et al. 2008: 16.

https://doi.org/10.1515/9783110719628-006

up under normal conditions.[5] Such a life history can occasionally be inferred through analysis of archaeological skeletal remains.[6] Flooding too leaves a long term imprint, both on the environment and on those who experience it, with appreciable negative impacts on the mental health of those affected.[7] Clearly, in addition to tangible impacts, the experience of disaster also has a profound long-term impact on the psyche of those who live through these events.

While these long-term impacts are apparent today, it is doubtful whether such considerations were readily appreciated during the medieval period. Memories of extreme natural phenomena, however, were preserved over the longer-term in other ways, albeit with different agendas in mind. Modern memorials to disasters often serve multiple purposes – these may encompass assisting survivors in coming to terms with loss, acting as reminders to avert similar future occurrences or for political and/or religious motives. In this chapter the different categories of evidence for disaster memorialisation during the medieval period are explored. In doing so, the role that any preservation and transmission of memory played in bolstering resilience from recurrent disasters, if any, is investigated. Particularly, any evidence for the existence of positive feedback loops or 'adaptive cycles' in amplifying medieval resilience to natural hazards is assessed.

6.1 Memory in the past

Natural disasters can have profound impacts: on the lives of individuals who experience them, the trajectories of settlement development as well as wider economic and demographic trends. As a result, their occurrence was usually highly memorable; forming distinct moments crystallised in the memories of those who lived through them. Over time, these would have become looser familial and communal anecdotes as they were passed down to subsequent generations. As oral traditions are repeatedly retold they eventually lose detail, are forgotten or mythologised – losing the ability to connect them with the historical event they recall, as conceptualised in Figure 6.1. An example may be provided by the belief, retold by Geoffrey of Monmouth,[8] that Stonehenge was constructed as a war memorial, reusing stones transported from an earlier structure located on the legendary mountain of Killaraus in Ireland – a story which has been interpreted as a partial, yet mythologised and confused, account of the transportation of the blue stones from the west (in actual fact from Wales) during the Bronze Age.[9] Work on oral history in traditional societies suggests that it is possible for memories to

5 Huang et al. 2010.

6 Annis et al. 2018: 136.

7 Fernandez et al. 2015.

8 Thompson 1999: 133–136.

9 Piggott 1941.

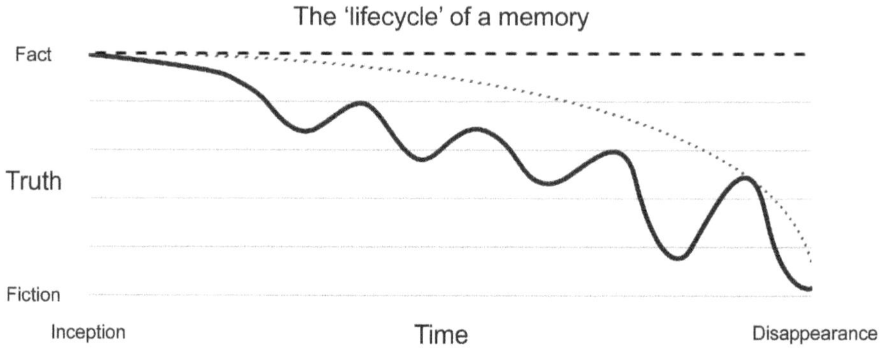

Fig. 6.1: The 'lifecycle' of a memory. The dashed line indicates an idealised transmission of a memory over time in which the memory is perfectly transmitted to future generations without loss of knowledge or the introduction of incorrect or fictional elements. The solid line presents a more realistic model in which the memory slowly deteriorates from a factual representation of real events to an account which, as a result of inaccuracies introduced over time, has become more and more mythologised and is eventually more or less a fictional account. The dotted line indicates the trend line. One problem with this model is that memories are not in fact absolute truth at their inception, as they are influenced by the viewpoint, biases and beliefs of the witness to the events they recall. Created by the author.

be accurately passed down for a maximum of approximately 200 years.[10] After this, as in the example of Stonehenge, events become mythologised and exaggerated with precise details forgotten. In the example of hurricanes in the Philippines, extreme storms are often forgotten after only *c.*100 years.[11] Of course, hurricanes are a common occurrence in the Philippines and this perhaps explains why even the highest magnitude events do not persist in the popular memory for longer, as more recent events surpass the stories of the past, in relevance and immediacy if not always magnitude. It is therefore plausible that in any given society, individual incidences of relatively routine and well-known hazards will be forgotten more rapidly than events which sit beyond what is considered 'normal'. It should be noted, however, that, while the occurrence of a specific disaster might be forgotten, this does not necessarily imply that ancestral wisdom concerning how to cope with that type of hazard also ceased to be remembered.

Usually, natural hazards directly affect only a limited number of people. During the medieval period, these were statistically most likely to be peasants – a group also associated with poor literacy and the relatively low survival of any documentary evidence they did create. In institutional and official records, which survive in far greater

10 Bradley 2002: 14.
11 Warren 2015: 214.

numbers, while the occurrence of natural disasters themselves was often recorded, although usually in low detail or for very specific reasons, the way in which these events were remembered by affected individuals in the years that followed was almost never a subject which merited documentation. As a result, exactly how memories of natural hazards were created, maintained and transmitted during the medieval period is a difficult subject to approach from the surviving textual sources.

In some respects, the medieval period has been conceptualised as a time when memory and the past were relatively unimportant, at least compared to those periods which preceded and followed. The lack of obvious war memorials from the medieval period compared to Roman antecedents and the equestrian statues which rapidly appeared following the Renaissance offers one such example.[12] Similarly, medieval material culture has been interpreted as signifying little interest in old things[13] and evidence for medieval interaction with artefacts from earlier periods has been thought to display "a rejection or suspicion of material links with the past".[14] New, therefore, seems to have generally been preferred to old. The importance of memories and the past to medieval people, however, is demonstrated by various strands of evidence including the curation of old objects, likely often family heirlooms, found among archaeological assemblages,[15] the foundation of churches and ecclesiastical institutions to commemorate past events[16] as well as the popularity of the cult of the saints which purported, and was widely believed, to commemorate historical events – the past clearly mattered.

That natural disasters were communally remembered during the medieval period also occasionally finds confirmation in the documentary record. Edward Hall, an early 16th century chronicler, for example, was well able to recount the devastating 1483 flood on the Severn, near Gloucester, which occurred 14 years before his birth, and 65 years before the publication of his chronicle, and was apparently well known amongst the local community.[17] Understandably, high magnitude events, such as the 1483 Severn flood, appear to have provided particular cause for comment and subsequent memorialisation. Such events, which transcended accepted norms, were frequently described with the adjective 'great'[18] – as indeed has been the case in many more recent nat-

---

12 Borg 1990: 1–3. Note, however, that some medieval battles did attract memorials such as the cross erected after the Battle of Neville's Cross, Durham, in 1346. See Holland 1835: 176–177. Another example is the 'Battle Stone' which marked the site of the 1388 battlefield at Otterburn, Northumberland. See Tyson 1992: 77.

13 Hinton 2005: 223.

14 Gerrard 2007a: 170.

15 Gilchrist 2012: 237–242.

16 Borg 1990: 4–6.

17 Hall 1809: 394; Rhodes 2006: 33.

18 Or its Latin equivalent *magna*.

ural disasters.[19] The extreme drought of 1540, for example, was described as a 'great drought' by English contemporaries[20] while the use of the word 'great' as a descriptor in a Papal letter written in 1410, describing the storm which inflicted damage to the chapel of St Martin-le-Grand, London,[21] likely indicates that the chapel was a victim of the storm of 1362;[22] which certainly caused severe damage in London and would undoubtedly have been remembered 50 years later.[23] The 15th century historian of the Abbey of Westminster, John Flete,[24] included a description of this storm, and the long period of repairs in its aftermath, in his history of the Abbey's estates.[25] Although Flete himself was not an eyewitness, he no doubt grew up hearing stories from the more senior monks of the ferocity of the 1362 storm. That laymen held similar recollections of that storm may be suggested by the inclusion of the St Maur's Day storm in a calendar of saints' days and historical events at the beginning of an early 15th century Bible thought to have been produced for a lay Londoner.[26] Furthermore, notes in the margins, or on otherwise blank pages, of ecclesiastical books from the 15th century commonly referenced the St Maur's Day storm,[27] highlighting its recollection, at least by monks in the *scriptorium* if not by the wider populace.

Documents which describe natural hazards quite tangentially also occasionally hint at the persistence of memory. In describing the flood of 1338 in Marshland, for example, the historical material gathered by DUGDALE records that "within the compasse of *Lx.* years then past, one Parish Church, with the Parsonage, was by the breaking in of the Sea, ruined: and within *xxx* years, a Mannour house ... with an hundred acres of Land, utterly lost for ever".[28] Another example comes from a Papal letter dated to 1404 or 1405 which mentions how a parish church "had been founded of old" near Newton, Cambridgeshire, at a place called "Saltmarsh" and "through storms of the sea and flooding tides ... been devastated and abandoned and transferred to a place more remote from the sea".[29] In both cases, exactly when the old parish church was abandoned is unknown but candidate events for the marine flooding include the storms

19 Examples include the 'Great' blizzard which affected the Atlantic coasts of the US in March 1888, the 31st January/1st February flood of 1953 which is commonly remembered as 'The Great Flood' in England (and the events of that night are thoroughly documented in Hilda Grieve's 1959 book 'The Great Tide: The Story of the 1953 flood disaster in Essex') as well as the 'Great Storm' of 1987 in England. BURT, MANSFIELD 1988.

20 FABYAN 1811: 701; BAILEY 1853: 411.

21 BLISS, TWEMLOW 1904a: 208.

22 CONYBEARE 1910: 248.

23 More detail on the storm is provided in Chapter 3.3.

24 B. *c.*1398 d. 1466.

25 ARMITAGE ROBINSON 1909: 135.

26 SOLOPOVA 2016: 211, 214.

27 See for example: DUL MS Cosin V.iii.19 f. 19r; HL HM 37539 f. 154; HL HM 28174 f. 143v.

28 DUGDALE 1662: 255.

29 BLISS, TWEMLOW 1904a: 24–25.

of 1287/88, which affected a number of nearby towns.[30] In the latter case, the marine inundation of 1338, which reportedly swept away the church of Walpole St Peter, Norfolk,[31] situated less than 8km from Newton, is another candidate which would fit well with the devastation described and would equate to the survival of memory of the event for some 67 years. Such cases make it apparent that relatively precise memories of disasters and information relating to their occurrence easily endured for 50 years and sometimes more.

In other cases, it appears that oral traditions recounting the occurrence of medieval disasters could survive for considerably longer. In the case of the 1345 Gauldalen landslide in Norway, local oral traditions of the event, some of which contradicted each other, persisted at least to the 1770s, over 400 years from the date of the landslide, when they were recorded by the historian Gerhard Schøning.[32] While in this case, a near contemporary written account of the disaster survives,[33] in the form of the Icelandic *Skålholt Annals*,[34] the oral tradition includes details that are not mentioned in the documentary account – some of which are now thought to be key to understanding how the disaster unfolded.[35] A similar survival of an oral tradition, recorded in the early 20th century, relates to the besanding of the medieval town of Kenfig, Bridgend, by aeolian sand. Reportedly, some of the town's houses were engulfed so rapidly by sand that their inhabitants had to be dug out and rescued from the structures buried by sand.[36]

The oral record of the storms of 1287/88 provides another case of the possible long-term survival of memories related to disasters. Knowledge of the storm certainly existed into the early 19th century as the remains of a medieval ship, unearthed in a stretch of the River Rother, were initially interpreted as a wreck damaged and sunk during the storms of the late 13th century.[37] Equally a local legend of unknown antiquity purports that an outcrop of holly trees near Lydd, Kent, are the only remains of trees planted by a Frenchman named Stepvans, the majority of which were buried by shingle on the night of a storm in 1287.[38] Whether an oral tradition of these storm events persisted uninterrupted from the 13th century until the recent past or was 're-kindled' by a more

30 Pawley 1984: 74.
31 Pevsner 1962: 745; see Chapter 4.
32 Helland, Steen 1895.
33 Perhaps as a result of this 'external storage of memory' the survival of the local oral tradition was assisted in a way that would not have been the case in completely non-literate societies.
34 Storm 1888: 211–212.
35 Rokoengen et al. 2001.
36 Morris 1907: 6. Note though that in this case it is unclear precisely when the sand movement that the local tradition relates to took place. Although problems with sand occurred at the site from the early 14th century, these persisted into the post-medieval period. See Wessex Archaeology 2012b: 3.
37 Wallis 1823: 84–85. This vessel is now thought to date to a significantly later period, *c.*1500. See Fenwick 1978: 258–260.
38 Scott 1965: 29.

recent discovery of documentary evidence is, however, difficult to gauge. All of these examples indicate the possibility that oral traditions recording medieval disasters survived for several hundred years, transmitting information which, theoretically, could have assisted future generations dealing with the same type of hazards. In the case of the Gauldalen landslide, knowledge of the oral tradition could conceivably have helped local populations interpret conditions likely to result in a comparable landslide – allowing a measure of forewarning. Knowledge that sand could envelop human habitation, as it reportedly did at Kenfig, might have encouraged active preventative measures while providing an example of how to act if such unfortunate circumstances did arise. The traditions relating to the 1287/88 storms, meanwhile, likely contributed to a more general awareness of the risk posed by storm events in the local area rather than offering preventative or protective advice.

When particularly severe events struck, it was common to compare them to those that could be remembered from the past. This provides another rare glimpse into how memories of disasters were, or were not, preserved. This is borne out by a detail mentioned by the anonymous chronicler from the Abbey of Melk, Austria, regarding the high magnitude flooding which occurred along the Danube in 1501: "within 100 years hardly anyone could remember such a high flood as a 107 year old woman … testified".[39] Although less specific, this can also be seen in chroniclers' descriptions of the 1362 gale which include descriptions of the storm as like nothing "seen or heard before … in England"[40] or "it was thought [nothing similar] had been seen in earlier times".[41] The inscription documenting the 1342 flood at Würzburg, Germany, meanwhile, describes how, during that event, water "reached as high as never before".[42] Comparably, the conditions experienced during the drought of 1540, by some estimates the driest summer across Europe in the last 1300 years,[43] "had scarcely ever been known in the kingdom".[44] Although most of these events were verifiably unprecedented – well explaining such assertions – these kinds of comments must be viewed critically. Such claims are routinely made today in the modern media's reporting of natural catastrophes, even when the events themselves are both routine and with recent precedents.[45] As the occurrence of natural disasters was frequently exploited by medieval populations to petition for concessions in taxation, or other dues owed,[46] describing the event as completely beyond human experience, even when this was not literally true, can only have enhanced their case. Therefore, such descriptions

39 ROHR 2007: 95.
40 SCOTT-STOKES, GIVEN-WILSON 2008: 119.
41 MARTIN 1995: 185.
42 DOTTERWEICH et al. 2003: 148, Fig. 7.
43 BÜNTGEN et al. 2006; WETTER, PFISTER 2013.
44 BAILEY 1853: 411.
45 FAVIER, GRANET-ABISSET 2012: 120–121.
46 As was the case following the Fenland floods of 1338; see Chapter 4.

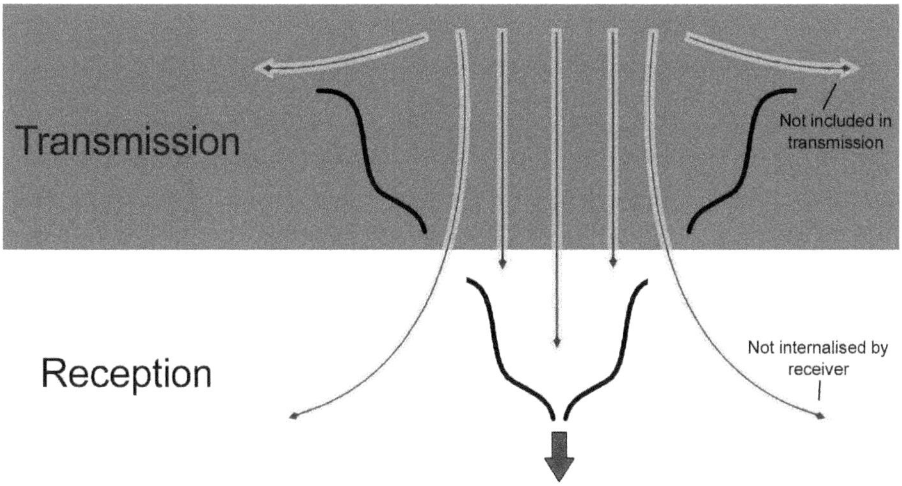

Fig. 6.2: The transmission and reception of memories. Where memories are transmitted orally they are 'filtered' iteratively at each retelling. As a result, details are lost as certain pieces of information fail to be transmitted while, of the data that is transmitted, some fails to be properly received. Each transmitter and receiver may also introduce their own embellishments which were never present in the version of the memory they originally received. Created by the author.

may have at times been exaggerated to amplify the potential afforded by a disaster to renegotiate existing customs, arrangements and obligations.

While, clearly, medieval memories of natural disasters were made, transmitted and preserved, they remain a complex topic. Memories which exist only in human consciousness are, in some regards, like ripples in a pond, beginning as quite pronounced disturbances which, over time, give way to more gentle movements before fading out altogether. At the same time, however, at any given point along a memory's life-cycle it could, quite suddenly, change its character or absorb new aspects as the memory is retransmitted and internalised through iterative retelling. Through each oral retransmission, a memory must pass through two 'filters', conceptualised in Figure 6.2 – what the teller chooses, consciously or not, to include and exclude from their 'transmission' and what the listener chooses, consciously or not, to hear and remember. Of course, in a complex, literate society – as existed across medieval Europe – there were many additional ways to encode and transmit memories which, to some extent, bypassed this 'filtration' system. The remainder of this chapter, therefore, considers the additional ways in which society created and preserved memories relating to natural disasters during the medieval period.

6.2 Popular verse

The transmission of popular verse was an important method of spreading and communicating news of events. In an age before newspapers, radio or modern digital media, news from the wider world was relayed largely by word of mouth. Poetry played an important role in this process, presumably from an early date, not only enlivening news from far and wide but also packaging it in a form that could easily be remembered and passed on to family, friends and neighbours. The occurrence of natural disasters, as in modern news reporting, was a topic which readily excited popular interest and was therefore commonly encoded in this way. In medieval Welsh poetry, for example, extreme weather events such as floods and anomalously cold seasons were frequent subjects.[47] Verse continued to be a key way in which memories were preserved into the post-medieval period, with a significantly larger corpus of surviving source material as the invention of printing facilitated the spread of such material.[48] Ballads were also an important method by which news was disseminated in post-medieval England – giving notice of notable events such as criminal executions as well as the occurrence of environmental hazards.[49] Medieval poetry operated in much the same way providing a source of news that did not require literacy and could easily be remembered and recounted and, as with early modern ballad performance,[50] musical accompaniment enhanced the popularity and, therefore, the longevity of these works.[51] As a medium for transmitting and encoding the occurrence of events, verse operated in a complex way. For poetry to pass from the mind of the poet into wider society – and survive into the present – it first had to be pondered on and composed by the poet and then performed in front of multiple audiences. At some point the poem had to be written down or taught orally to others in order for it to spread. Where textual copies survived these were, over time, collected, read and examined by scholars and interested parties. In some cases, poems may have been completely forgotten in the oral tradition only to be rediscovered in manuscript form. In this way, knowledge of the events discussed in a poem would have "waxed and waned across the centuries" and would be "recalled and recycled at particular times depending on access and interest".[52]

Popular verse relating to natural disasters presumably flourished, and reached a large audience, only when hazards affected a sufficiently high number of people

47 Griffiths et al. 2017.

48 Pfister 2011: 8.

49 McIlvenna 2016.

50 McIlvenna 2016: 290–291.

51 Examples which seem to date from the medieval period include *Sir Patrick Spens*, which Walter Scott interpreted as a mythologised memory of a voyage undertaken in the late 13th century, and *The Battle of Otterbourne*, which preserves memory of the English defeat to the Scots at Otterburn, Northumberland, in 1388. Scott 1807: 295–298, 345–353.

52 Griffiths et al. 2017: 104–106.

across a wide enough area, allowing the promulgation of verses that were meaningful to a large number of people. Regional and local disasters on the other hand, affected much smaller numbers and, as such, although verses may have been composed and known locally, it was less likely that these would survive over a number of years or be recorded by contemporary chroniclers. The St Maur's Day storm of 1362 fits into the former category and, as such, offers a rare glimpse into the memorialisation of a medieval natural disaster through poetical verse. As it affected the whole of southern England, the majority of the population could find interest in recalling an experience that many had directly experienced and all knew something about. A variety of strands of evidence suggest that verse was an important way in which these memories were transmitted, both in the immediate aftermath of the storm and over a much longer period.

Perhaps most famously, the St Maur's Day storm appears in the allegorical poem *Piers Plowman*; its inclusion in which provides a *terminus post quem* for the composition of the poem itself – which probably occurred within a decade of the storm.[53] *Piers Plowman* went on to become a highly influential and popular work in late medieval England[54] so the storm's inclusion in the poem would have done much to perpetuate and spread knowledge of the event. Two other near-contemporary verses, which seem to have gained popularity in the area affected by the storm, were recorded by the anonymous chronicler who continued Adam Murimeth's chronicle following his death. It appears that from 1359 this anonymous chronicler was a contemporary of the events they describe,[55] meaning that they were presumably directly aware of these verses being recited among communities that had lived through the storm. The two verses documented are given below:

Contemporary verses documenting the 1362 storm

Latin Original
C. ter erant mille decies sex unus et ille
Luce tua, Maure, vehemens fuit impetus auræ[56]

Ecce flat hoc anno Maurus in orbe tonans[57]

English Translation
One Thousand, three hundred, sixty and one
Your light, Maur, vehemently attacks the air
(Translated by the author)

53 Bloomfield 1962: 114, V: 14.
54 Hudson 1988.
55 Hog 1846: xviii.
56 Hog 1846: 196.
57 Hog 1846: 196.

> Behold! in this year, Maurus thunders in the world
> (Translated by the author)

It can easily be imagined that these two short phrases – which are concise and low in detail, conveying only the fact that a storm occurred in a particular year – became widely known and were frequently repeated by people across the area affected by the disaster. This would have provided a means of commemorating an unprecedented event that probably touched nearly everyone in some way. The widespread popularity of at least the latter of these phrases is confirmed by the fact that it was engraved at the base of the tower of the church of St Mary, Ashwell, Hertfordshire (Fig. 6.3). The last line of this graffito, transcribed below, excluding the first two words, and first letter of the third word, is almost a perfect match for the verse given in the continuation of the Chronicle of Adam Murimeth.[58]

The inscription at St Mary's Church, Ashwell, documenting the 1362 storm

Latin Original
[mccc]xlix
pestilencia
M C T Xpenta miseranda ferox violenta
mcccl
Supest plebs pessima testis in fine que ventus validus
[h]oc anno maurus in orbe tonat mccclxi
(see Fig. 6.3)

English Translation
[13]49
There was a plague
One thousand, three times one hundred, five times ten, miserable, wild and violent
1350
The dregs of the populace survive to witness, In the end a strong wind
Behold! in this year, Maurus thunders in the world 1361
(Translated by the author)

Although the inscription of the graffito cannot be dated precisely, the church was under construction at the time of the storm – the chancel was completed in 1368 while the tower remained under construction until 1381[59] with structural timbers from the tower dating to the decade immediately after the storm.[60] It is likely, therefore, that the graffito was carved by a literate mason working on the church at this time.[61] This

58 Hog 1846: 196.
59 Pevsner 1977: 74–75.
60 The timbers were dated through dendrochronology to 1365–1376. Miles et al. 2003: 110–111.
61 In relation to memorials, including popular verses, which recalled the famine of 1315–1322, Jordan notes that these were most commonly penned by those below the upper echelons of society; by people

Fig. 6.3: The graffito at St Mary's, Ashwell, Hertfordshire, engraved at the base of the tower. The text, highlighted, describes the outbreak of plague in 1361 and goes on to mention the storm of St Maur's Day 1362. The last line is nearly identical to a popular verse given by a contemporary chronicler. Photograph by the author. Inscription artificially highlighted.

possibility is bolstered by the fact that the inscription is at least 2m above ground level, suggesting that scaffolding may have been in place when it was inscribed – as it would have been both difficult and unnatural to engrave it at this height from the current floor level. Furthermore, a double column of figures carved on the walls nearby has been interpreted as the masons' wage sheet[62] – indicating that they had no qualms about marking the walls. One possibility, therefore, is that the unfinished church was damaged, or work was disrupted, by the storm and a literate stone mason decided to commemorate this event by carving an inscription into the wall of the tower where they were working.

Beyond the verses that circulated in the months and years that immediately followed the occurrence of the storm, there is some evidence that memory persisted through poetical compositions. To the examples given above can be added a poetical verse included in John Harding's *Chronicle of England*, composed no earlier than 1436, which seems to have been an original composition made at least 70 years after the storm's occurrence.[63] The verse recalls the magnitude of the wind and the high degree of damage caused across England but also, interestingly, introduces a falsehood – that the storm was accompanied by an earthquake. This fact goes unmentioned in any

who were literate but in touch with the common lay population such as friars, University masters or parish clergy. Jordan 1996: 183.

62 Champion 2015a: 209.

63 Ellis 1812: 330.

contemporary source and probably stems from the Aristotelian belief, transmitted to medieval European scholars through texts such as Isidore of Seville's *De rerum natura*, that earthquakes were caused by subterranean winds and eruptions released these winds into the air causing storms.[64] The inclusion of this detail in the verse can therefore only be a later embroidery to the memory of the storm. Though the evidence is fragmentary, we can see to some extent that the initial verses which recalled the storm, though low in detail, provided the essential information – that a severe storm had occurred in a particular year – while, after the passing of time, though the verse still relayed these basic facts, the memory had begun to be embroidered with the addition of incorrect and exaggerated elements.

Verse documenting the 1362 storm from John Harding's *Chronicle of England*

In that same yere was on sainct Maury's day,
The greate winde and earth quake' meruelous,
That greatly gan the people all affraye,
So dreadfull was it then and perelous,
Specially the wind was so boistous,
The stone walles, steples, houses, and trees,
Were blow doune in diuerse ferre coũtrees.[65]

6.3 Inscriptions

Although not all were metrical in their composition, inscriptions which commemorated the occurrence of extreme natural events – like the graffito from St Mary's, Ashwell – were a relatively common method of creating a lasting memorial and these were made to record the occurrence of all kinds of hazard. An inscription carved in in 1317 at the church of Saint Katherine, Oppenheim, Germany, for example, commemorates the famine of 1315–1322 – reminding future church goers of the high price of bread during this time of dearth.[66] The inscription dates the construction of the church to 1317 and in this case, as with the Ashwell graffito, it seems almost certain that it was inscribed by one of the masons working on the church. In seismic regions, comparable inscriptions are also commonly found in ecclesiastical settings memorialising repairs and restorations necessitated by earthquake damage – as in medieval Italy. One such example, at the church of San Mauro of Costozza, in Vincenza, is a 14th century inscription recording the earthquakes of 1117, 1222 and 1348.[67]

64 Kendall, Wallis 2016: 173.
65 Ellis 1812: 330.
66 Jordan 1996: 184.
67 Forlin, Gerrard 2017: 104–105.

Commemorative inscriptions were also widely created to recall destructive floods. The 1342 flood in central Europe, for example, which was one of the region's most severe recorded floods, provoked the creation of a number of inscriptions detailing the ferocity of the inundation and the extent of flooding.[68] One example, which originally occupied the wall of a house in Würzburg, Bavaria, records the damage wrought by the flood as well as the name of the mason who built the house.[69] Once again, it appears that a mason was the author of this inscription – though whether masons acted independently in the creation of these memorials, or were compelled to do so by their clients, is unknown. Some inscriptions, such as the one at Würzburg, display an attention to detail that suggests they are unlikely to have been haphazard unplanned additions – a scenario which is more likely in the case of a relatively untidy *graffito* such as the one from St Mary's, Ashwell.

The St Elizabeth's Day flood of 1421, which devastated settlements in South Holland, is another example memorialised by a number of inscriptions. One such, transcribed below, comes from inside the Grote Kirk, Dordrecht, from the south wall by a door underneath the organ.[70] Another was installed above the Spuipoort, one of the city gates. Whilst the former appears to have been installed shortly after the event it recalls, the latter inscription only appeared in 1609, almost 200 years after the event it commemorated.[71] Clearly, memory of the event in the town survived for a long period although, as other strands of evidence make clear, this was not without fictional embroidery of the historical facts. The flood destroyed a number of villages located in the Grote Ward and, as Dordrecht was one of the towns to which the survivors relocated, memory of this event must have been particularly strong within the town. The survival of local memory was reinforced by the fact that, at least into the early 16th century, the remains of the inundated villages could still be seen when sailing in the area.[72] Local folklore surrounding the flood, however, soon developed.[73] One tale, frequently retold centuries later, concerned a baby who had apparently been found during the flood floating in a basket accompanied only by a black cat. This story must have surfaced relatively soon after the flood's occurrence as it is depicted on the left panel of the St Elizabeth's Flood Altarpiece (see Fig. 6.4), which was painted in the early 1490s to commemorate the flood. For centuries afterwards, certain families in Dordrecht claimed descent from this child, who was either male or female depending on who was asked. The number of villages destroyed by the flood also swelled over time to 72, an auspicious number in

68 See for example DI 66, Lkr. Göttingen, Nr. 9 [Online] http://www.inschriften.net/landkreis-goettingen/inschrift/nr/di066-0009.html [Accessed 01 February 2021].

69 Campbell 2016b: 283, note 64.

70 This information comes from the website of the Regional Archief Dordrecht, inventory number: 551_30149. http://beeldbank.regionaalarchiefdordrecht.nl/ [Accessed 01 February 2021].

71 Balen 1677: 770.

72 Junius 1588: 182–183.

73 Pollmann 2017.

Christian theology, and this inaccurate detail was included in both of the inscriptions that recorded the flood in Dordrecht. Clearly, in this case, the memory of the event was embroidered relatively quickly after the event's occurrence and perpetuated through these inscriptions as well as local folklore. Medieval inscriptions of this type are usually concise – ensuring the preservation of some knowledge of the event – but, as a result, they fail to prevent the embroidery and fabrication inherent in the long-term oral transmission of memories. Where such invented and mythologised elements are included in the inscription they can even perpetuate the survival of inaccurate information.

Inscription commemorating the flood of 1421 in the Grote Kirk, Dordrecht, South Holland

Latin Original
A[NNO] XP̄I: 1421 NOVĒḂ 18 SV[BMERSAE]
MĀ[N]SERV̄[N]T 72 VILLE C[ON]TINVE
[DORDRACO][74]

English Translation
In the year of Christ: 1421 November 18 72 towns
were permanently submerged around Dordrecht
(Translated by the author)

Inscription commemorating the flood of 1421 originally installed above the Spuipoort, Dordrecht, South Holland

Dutch Original
T Landt en Water dat men hier siet ware[n] 72 proch[y]en na geinundeert devrt water crachtich int iaer 1421 waer achtich[75]

English Translation
On the land and water seen here, 72 parishes, as the chronicle says,
were inundated by the force of the water in the year 1421[76]

Flood marks

In addition to providing qualitative documentation of the impacts of particular inundations, in the case of floods, and more rarely droughts,[77] inscriptions presented a method of preserving memory of the height reached by floodwaters in a given event at a particular location. The height reached by floodwaters is an important piece of information which both reflects the severity of the inundation and assists in measures

74 van Gijn 1908: 197.
75 Balen 1677: 770.
76 Pollmann 2017: 124.
77 Macdonald 2007: 136.

to prevent such an event recurring through future planning; erecting flood defences or zoning (re)construction away from areas which are commonly flooded. Evidence that this information was frequently noted and remembered can at times be discerned from other forms of evidence. John of Oxnead, for example, recorded that the flooding caused by the December 1287 storm reached more than a foot above the high altar of the parish church at Hickling, Norfolk[78] while in New Romney, Kent, oral memory persisted that the pillars of the church of St Nicholas were stained at the level reached by storm tides during the floods of the same season.[79] While such documented or orally transmitted memories of the levels reached by floodwaters were useful, flood marks, also called Noah's marks, represent a permanent material memorialisation of this information which were relatively commonly erected in towns exposed to flooding, although not in all areas.[80] Examples are found across Europe – for example at Hannoversch Münden, Lower Saxony, an inscription on the parish church was set at the level reached by floodwaters during the high magnitude flood of 1342.[81] A later example comes from Krems, Austria, where a number of inscriptions were affixed around the town, including to the city gates, recording the level reached by flooding as a result of ice on the Danube in 1573.[82] Some cities, which were continually affected by flooding, such as Rome, are littered widely with inscribed river level markers dating to the medieval period.[83] A rare documentary reference to such marks is found in a 1341 confirmation of land relating to the town of Saint-Émilion, in the English possession of Gascony, in which the boundaries of the town's jurisdiction are described in reference to local landmarks including two flood marks.[84]

Flood marks provide a valuable means of preserving knowledge of the levels reached along particular catchments during past flood events – extraordinarily useful information for increasing resilience in the face of flooding. Such data can inform future decisions and enhance efforts to protect areas already exposed to flood risk. However, one major problem with flood marks is their impermanence. Most frequently, flood marks were inscribed on bridges or walls close to river banks – areas which were especially vulnerable to damage from the hazard they memorialised.[85] Many medieval flood marks, it can therefore be assumed, have been lost while, of those examples which have survived, their present location cannot always be assumed to be where they were originally erected or inscribed. Repairs, reconstruction and the reuse of

---

78 Ellis 1859: 270–271.

79 Grimson 1978: 34; visible in Fig. 5.24.

80 Griffiths et al. 2017: 103.

81 Hoffman 2014: 325–326.

82 Floes of ice created a dam across the river causing flooding at Krems which affected two thirds of the town. Rohr 2013: 137–138.

83 Aldrete 2007: 55–56.

84 Maxwell Lyte 1900b: 161.

85 Macdonald 2007: 138.

building material over the centuries has frequently resulted in medieval stonework being repositioned from the setting in which it was originally installed. As relocation usually raised or lowered the flood mark from its intended position, many medieval examples can therefore no longer be trusted as reliable indicators of the flood levels they purport to represent. Even so, the presence of inscribed flood levels, in their original location or not, continue to serve as reminders of the possibility of high magnitude flooding and the omnipresent need to protect local populations in areas where flooding has occurred in the past.

6.4 Commemorative activities, material culture and other manifestations of memories

6.4.1 Commemorative events

In the modern world, anniversaries provide opportunities for reflection and remembrance and this is particularly the case with regards to natural disasters.[86] Some modern studies have even demonstrated the existence of 'anniversary reactions' in which people who have experienced highly stressful situations, such as a natural disaster or a wartime experience, are particularly affected emotionally and mentally at or around the anniversary of the event they experienced.[87] Anniversaries were of great importance during the medieval period – the majority of Church festivals were anniversaries of some kind; saints' days were usually the date of martyrdom or another important moment in a particular saint's life and the major festivals of Easter and Christmas commemorated dates of particular events in the life of Christ. The annual custom of Rogation processions, observed throughout medieval Europe to bless crops, fields and the parish, arose out of commemorative activities reported to have been instigated by Pope Gregory I to combat natural disasters that had afflicted the city of Rome in the 6th century.[88]

While anniversaries commemorating events known through scripture were of paramount importance to the medieval Church, it is also evident that, in a number of cases, the anniversaries of more recent events were cause for commemoration. An example from the Mediterranean reveals that following the severe flood of 1330 in Nicosia, Cyprus, a procession was conducted on the anniversary of the flood to commemorate those who had lost their lives and preserve memory of the event. That this may have contributed to ensuring that memory of the flood survived may be indicated by the fact that travellers to the area were aware of the risk from the river into the 15th century.[89]

86 See for example Blanchard-Boehm, Cook 2004.
87 Assanangkornchai et al. 2007.
88 Latham 2015.
89 Charalambous et al. 2016: 196.

Another example of a commemorative anniversary event occurred in 1341 when Pope Benedict XII issued an indulgence to encourage attendance at a memorial for victims of famine, presumed to relate to the agrarian crisis of 1315–22, although the documentary evidence does not specify which famine was commemorated.[90]

The lack of obvious forms of disaster commemoration in the medieval period may be misleading. Although perhaps not restricted to anniversaries of the event, it seems likely that commemoration of disasters would have been included in regular church processions, such as those of Rogation, in which, while carrying out blessings for the future prosperity of the town or parish, negative past occurrences, such as natural disasters – and how they could be prevented in the future – would naturally have been brought to mind. This kind of intention might have motivated the burial of a medieval crucifix, discovered through archaeological excavations in New Romney, Kent, along a known procession route through the town. Given the recurrent storms and flooding of the late 13th century,[91] the deliberate deposition of the crucifix might have been believed to protect the town against future floods. If this was the true intention, such an act, while seeking protection in the future, simultaneously commemorated the disasters of the past.[92]

6.4.2 Individual commemoration

In addition to the anniversaries of disasters, in the present, memory is also preserved in relation to individuals who lost their lives as a result of a disaster. Today, memorials are often inscribed with the names of the dead, commemorative messages are often written in 'books of remembrance' and informal 'shrines' are created at the site of the disaster or another meaningful location.[93] This kind of commemoration of individual disaster victims, in the modern sense, does not appear to have been common in the medieval world. A rare exception may be the grave of a chaplain named Ralph, who was killed when he was struck by a beam blown down into the interior of the chapel of St Pancras at St Augustine's Abbey, Canterbury, during the 1362 St Maur's Day storm.[94] Afterwards, we are told that his body was buried inside the chapel beneath a marble gravestone – so he was buried at the site of the disaster that killed him. Archaeological excavations did uncover 14th century burials in the chancel[95] but later interventions had removed the medieval floor surface[96] along with any grave slabs, although a slab of

90 Lucas 1930: 362.
91 Discussed fully in Chapter 3.
92 Standley 2020.
93 Brennan 2008; Maynor 2016.
94 Davis 1934: 564.
95 Blockley 2000: 68.
96 Blockley 2000: 60–61.

Bethersden 'marble'[97] was recorded in a doorway along with 15th century stonework.[98] Graves in the nave and entranceway, thought to be medieval in date, were covered with slabs of Portland limestone.[99] Whether any of these inhumations or grave slabs relates to the burial of Ralph the chaplain, however, is impossible to confirm.

Grave markers or commemorative plaques, with or without inscribed names of victims, as would be common today at the site of a disaster, were the exception rather than the rule. It seems unlikely, however, that memories of those no longer present – either those who had been killed as a direct result of a catastrophe or those forced to relocate and move out of an affected area – did not persist. An example comes from the Great Famine in Germany which seems to have driven many inland dwellers from their homes to the coast in search of food. Documentary evidence suggests that memory of this mass migration, from the countryside to the coast, was remembered at least up until the Reformation.[100] A rare example of a memorial to an individual who became a casualty of a natural disaster, on the other hand, comes from Göttingen, Lower Saxony, where a commemorative plaque attached to the base of a crucifix records the unfortunate death of one Hermen Goltsmet who drowned during the "great flood of Saint Margaret's Day [20th July]" in 1342.[101]

The reasons for the scarcity of overt individual commemoration are likely twofold. Firstly, low literacy rates reduced the value of erecting commemorative markers in this way, a point which is reinforced by the fact that individual grave markers bearing the names of the deceased only became common from the 18th century.[102] Secondly, individuals were remembered in other ways; primarily through oral memory as well as friends' and family members' thoughts and prayers – which have left little, if any, material evidence. One rare manifestation may be the belfry of the Augustinian friary at Clare, Suffolk. As already recounted in Chapter 4, this friary received a large donation in 1363 to fund the construction of a new bell-tower, likely necessitated by storm damage the preceding January. The sizeable donation however, led to the appointment of priests to pray for the souls of the benefactor, his family and "any to which he is obliged".[103] Such prayers – although more often from family rather than Augustinian friars – were probably the primary way through which individuals were remembered after death and disaster victims were no exception.

---

97 This is in fact a type of limestone although this distinction was probably not made in the Middle Ages.

98 Routledge 1882: 104.

99 Blockley 2000: 59.

100 Diemar 1909: 238.

101 Arnold 1980. Although the plaque and crucifix are now affixed at the top of the tower of Göttingen's St Albani Kirche this is unlikely to have been the originally intended location.

102 Renshaw, Powers 2016: 160.

103 Harper-Bill 1991: 85.

6.4.3 Commemorative material culture

In some cases, memories became intertwined with objects. For example, in 1315, during the Great Famine, the pressure placed on the Cistercian Abbey of Aduard, in Groningen, the Netherlands, by those seeking alms, led Abbot Eylard to order a huge cauldron to be made in order to prepare enough food for the needy. After the famine, the enormous cauldron was displayed to visitors to the abbey, even until the 18th century, with its size and the embroidered memory of its provision for the hungry affording it comparable status to a holy relic.[104] Somewhat analogous was the wine produced from grapes which matured during the 'mega-drought' of 1540. The long summer drought caused the grapes on the vines to become raisin-like and delayed the grape harvest until rains arrived in late August of that year. The delayed harvest resulted in a particularly prized, sweet and sherry-like vintage as a result of the grapes' extra sugar content. Reserved for special occasions, this wine was preserved for many years and was still well known in 1631 when invading Swedish soldiers searched Würzburg for barrels containing the 1540 vintage – although they were unsuccessful in locating them.[105] Similarly, up until the late 18th century, handfuls of (durum) wheat were preserved in Zürich from the 1540 harvest as a reminder of the terrible drought of that year.[106] In all of these cases, relatively everyday objects; a cauldron in the case of the Great Famine, and wine and wheat in the case of the 1540 drought, became imbued with special significance as a result of their relationship with anomalous natural events. As a result, they became vehicles which both visually (the great size of Abbot Eylard's cauldron, or the notably parched nature of the sheaves of wheat preserved in Zürich), and through oral tradition, transferred memory of these events down to future generations.

In all probability, certain types of artefact served to propagate memories relating to disasters in ways that will always allude the archaeologist. The occurrence of heirlooms in medieval domestic settings, for example, are one such class of artefact.[107] The archaeological record attests to the continued use of a wide variety of objects, including ceramic vessels, dress accessories and other miscellaneous artefacts, for two or three generations and sometimes longer. Heirlooms were kept either for their intrinsic or practical value as objects or because of their personal associations. Objects which "constructed family biographies"[108] such as those associated with life course rituals – birth and baptism, for example–were often kept as heirlooms. Although no medieval examples are known, it is easy to imagine how objects which had been saved from a home in the face of an impending natural hazard – as was the case as the flood-

104 Jordan 1996: 110, 230.
105 Glaser et al. 1999: 192.
106 Pfister 2017: 185.
107 Heirlooms have been foci for archaeological reserch in recent years. See Gerrard 2007a: 179; Gilchrist 2012: 237–242.
108 Gilchrist 2012: 248–249.

waters arrived in the Grote Ward in November 1421[109] – might have garnered particular significance, "gaining new meaning and status",[110] in the memories of survivors and subsequent generations. Certainly, in the case of town fires in medieval Bergen, the lack of ironwork among the archaeological assemblage suggests that material was salvaged from structures damaged or destroyed by fire.[111] A more recent example comes from the devastating explosion of a ship in the centre of Leiden, the Netherlands, in 1807 which resulted in over 150 fatalities and caused thousands of injuries. After the disaster, however, the wooden and metal debris from the ship was collected and curated by survivors and passed down to future generations.[112] Such objects, through their connection to a family member who had survived, and come through, great danger accumulated a distinct object biography[113] and might have been perceived as having luck-giving or protective powers for their descendants. Of course, such sentimental and metaphysical importance attached to heirlooms can only have persisted as long as memory of their connection to the event and family member persisted – "their social currency as heirlooms … diminished as oral memory of ancestors faded".[114] Heirlooms associated with disasters were probably not common, and are unlikely to be identifiable archaeologically, but their existence is perhaps less fanciful than it may at first seem.

6.4.4 Ex votos

The ex voto is a category of commemoration which fulfils a vow by thanking a saint, saints or other holy intercessor for their help in the past. These can take the form of paintings, other types of artistic depictions or votive artefacts[115] and are found throughout post-medieval Europe[116] but were similarly common during the Middle Ages. An ex voto might be created for a wide variety of reasons; safely coming through childbirth or recovering from illness for example, but some memorialise and commemorate disasters and, in doing so, preserve memory in a physical form which would have been visible and meaningful to future generations. They are most frequently found in church settings – as the shrine of the saint to which one made the vow was the most appropriate place to leave the ex voto. While most ex votos were cleared away at the Reformation, some examples have survived such as the assemblage of wax effigies

109 Kleinhans et al. 2010: 70.
110 Gerrard 2007a: 179.
111 Hansen 2015: 169.
112 Reitsma, Ponsen 2007: 7.
113 Gosden, Marshall 1999.
114 Gilchrist 2012: 241.
115 As discussed in relation to folded coins in Chapter 5.
116 Pfister 2011: 10.

discovered at Exeter Cathedral[117] following air raid damage during the Second World War. These are interpreted as ex voto offerings originally hung by the tomb of Bishop Edmund Lacey who, although uncanonised, attracted pilgrims after his death and was held responsible for a number of miracles.[118] Ex votos relating to sailors are particularly common, with many referencing St Nicholas – a renowned protector of sailors. Similarly, graffiti depictions of ships are found across medieval Europe, with examples found adorning buildings in Spain, Britain and Scandinavia,[119] though their function is not always clear. Some seem to be depictions commemorating real life events, such as battles at sea, while others are more likely to be manifestations of prayers, to saintly intercessors, for protection for particular vessels, or to give thanks after a successful journey.

One of the most impressive examples of an ex voto related to a medieval disaster comes from Dordrecht, the Netherlands, where in the aftermath of the Saint Elizabeth's Day Flood of 19th November 1421,[120] the survivors commissioned an altarpiece to thank St Elizabeth for their survival. The altarpiece is composed of two panels which depict the flooded landscape around Wieldrecht (see Fig. 6.4). These panels acted as doors to the altar which, when opened, revealed scenes from the life of St Elizabeth. The town of Wieldrecht was lost in the 1421 flood and its surviving residents relocated to Dordrecht where, in 1438 the churchwardens provided an altar for their use. After some time, the survivors furnished this altar with the St Elizabeth altarpiece commemorating the terrible disaster that had swept away their homes – and perhaps some of their friends and family members.[121] The survivors were clearly grateful for their own survival which they attributed to the intervention of St Elizabeth on their behalf. As a result, memorialization of the event was probably intimately connected to protection from any re-occurrence in the future.

Another method of fulfilling a vow to a saint was the foundation of chapels or religious houses as thanks for survival during a hazard. Such foundations became memorials to the event as well as the saint who had provided protection. These dedications were, by necessity, made by those with means who could afford to fund the establishment of religious houses. Some royal dedications such as the Abbeys of Hailes[122] and Vale Royal[123] provide examples of houses founded as thanks for their rich patrons surviving natural hazards[124] – in this case storms at sea. Comparably, local folklore

---

117 Discussed more fully in Chapter 5.

118 Radford 1949.

119 Gerrard 2003: 336–340; Champion 2015b; Westerdahl 2013.

120 This event is discussed above in relation to the later folklore which grew up around the flood.

121 Helmus 1991: 139.

122 Page 1907: 96.

123 Denton 1992: 124.

124 Although note that in the case of Vale Royal, which legend suggests was founded as a result of a vow made by Edward I during a storm on his return from the Holy Land, there is evidence that the

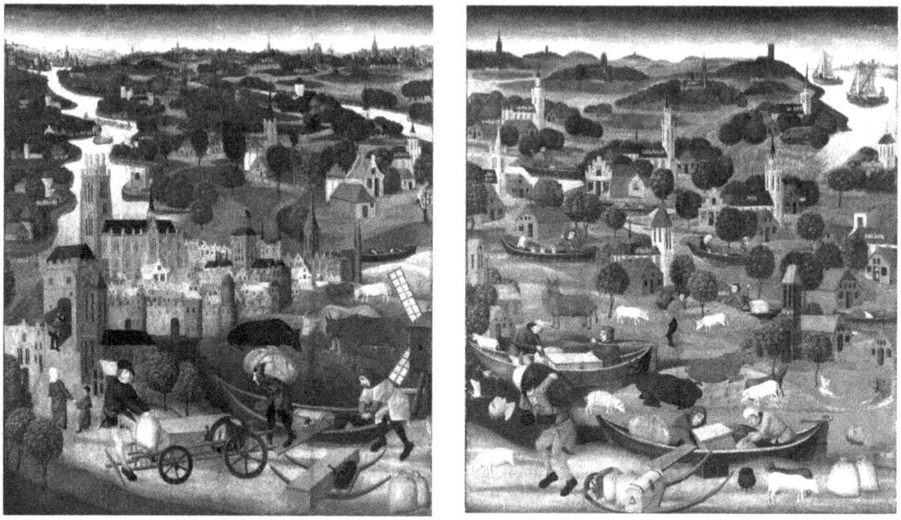

Fig. 6.4: The left and right panels of the St Elizabeth's Day Flood Altarpiece. Originally these formed the doors to an altar which, when opened, revealed scenes of the life of St Elizabeth. Rijksmuseum. Object numbers: SK-A-3147-A and SK-A-3147-B. Public Domain.

from Fingal, Ireland,[125] holds that a chapel dedicated to St Maur was founded at Rush by knights caught at sea during a storm in thanks for their safety.[126] Although in Italian towns many late medieval churches and chapels were founded as ex voto dedications in return for protection from plague,[127] this practice does not seem to have been common in northern Europe during the same period although monuments, especially crosses and columns, commemorating later post-medieval plagues, especially in the 17th century, are relatively common across Europe.

Ex votos – including pictorial representations, votive figures and dedications – did not generally preserve particularly useful information for future populations in how to deal with natural hazards in a practical sense. However, within the medieval world view, they served a very important purpose. By providing a lasting manifestation of prayers and vows to the saints for protection, old ex votos, which were visible in local parish churches, exemplified a method to guard against the danger posed by natural

Abbey was founded prior to the storm suggesting that the Abbey's foundation may have been intended to provide protection rather than as thanks for survival.

125 See: National Folklore Collection, UCD: The School's Collection, Volume 0786, pp. 56; 130–131. https://www.duchas.ie/en/cbes [Accessed 26 January 2018].

126 Given the dedication, it is tempting to connect this foundation with the 1362 St Maur's Day storm but the architecture of the chapel and the fact that the local legend specifies that the knights were 'crusaders' seems to favour an earlier date.

127 Avery 1966.

hazards which had seemingly been successful in the past. As such, although from a modern scientific viewpoint vows to the saints and ex voto inscriptions or paintings are powerless, to medieval populations they provided a reliable means to protect against worldly dangers. Ex voto ship inscriptions and altarpiece paintings, therefore, provided subsequent generations with demonstrations of the efficacy of such prayers and vows in the face of disaster.

6.4.5 Memory and structural repairs

Repairs to structures offer another alley down which memories must have been focused. Buildings and infrastructure damaged by extreme events must have found a prominent place within communal memory. Structural damage inflicted as a result of a natural disaster caused a sudden disjuncture which was often visible in the fabric of the building long after the subsequent repair work had been completed which, in the case of large buildings such as cathedrals, could be ongoing for decades. New repairs and additions usually differed stylistically from what had previously existed – as in the case of the church of St Nicholas, New Romney, repaired after the storms of 1287/88 or the church of Walpole St Peter, Norfolk, repaired after damage from flooding in 1338 – meaning that the structure, although repaired, was fundamentally altered as a result of the natural hazard. Repairs themselves then must have become points by which memories of extreme events were repeatedly retold, preserving and reinforcing communal memory of the event. This was often manifested simply by the addition of the prefix 'new'[128] to the name of the settlement or structure; for example New Winchelsea, relocated after Winchelsea was lost to sea floods, or the 'new' tower built at Stowmarket after 1362. This communal memory attached to repairs may be seen following the 1356 earthquake in Basel, Switzerland. Repairs necessitated by this event could still be identified and commented upon by visitors and local antiquaries even into the 16th century[129] illustrating how well known the story of their damage and subsequent repair had become. Another example comes from Ullensaker Church, Akershus County, Norway, the original stone structure of which was destroyed by a landslide in the late 15th century. In the aftermath, however, a new wooden church was constructed which incorporated structural elements, most notably a large door, as well as fittings such as the stone font and a bell, from the original church.[130] In this case, especially as it relates to a holy building, it is hard to imagine that the reuse of structural fragments from the earlier church would not have afforded frequent opportunities for remembering the antecedent structure and the event which had swept it away. At

128 Although note that the prefix 'new' did not necessarily imply a natural catastrophe had occurred.
129 Hoffman 2014: 307.
130 This information comes from http://www.norgeskirker.no/wiki/Ullensaker_kirke [Accessed 01 February 2021].

Bergen the continued reuse of structural timber after multiple town fires is attested through dendrochronology which demonstrates that high volumes of timber were reused in the rebuilding of structures razed by fire – creating tangible links with the structures that had previously stood in their place.[131] These examples demonstrate that repairs and the structural alterations necessitated by hazards, were themselves cause for the creation and propagation of memory.

In some respects, the memory of what had been damaged and required repair in the past could increase future resilience. In the case of earthquakes in medieval Cyprus, by taking note of what structural features withstood seismic damage and which elements failed, medieval cathedral builders appear to have been able, over time, to move toward more resilient designs.[132] More earthquakes provided more opportunities to learn and improve designs so that they were less likely to collapse. Cypriot churches, therefore, gradually became better adapted to earthquake damage increasingly favouring features such as smaller windows and doorways, reduced vertical height, more massive buttresses and increased symmetry – all of which served to reduce the chance of collapse in an earthquake. Similar processes may also have applied in the face of other hazards. Windstorms, for example, slowly led to designs becoming more aerodynamic; pyramidal spires gave way to conical forms which experienced reduced wind loads, and the development of lead roofing provided a stronger roofing material.[133] Comparable developments can, likewise, be seen in bridge design as some bridges were swept away by floods while others were able to withstand them.[134] All of these structural developments required the preservation of memory of what had been damaged and what had survived during previous disasters – something which could be deduced from standing structures and visible patterns of repair which local oral memory linked to particular events.

6.5 Summary

The memorialisation of disasters is clearly a complex topic which has many facets. At one end of the spectrum it relates to highly personal experiences which have the power to shape or destroy lives while at the other it relates to informing public policy and methods of mitigating the impact of future hazards. As such, during the medieval period and at any time, disasters were memorialised in many different ways, not all of which, and in fact the minority, persist in either the material or textual records. Since the early modern period, and especially the advent of computing, our ability to collate data with the aim of forecasting the future has greatly increased. In the present, this means

131 Hansen 2015: 165–166.
132 O'Neill 2020.
133 Baker 2007.
134 As discussed in Chapter 4.

that maintaining detailed and accurate information concerning past occurrences of natural hazards is vital to how we conceptualise the mitigation of their future impact. For a number of reasons, this was less important during the medieval period. Firstly, due to the lack of scientific and technological understanding, the type of data that is useful today – such as precise measurements of wind speed, precipitation or wave height – were not of particular interest or efficacy, even if they could be measured, during the Middle Ages. The exception were flood levels which were either inscribed onto structures or communally remembered in relation to a well known and permanent local feature. Secondly, because such high resolution information served little purpose, it was unnecessary to maintain communal memory of extreme events in high detail; it was enough for a coastal community to know that the sea had flooded their town before, and no doubt the particular streets and buildings that had been damaged or affected would have been remembered, without knowing the detailed characteristics of the storm that caused the flooding. Additionally, no sweeping generalisations can be made about the length of time over which memories survived or how accurately they were transmitted. In some cases, such as the Gauldalen landslide in Norway, relatively accurate information seems to have been conveyed over hundreds of years while in others, such as the 1421 St Elizabeth's Flood, inaccuracies seem to have crept in almost immediately after the event. Communal memory seems to have operated on a case by case basis making it hard to predict what survived and what fell by the wayside.

Most of the categories of evidence described above commemorating disasters do not provide much information which would have been practically useful to subsequent generations. The majority simply acted as reminders that, in the past, perhaps with a precise date, a particular hazard occurred in a particular area. Some, such as ex voto ship carvings, do not even convey this level of information. As such, turning to the concept of an adaptive cycle, it initially appears that most of these methods of encoding and transmitting memories did not have a significant impact on any 'adaptive cycle'. Two of the most obvious exceptions are flood marks and structural repairs. The former readily communicated the levels reached by floodwaters to future generations, providing the information necessary to mitigate the impact of flooding of that magnitude in the future.[135] In the case of the latter, visible structural repairs, and sometimes accompanying inscriptions, informed what could withstand natural forces and what could not, helping to build more resilient structures through an iterative process of trial and error over time.

Although the other categories of evidence examined above do not appear to have actively contributed to increasing resilience through an adaptive cycle, this is not wholly true. One of the most fundamental steps to improving resilience against hazards is first accepting that their occurrence is possible and presents a risk to society. By

135 Although note that it is unclear to what extent the information provided by flood marks was actually heeded.

attesting to the past occurrence of hazards, even in such low detail, this is exactly what the majority of these types of memorial did – either within kin groups (through oral memory, perhaps of ancestors, and the biographies of heirlooms) or wider society (as in the case of poetical verses, inscriptions, commemorative events and structural repairs). Additionally, some types of commemoration contributed to a peculiarly medieval adaptive cycle. Ex votos provided a material manifestation of a prayer to a holy intercessor for help, often against hazards, which had been believed to be effective in the past. In so doing, they provided an exemplar to future generations of a method to deal with worldly dangers that could be relied upon – and the evidence suggests that, in turn, many did.

7 Discussion

The preceding chapters have presented a wide array of evidence related to the occurrence and impact of disasters over the short- and long- terms. Chapter 3 presented a number of high-resolution reconstructions of catastrophic disasters which impacted northern European populations during the medieval period. Chapters 4 and 5, on the other hand, illustrate the dichotomy between two seemingly opposing response strategies. On the one hand, practical and physical responses, which provided tangible protection from the impact of a future recurrence of a natural hazard, or relief in its aftermath, while on the other, religious and superstitious responses which were widely held to provide efficacious protection against these unpredictable natural events – either in isolation or in concert with other practical measures. This chapter draws these strands of evidence together, setting them in a wider geographical and historical context. This provides a renewed understanding of the occurrence and impacts of the types of disaster considered in Chapter 3 and, in so doing, draws conclusions regarding the implications of these events – and disasters more generally – for medieval society. Additionally, the nature of the surviving evidence, both documentary and material, and how this might inform future research, is explored.

7.1 The causes of disaster

Natural hazards are, of course, the primary catalysts behind any natural disaster but, as discussed in Chapter 1, human decisions – such as the establishment of settlements in hazardous locations (low lying wetlands being one example) – are of equal importance in transforming the occurrence of these hazards into true disasters. Over the *longue durée*, human populations at a given location become familiar with the natural rhythm of their environment, including what natural hazards could reasonably be expected to occur, and what measures were appropriate to guard against these risks (as discussed in Chapter 2.1). As explored in Chapter 6, medieval society actively perpetuated communal memory of these events, although generally only preserving relatively low resolution information about the occurrence of natural hazards, through a diverse range of methods . A deviation from accepted norms in the way in which natural hazards occurred at a given location, or the occurrence of an unprecedented hazard, for which no collective memory exists, however, can have severe implications for human populations unprepared for these sudden changes. Medieval populations were, therefore, more likely to be afflicted by disasters when natural hazards occurred outwith patterns established over timescales that were meaningful to human society.[1]

1 As discussed in Chapter 6, this was generally not more than *c.*200 years.

https://doi.org/10.1515/9783110719628-007

7.1.1 Climate

Different types of hazard are the result of different natural processes but meteorological hazards are all influenced by climate – the long-term average weather experienced at a specific location. A change from an established climatic regime, therefore, especially one that has persisted for a lengthy period, can result in new and, for human populations, unknown weather patterns. Such a picture is gleaned from climatic reconstructions based on environmental proxies for northern European climate over the medieval period.[2] At the heart of the matter is the shift from the the so-called 'Medieval Climate Anomaly', characterised by generally warm and stable conditions from *c.*900–*c.*1300, to the 'Little Ice Age', which is thought to have typically consisted of colder and drier conditions between *c.*1300–*c.*1900.[3] Focussed on the 14th century, the transition between these two climatic epochs appears to have been marked by pronounced climatic instability with heightened storm activity[4] evidenced by a wide array of proxies including cliff top storm deposits from Shetland[5] and sediment cores from the North Sea.[6] The evidence obtained from such palaeoenvironmental proxies, therefore, supports the occurrence of meteorological hazards of a higher magnitude and/or greater frequency than had typically been seen in preceding centuries – a conclusion which is readily supported by the historical evidence.[7]

In concert with these climatic fluctuations, contemporary human society experienced particular challenges. Until the late 13th century, while climatic conditions were generally favourable, economic activity and population generally expanded whilst population density remained insufficient to pose major demographic problems. From the late 13th century, but especially in the early 14th century, population expansion combined with the universal desire to hold land meant that, in densely populated areas such as Norfolk, the average size of parcels of land available to tenant farmers gradually decreased – squeezing the ability of tenants to support themselves. Over time, this resulted in reduced living standards, increased inequality and, when crises forced some to sell their land, many were pushed into poverty.[8] From the late 13th century, therefore, society "found itself living in seriously reduced living conditions" and becoming "ever more prone to crises of subsistence".[9] By the early 14th century, interactions between this increasingly vulnerable population and the deteriorating climate resulted in acute crises such as the Agrarian Crisis of 1315–1322 – the catalyst

2 MANN et al. 2009; BÜNTGEN et al. 2011: 580–581.
3 DAWSON et al. 2007.
4 TROUET et al. 2012.
5 HANSOM, HALL 2009: 49.
6 HASS 1996: 139.
7 See for example BAILEY 1991.
8 CAMPBELL 2016b: 193–196.
9 CAMPBELL 2005: 4.

for which were rains of unprecedented intensity. Not only, therefore, did the fluctuating climate of the late 13th and 14th centuries favour the occurrence of hazards such as windstorms, heavy rains and consequent flooding but, in many places, society was reaching a point which made it peculiarly vulnerable to the impacts of these hazards.

The global changes in atmospheric circulation during this period also played a key role in enabling the occurrence of the greatest disaster of the period – the Black Death. Plague (*Yersinia pestis*) naturally occurs among central Asian rodent populations and, under favourable environmental conditions – notably warm springs and wet summers – the prevalence of plague in these populations drastically increases.[10] Positive climatic anomalies which resulted in conditions conducive to rodent population expansion in central Asia during the medieval period, therefore, would have caused natural reservoirs of plague to expand while subsequent negative climatic fluctuations caused these plague-bearing populations to collapse. Such a collapse would drive the fleas, which spread the disease among these populations, to search for alternative hosts – including livestock, beasts of burden and humans.[11] Swings in climate in central Asia, therefore, promoted booming populations of plague-bearing rodents from which, during negative climatic anomalies, the disease could make the transition to other hosts – who carried the pathogen to Europe along the 'silk roads' – the end result of which was perhaps the most consequential event of the Middle Ages.

Archaeological evidence, both direct and indirect, for the impacts of these climatic changes on contemporary European society can be found widely. An obvious indicator comes from the increased number of coastal sites affected by wind-blown sand inundation from *c.*1300, presumed to be a result of increased storm activity mobilising bodies of sand.[12] The extent to which climatic fluctuations played a role in social changes at this time, however, is significantly more debatable,[13] though evidence from a number of sites across northern Europe have been interpreted in this light. According to Buckland et al., for example, the climatic decline of the mid-14th century was a contributing factor to the abandonment of the Norse colonies in Greenland – as a run of anomalously cold years made continued settlement untenable.[14] In the Hebrides too, climatic decline may have prompted a shift in settlement. On South Uist, during the 14th and 15th centuries, the medieval population relocated from the coastal grassy dune plains, known as machair, to rocky outcrops to the east – perhaps as a result of anomalous weather patterns which caused instability of the machair environment.[15] Perhaps relatedly, archaeological evidence points to a dramatic decline in the availability and

---

10 Stenseth et al. 2006.
11 Schmid et al. 2015.
12 See for example Brown 2015.
13 As discussed in Chapter 1.3.
14 Buckland et al. 1996: 94–95.
15 Sharples 2005: 195–196.

consumption of marine fish in the area during the 14th century.[16] Although this has traditionally been interpreted as a result of regional political turmoil, the sharp decline could relate to fluctuations in sea temperatures, which altered where fish stocks could be found, and increased storminess, which made fishing a more hazardous pursuit.[17] In this vein, medieval ship builders appear to have made their vessels more sturdy and robust in tandem with the climatic downturn during the transition into the Little Ice Age.[18]

In Sweden, meanwhile, pollen sequences from upland medieval settlements indicate that the mid-14th century coincides with a drop in cereal crop coverage of roughly 50%.[19] While this is primarily interpreted as a product of farmers' migration, driven by plague mortality, to more productive southerly farmland, the spread of the disease, as discussed above, held a significant climatic component and poor growing conditions during the years of the pandemic itself likely exacerbated the impact of the crisis.[20] A similar interpretation has been advanced for cultivation in upland areas of Britain as a result of the climatic changes of the medieval period,[21] although this picture has more recently been significantly qualified.[22] Potential impacts, though, are also frequently theorised to have affected lowland sites. For example, at a moated site excavated at Cedars Field, Stowmarket, Suffolk, the raising of the height of the moat platform in the early 14th century has been interpreted as a response to the increased incidence of flooding caused by the contemporary climatic fluctuations[23] while evidence for pastoralism at nearby Cedars Park is thought to indicate a switch away from arable agriculture influenced by the difficulties of working clay soils under unfavourable climatic conditions.[24]

In the archaeology of British medieval settlements, such a narrative of decline over the course of the 14th and 15th centuries is almost ubiquitous[25] and is supported by

16 Serjeantson 2013: 78–9.
17 Oram 2014: 232.
18 Dhoop 2016: 208–209.
19 Lagerås et al. 2016: 43.
20 Campbell 2010b: 300–303.
21 Parry 1975: 11.
22 Tipping 2002: 21.
23 Anderson 2004: 28.
24 Woolhouse 2016: 122. In this case, although, as with the other examples given here, climatic change is likely to have played a role, social and economic factors are more likely to have been decisive in the shift from arable to pastoral agriculture. As with the criticism attracted by a similar interpretation applied to the desertion of the medieval villages at Barton Blount and Goltho in the 1970s (see Beresford, Hurst 1971: 21), while climatic fluctuations may have made arable agriculture a less attractive proposition, the catalyst for the switch to pastoralism is more likely to have been a reflection of changing market demands and the reduced labour requirements of pastoral compared to arable agriculture. See Rippon 2001: 31–33.
25 Dyer 1990: 115–116; Rippon 2001: 16–17; Gerrard 2007b: 987; Everett, Boulter 2010: 59; Hindmarch, Oram 2012: 283–293; Dyer 2018: 204; Young 2020.

systematic test-pitting across a wide area.[26] Importantly, however, there are many plausible factors which may explain the contraction of settlements during this period which do not relate to climate and, notably, contraction and decline was neither synchronous nor universal. At many sites, the onset of decline did not obviously occur until the 15th century while the majority of settlements remained continually occupied – and some even experienced expansion throughout the centuries of demographic decline and stagnation.[27] Many of these indicators of decline are likely to have been affected by a climatic component but, equally, a wide array of social and economic push and pull factors exerted significant influences in shaping the dynamics of population, settlement and economic activity. Undeniably, significant disruption coincided with the climatic fluctuations of the 14th century but there is no linear relationship between climatic change and human society – climate did not determine what people did but it affected the viability of possible options.

The climatic fluctuations centred on the 14th century clearly played a significant role in the causation of many of the disasters which characterised this period. The case studies considered in this book mostly occurred during this period of climatic transition in the 13th and 14th centuries. Although they were primarily selected based on the availability of source material, rather than their chronological clustering, the occurrence of disasters for which particularly large quantities of source material survive during these centuries might be more than a simple coincidence. As meteorological hazards demonstrably occurred with greater severity during this period and, at the same time, a variety of considerations may have reduced society's ability to cope with environmental shocks, it is reasonable to suggest that high-impact disasters – those which were more likely to leave a significant trace in the historical and archaeological records – were also more likely to occur. In the case of any individual hazard, however, it is exceedingly difficult to isolate the specific contribution climatic change may have played in the occurrence and magnitude of the event. Taking the storms of 1287/88 and 1362 as case studies, climatic proxies provide only uncertain indicators that wider climatic trends exerted an appreciable influence on the occurrence of these storm events.

7.1.2 The late 1280s

The series of storms that affected south eastern England and the North Sea coasts of continental Europe in the late 1280s were certainly anomalous. To what extent these were random weather events or manifestations of wider climatic trends, however,

26 Lewis 2016; Lewis 2020.

27 Dendrochronological dating of timbers from standing buildings, for example, suggests that a renewed phase of rebuilding occurred in the late 14th and early 15th centuries. See Dyer 2013 and Silvester 2010: 142–151.

is difficult to gauge – and, in fact, it is currently impossible to provide a definitive assessment. Through contextualising the situation in reference to climatic proxies, however, a tentative relationship may be suggested. Available reconstructions indicate that summer temperatures in Europe cooled dramatically from 1286 to 1287 with only gradual warming over 1288 and 1289.[28] This period of cooling coincides with a spike in volcanic sulphates detected in Arctic and Antarctic ice cores in 1286.[29] The volcano responsible for the output of these sulphates in 1286 is currently unknown but must have been either a large southern hemisphere, low latitude, eruption or at least two simultaneous eruptions in both hemispheres.[30] The cooling associated with this event can be traced widely, for example in the tree ring growth patterns of Siberian juniper and larch.[31] In addition to the atmospheric sulphates generated by the 1286 eruption, the late 13th century also coincides with significantly elevated volcanic sulphate levels in the atmosphere – between 2 and 10 times greater than at any other time in the last millennium.[32] To this abnormal level of volcanic forcing can be added the effects of sunspot activity and overall solar irradiance which also entered a depression during the late 13th century, in what is now known as the Wolf Minimum.[33] From the above evidence, it is apparent that manifold factors were influencing global climatic circulation during the 1280s, bringing about significant cooling in Europe and the North Atlantic. Such cooling could have enhanced the meridional temperature gradient leading to storms of greater intensity than might have been expected in the absence of abnormal cooling.[34] In turn, this may have theoretically increased the intensity of the storms of 1287/88 and their ability to cause the kind of damage attested by the documentary accounts from south-east England and the archaeological record across New Romney.[35]

7.1.3 1361/1362

In the case of the storm of 1362, the climatic and meteorological conditions prior to the storm's occurrence can be reconstructed in vague terms from historical descriptions of the weather over the preceding months. For the second half of 1361 the evidence suggests that, following a summer of drought in England,[36] sea ice was present in Iceland during the autumn[37] yet, by Christmas time, fruit trees were in bloom near

28 Luterbacher et al. 2016: 1–12; see Fig. 7.1.
29 Sigl et al. 2015: Supplementary data 5.
30 Zhou et al. 2006: 2779.
31 Hantemirov et al. 2004: 161.
32 Gao et al. 2008.
33 Named after the noted sunspot researcher Rudolf Wolf, b. 1816 d. 1893.
34 Trouet et al. 2012: 53.
35 As discussed in Chapter 3.1.3.
36 Brandon 1971: 3, Ogilvie, Farmer 1997: 127.
37 Storm 1888: 359.

Paris[38] which enjoyed a mild, if wet, winter.[39] The winter in Germany and northern Austria, meanwhile, was reportedly harsh, although little snow fell in the latter and there was more rain than snow in Bavaria.[40] This unusual winter season was followed by a 'great' drought in Croatia in spring 1362.[41] The presence of sea ice, meanwhile, suggests cold conditions in the Arctic though flowering trees indicate an unusually warm winter in northern France. Winter warming in the northern hemisphere is a documented result of high-sulphur volcanism in the tropics[42] and, interestingly, a strikingly similar description of a mild winter in northern France followed the 1257 Samalas 'mega-eruption'. No comparable eruption, however, is known for the 1360s – Mount Yakeyama, Niigata Prefecture, Japan, erupted in 1361[43] but this was a relatively low magnitude eruption which does not register in Arctic or Antarctic ice cores[44] and is therefore unlikely to have exerted a decisive influence on European climate. More locally, another possibility is the 1362 Icelandic eruption of Öræfajökull but the historical evidence suggests it only began erupting in June 1362,[45] after the occurrence of the St Maur's Day windstorm. Furthermore its ejection of only ~1.7 Mt of sulphate aerosol into the atmosphere was probably insufficient to generate a significant climatic impact.[46]

The patchy and uncertain picture provided by descriptions in the historical sources can be augmented with the addition of climatic proxy evidence. These indicate a low-level spike in sea surface temperatures during the early 1360s which was sharply followed by cooling[47] with a peak in sea ice coverage in 1364.[48] This picture is echoed by reconstructed European summer temperatures – with 1361 and 1362 standing out as relatively temperate years against the extreme cold of 1360 and the onset of renewed cooling in 1363,[49] with anomalously cold summers in Slovakia during both years.[50] Meanwhile, the phase of the North Atlantic Oscillation appears to have favoured higher magnitude storms at this time.[51] These proxies provide quantitative evidence for the dramatic fluctuations in global atmospheric circulation which characterized the period from the late 13th century through to the end of the 14th century, with the late 1350s and early 1360s registering as one of the peaks of environmental instability between 1300

38 Birdshall, Newhall 1953: 108.
39 Kiss 2016: 44.
40 Kiss 2016: 44.
41 Kiss, Nikolić 2015: 13–14.
42 Lavigne et al. 2013.
43 Hayatsu 1994.
44 Sigl et al. 2015: See supplementary data 5.
45 Thorarinsson 1958: 29.
46 Sharma et al. 2008: 736.
47 Dawson et al. 2007: 431.
48 Massé et al. 2008: 567.
49 Luterbacher et al. 2016: 1–12; see Fig. 7.1.
50 Büntgen et al. 2013. http://www.ncdc.noaa.gov/paleo/treering.html [Accessed 31 August 2016].
51 Trouet et al. 2012: 53.

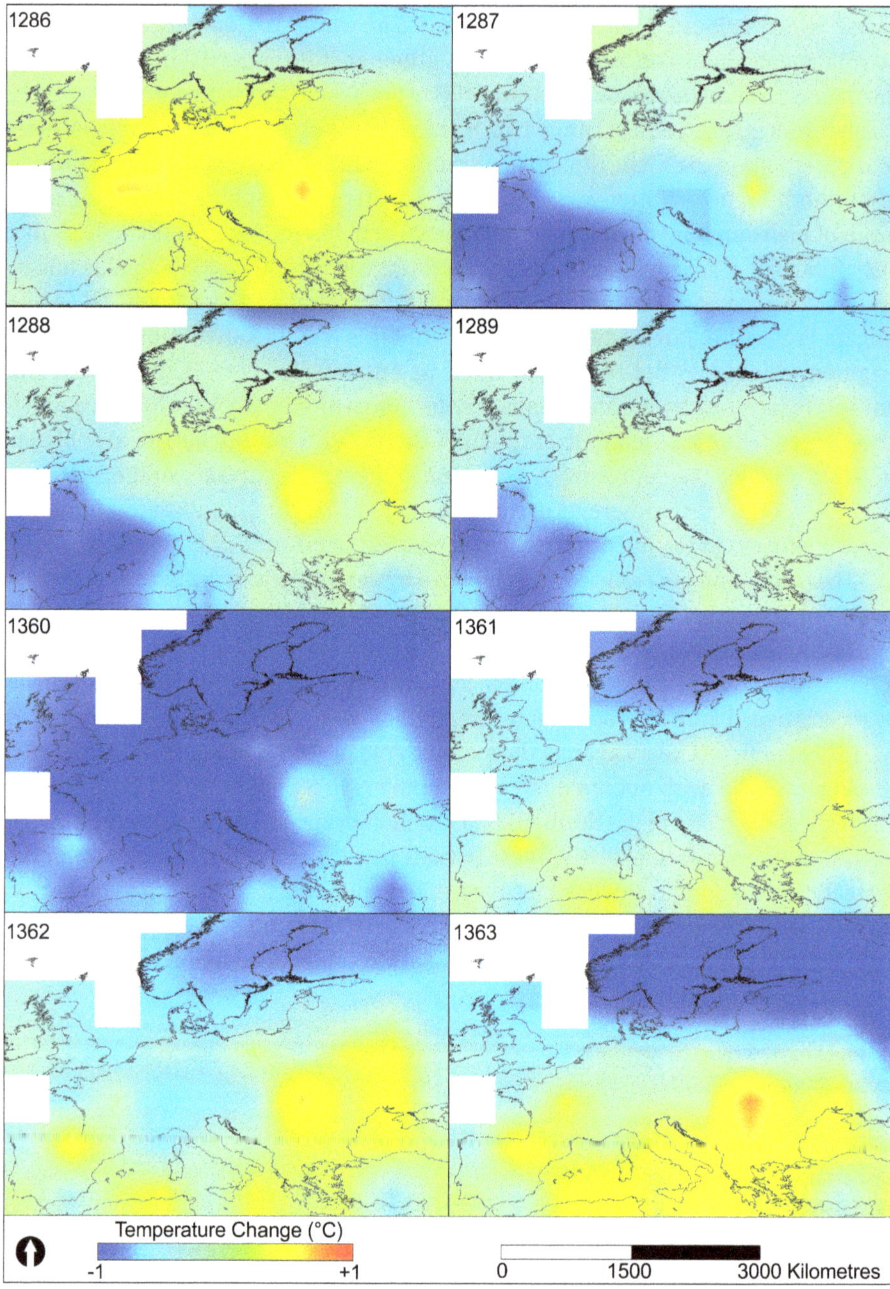

Fig. 7.1: Summer temperature variance reconstructed from proxy evidence across the European continent between 1286–1289 and, below, 1360–1363. Created by the author using data from LUTERBACHER et al. 2016.

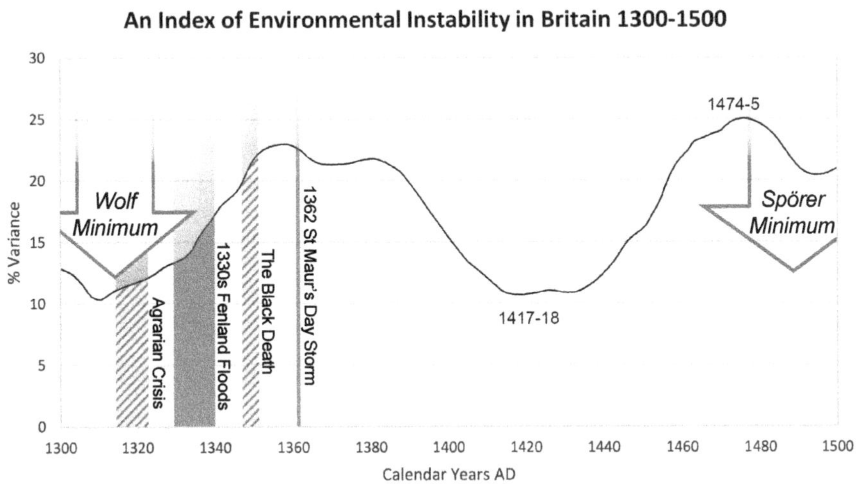

Fig. 7.2: An index of environmental instability in Britain 1300–1500. Adapted by the author after CAMPBELL 2016b: 339.

and 1500.[52] As there is limited evidence to suggest this climatic variability was caused by volcanic forcing, the most likely explanation is increased sunspot activity and solar irradiance – which peaked sharply between 1355 and 1375.[53] Again, under this scenario, the increased thermal energy input into Earth's atmosphere could have played an element in the genesis of a particularly high magnitude extratropical cyclone.

7.1.4 Summary

Although both the storms of 1287/88 and 1362 appear to have occurred during years of abnormal climatic variation, it is important to emphasise that there are great difficulties in attributing single events to climatic change. Windstorms are a particularly problematic phenomena in this regard meaning that any estimates as to the climatic contribution behind these events are extremely uncertain.[54] That being the case, the cooling which characterised the North Atlantic and western Europe in the late 1280s may well have heightened the intensity of the winter storms which struck England and the Low Countries in 1287 and 1288. Equally, the storm of January 1362 demonstrably occurred during a period of heightened climatic instability when the North

52 CAMPBELL 2016b: 339; see Fig. 7.2.

53 SOLANKI et al. 2005; DELAYGUE, BARD 2010: See supplementary data.

54 NATIONAL ACADEMIES OF SCIENCES, ENGINEERING, AND MEDICINE 2016: 111–114.

Atlantic Oscillation favoured storms of increased magnitude. Such linkages may also have played a part in the other disasters discussed in Chapter 3. The clustering of flood events in Norfolk in the 1330s, therefore, must be viewed alongside many of the highest magnitude disasters of the medieval period which struck during this period of heightened climatic instability (See Fig.7.2).[55] Central Europe's highest magnitude flood in 1342,[56] Norway's deadliest landslide in 1345[57] and even the floods, epidemics, drought and famine which characterise the early 1340s in Yuan China[58] may, therefore, be manifestations of this global trend. While it is difficult to make any unqualified statements, it is likely that the global climatic shifts taking place when the hazards discussed above occurred affected their magnitude and, therefore, the severity with which they impacted contemporary human society.

7.1.5 Other factors

While climate was certainly an important consideration in the multitude of weather-related disasters which occurred over the 13th and 14th centuries, other factors affected the impact that these hazards had on contemporary society. In the case of the 1287/88 storms, the occurrence of earlier storms in the mid 13th century and the ongoing alterations to the drainage of Romney Marsh during this period,[59] which made New Romney a less attractive port and eventually removed its access to the sea altogether, must have exacerbated the economic downturn which this series of storms initiated (see Chapters 3.1.3 and 4.7).

Turning to the flooding in the Fens in 1338, a number of factors likely also exacerbated the effect of this disaster. Widespread poverty among the communities affected by flooding, reported in the documentary sources[60] – which would have increased the vulnerability of populations to the floods when they struck – may have been linked to increased tax demands placed on the English populace as a result of the outbreak of war with France in 1337. According to Ormrod, the additional taxation levied on the laity in 1337 and 1338 was significantly higher than might have been expected given recent precedents.[61] One indicator that this caused hardship and a degree of civil unrest amongst the peasantry elsewhere in East Anglia comes from the relative spike in trespass on the Lord's demesne reported to the manorial court at Walsham-le-Willows,

55 Campbell 2016b: 277–289.
56 Dotterweich et al. 2003.
57 Rokoengen et al. 2001.
58 Brook 2010: 72.
59 Rippon 2002b: 90–96.
60 Maxwell Lyte 1915: 62.
61 Ormrod 1991: 183.

Suffolk.[62] As well as heavy taxation, the widespread failure of the harvest the following year must have made the situation in Marshland particularly challenging.[63] The Fens were also particularly densely populated at this time,[64] meaning that scarcity would have been amplified and the local people would have been especially vulnerable to the impacts of harvest failure and famine. Earlier disasters too may have still been exerting an influence – the panzootic of the late 1310s and early 1320s resulted in a sharp decline in livestock herds in England and recovery was a gradual process. In Norfolk, however, the reduced reliance on oxen for ploughing, compared to neighbouring counties,[65] as well as the preference for sheep over cattle, at least in Marshland,[66] may have reduced local exposure to the panzootic and its after effects. Elsewhere, even by the late 1330s, herd numbers and prices had not fully returned to their pre-panzootic norms[67]

In both of the episodes of flooding referenced above, the storms of 1287/88 and the inundation of 1338, a further consideration may have been the transformation of coastal wetlands by human exploitation. A direct result of agricultural expansion into reclaimed marshland environments, especially those in which peat was present, was soil compaction and peat shrinkage.[68] As a result, by *c.*1300, agricultural land in reclaimed marshlands around the North Sea littoral was topographically lower and thus at an increased risk of flooding. By the late 13th and early 14th centuries these environments, all around the coasts of the North Sea basin, had grown increasingly vulnerable as an unintended consequence of human activity. Although in these cases, the worst of the demographic decline was yet to come, Rippon also points out that reduced population levels would have suppressed landowners rental returns leaving the maintenance of the flood defence infrastructure to a diminished pool of manpower.[69] At the same time, the increased occurrence of storms and flooding as a result of climatic fluctuations increased the risks to coastal wetland landscapes as well as the costs of protecting them through maintaining dikes and drainage infrastructure.

The impact of the storm of 1362 was similarly exacerbated by contemporary conditions. The windstorm came in the midst of a renewed outbreak of plague which had arrived in England in 1361,[70] a fact bemoaned by the graffito at Ashwell St Mary (see Fig. 6.3). Campbell regards this outbreak as "probably … the second greatest mortality crisis of the later Middle Ages"[71] which, in wiping out many of the children born in the

62 Kilby 2015: 77.
63 Campbell 2016b: 269–270.
64 Campbell, Barry 2014: 65, Fig. 2.2.
65 Slavin 2009: 20–22.
66 Silvester 1988: 165.
67 Slavin 2009: 50, 57–58; Campbell 2016b: 214.
68 Soens 2013: 211.
69 Rippon 2001: 18.
70 Horrox 1994: 85–88.
71 Campbell 2016b: 314–315.

decade after the Black Death, prevented any demographic recovery from the earlier bout of plague. These circumstances undoubtedly made the destruction caused by the windstorm more difficult to deal with, as workers and labourers were scarce and the sums they could demand, in spite of official price fixing, were high (see Chapter 4.1). Society was also likely still experiencing a 'hangover' from the demographic crash of the Black Death – which had struck only a decade prior and triggered significant adjustments in the established labour model. The resulting shortage meant that labour intensive tasks were increasingly abandoned or put on hold in the post-Black Death era.[72] This not only impacted the ability of landowners to restore the effects of natural hazards when these struck their estates but, in some cases, it also meant that maintenance and repairs were less likely to be performed which would have increased the vulnerability of communities to various types of hazard throughout this period.

The discussion of the issues presented above makes it clear that natural disasters do not occur in isolation. The most severe disasters arise when natural hazards are exacerbated by existing and ongoing factors. These include both the environment, with climatic fluctuations affecting the occurrence and magnitude of natural hazards, as well as endogenous considerations, such as wealth inequality, the distribution and availability of resources and the ongoing repercussions of earlier disasters. Climate was certainly not the only factor which governed the occurrence of disasters, therefore, but subtle changes could affect their occurrence. This is where BRAUDEL's overlapping layers of historical time,[73] encompassing gradual environmental changes, fluctuations in human society at the macro scale and the short-term agency of individuals, offers an effective model to conceive of the factors which caused disasters to arise and their consequences over the short- and long- terms. While changes in the environment, including climatic fluctuations, did not on their own determine when or where natural disasters occurred, they did influence, to a lesser or greater degree, the locations and periods during which natural hazards were more likely to occur. This, combined with social and economic forces and individual agency, expanded or constrained the risk to which medieval populations inhabiting a given location were exposed. Wider trends in both the environment and human society, therefore, played a considerable role in influencing to what extent disasters affected human society and how deeply their impacts were felt – as we shall see in the following section.

7.2 The impacts of disasters and the responses of society

The direct and long-term impacts of the disasters discussed in Chapter 3 have been discussed in detail in the preceding chapters. A number of aspects, however, warrant

72 CAMPBELL 2012: 124.

73 BRAUDEL 1972: 20–21.

further consideration and comparison. These include the economic impacts of disasters – while the damage and destruction commonly caused by disasters, sometimes across a large area, provides obvious evidence for the economically damaging effects brought about by disasters, there is also evidence to suggest that these events were, counter-intuitively, sometimes positive economic stimuli. Related to this theme is the way in which contemporary authorities managed the disruption and trauma caused by disasters and to what extent, if at all, disaster management strategies adopted by authorities developed and evolved over time. In addition, it is worth considering the intangible impacts of disasters. Through comparison with the impacts of natural hazards on modern populations, it is clear that disasters generate short- and long-term psychological and mental-health effects on individuals and populations. It can be assumed that such effects were similarly felt by medieval populations but there is a near complete absence of contemporary evidence – either historical or archaeological. This is concluded by a holistic review of the responses and mitigation strategies adopted by medieval populations in the face of the case studies considered in Chapter 3.

7.2.1 Economic impacts

In both the storms of 1287/88 and the storm of 1362, commodities, their production and distribution, were affected by the occurrence of natural hazards. In the case of the storms of 1287/88 the evidence suggests that pottery production, and/or distribution, saw a sudden shift in tandem with the occurrence of the storms. Specifically, the 1287/88 storms coincide with the replacement of sand and shell tempered wares, which had been common in the Romney Marsh area in preceding periods, with wares tempered only with sand.[74] Possible explanations for this change in the ceramics produced and consumed in New Romney, therefore, are that a coastal pottery production industry was taken out of operation, either through direct destruction or, perhaps more likely, supply and distribution networks were disrupted by the occurrence of the storms. The precise location of the pre-storm kiln responsible for the production of the sand and shell-tempered wares which fade away after the time of the storms is unknown. However, in the aftermath, pottery production seems to have shifted to the kiln identified *c.*18 km away at Ashford Potters Corner.[75] While such a change would have presumably hit the former producers of the sand and shell-tempered wares hard, potters in Ashford presumably saw a proportionate surge in business. Similarly, the damage that such flood events caused – particularly the repairs to the flood defences – may have provided an important source of employment for those on the margins of poverty in the affected regions.[76] Notably, in this case, however, quite outside the impact of natural hazards,

74 Barber 2008: 172; Canterbury Archaeological Trust 2010: appendix 5.
75 Grove, Warhurst 1952; Canterbury Archaeological Trust 2010: Appendix 5.
76 Bailey 1991: 207.

the entire region experienced population decline in the centuries which followed as a result of a combination of environmental conditions – the siltation of previously important watercourses and outbreaks of plague among the populace – political concerns – war with France which had formerly been an important trading partner – and market forces – which shifted demand from labour intensive grain to meat and wool.[77] In this case, therefore, while the storms may have exacerbated decline, the overall trajectory of human activity in the affected locale appears to have been governed by other factors.

Following the January storm in 1362, roofing materials, most notably tiles but also the labour of roofers and thatchers, were affected. The storm precipitated a spike in demand which, despite official regulation, sparked price rises. As a result, although a vast number were ill-affected when the storm caused damage to structures across England, for those with the skills and necessary resources (timber, roof tiles *etc.*) to mend the damage, the storm was good for business. The scale of demand is demonstrated by the high volume of tiles purchased; 123,500 tiles were bought from a tiler in Kings Langley, Hertfordshire,[78] as well as the fact that, around the years of the storm, labourers' wages in London jumped to an all time high.[79] In addition, the storm seems to have had an impact on the availability and value of land and property. Not only can this be seen in the administration of the estates of the Black Prince – the prince's Keeper of Fees in the county of Suffolk, Thomas de Stanydelf, proposed to sell houses at Icklingham, Suffolk, in the storm's aftermath,[80] perhaps to raise funds for other repairs[81] – but also in the activities of the Archbishop of Canterbury, Simon Islip.[82] His acquisition of land in the centre of Oxford in the aftermath of the storm, for the foundation of Canterbury Hall, a new college of the University, seems to have been motivated either by the availability of the land or the current owner's eagerness to sell as a result of the storm's impact.[83] Such transactions suggest that, while some needed to liquidate assets, others were keen to capitalise on the opportunity presented by the storm to acquire assets below their usual market rate.

As awareness of natural disasters grows, so analogies for the types of activities discussed above are emerging elsewhere across Europe. Another case of a disjuncture in ceramic production, akin to the aftermath of the 1287/88 storms, can be seen at Vila Franca do Campo, Azores, Portugal, following the 1522 earthquake and landslide which buried the town. This event appears to have ignited the local economy, forcing the instigation of a local ceramic industry to meet the sharp spike in demand. Previously,

77 Rippon 2001: 31–33.
78 Page 1914: 265.
79 Keene 2011: 49, Fig. 3.1.
80 Dawes 1933: 426.
81 Although note that the Black Prince ultimately ordered against this proposed sale.
82 D. 1366.
83 Wharton 1691: 415.

the ceramics in use across the town appear to have been entirely imported from the Portuguese mainland.[84] Exploiting disasters opportunistically, as the Archbishop of Canterbury seems to have done after the 1362 storm, also finds numerous comparisons. In the North Sea basin for example, it was frequently enshrined in local law codes that, if a landowner was unable to afford the maintenance costs of the flood defences protecting their lands, which were required to safely protect them, their ownership of the land was void and could be claimed by other parties. While initially intended to protect the integrity of communal flood defence systems, elites were often able to exploit these regulations to take control of large areas of land where the owner had failed in some way to abide by local maintenance regulations.[85] Costs would have been highest, and poor maintenance most obvious, in the immediate aftermath of a flood or storm event meaning that the occurrence of these hazards could actively assist the seizure and consolidation of land by wealthy elites at the expense of local landowners. Another example can be found in the case of the flash flood which destroyed the village of San Romano in Asturias, north-west Spain, in the 14th century. Here, devastation of the village led to its re-foundation with the new name of Villanueva[86] on the opposite side of the River Trubia.[87] However, the site of the destroyed village on the other side of the river, covered over by alluvial soils, seems to have quickly been acquired by local landowning elites,[88] with the area approximately corresponding to the zone affected by the flash-flood still owned by their descendants until the 1960s. Clearly, wealthy parties with sufficient financial capital were frequently able to exploit the situations generated by natural disasters to acquire land which would, normally, have been either unavailable or significantly more expensive. This accords with KLEIN's concept of disaster capitalism in which modern day disasters, both man-made and natural, are exploited for financial gain, usually by governments and corporations.[89] The general picture obtained from this discussion is that while disasters generated short-term economic crises they rarely had much impact on longer term trends. This being the case however, they did facilitate transactions which shaped affairs at the local level which could have extremely long-lasting consequences – as in the case of the village of San Romano, Asturias.

The precise economic impact of any given historical disaster is very difficult to gauge. While, in rare cases, detailed economic data from medieval Europe is available, untangling the specific contribution made by a particular natural disaster to regional prices, wages or profits, either positive or negative, is an extremely challenging task. From the examples considered above, it appears that rapid-onset disasters, while acting

84 FORLIN, GERRARD 2017: 104.
85 SOENS 2013: 228–229.
86 In English this translates as: New Settlement.
87 FERNÁNDEZ et al. 2019: 9.
88 FERNÁNDEZ et al. 2019: 10.
89 KLEIN 2007.

as short-term stimuli, were quickly overtaken by macro-scale economic trends to shape longer-term trajectories of change – as is exemplified by the population decline seen on Romney Marsh discussed above. A similar conclusion has been reached in reference to the occurrence and impact of the Black Death on the economy of Holland.[90] The existing social and economic make-up of Holland is thought to have been uniquely well placed to thrive under the new conditions brought about by the decisive changes of the Black Death. While the results of its occurrence, therefore, were fundamental in creating the right circumstances for a boom in the economy of Holland to occur, it was existing underlying qualities – weak feudal control and weak influence of guilds, the strength of the middle classes and relative freedom of the peasantry – which allowed economic growth to occur in these new circumstances. Perhaps the most significant impact of disasters, therefore, was causing changes that otherwise may not have occurred. In the case studies considered in this monograph, these included the transfer of land and property between different groups as well as alterations in industries and production, as seen in the pottery industry on Romney Marsh after 1287/88. These changes, although not significant enough to alter overall economic trends at the regional scale, did exert an influence on the futures of the local areas they affected. The occurrence of such events, therefore, created winners and losers – with some able to capitalise on opportunities at the expense of the difficulties experienced by others. Clearly, disaster capitalism[91] has a longer history within society which is readily reflected in the medieval evidence.

7.2.2 The management of disasters

One of the criteria by which previous scholars have compared the responses of past societies to natural disasters are the involvement of a society's power structures in the management and co-ordination of responses in the aftermath of these events.[92] In the modern world, government is widely expected to plan for, and implement, disaster mitigation strategies as well as facilitating the deployment of rapid-response emergency services to save lives and protect the public during the emergency phase. As SCHENK has examined in the case of flooding in 14th century Florence, such a role for authorities was certainly not an entirely alien concept to medieval Europeans[93] and this can be seen in the case studies examined in Chapter 3 to a varying degree. WHITE proposed a tripartite categorisation, pre-industrial, industrial and post-industrial (summarised in Table 7.1), of the ways in which different types of society respond to disaster – including the nature of social organisation and power structures involved.[94] This model assumes

90 VAN BAVEL, VAN ZANDEN 2004: 526–529.

91 KLEIN 2007.

92 See for example WHITE 1974: 5; CHESTER et al. 2012.

93 SCHENK 2017: 142–143.

94 WHITE 1974: 5.

Tab. 7.1: The tripartite categorisation of responses to disaster by societies at different stages of 'development'. Reformulated by the author after White 1974.

	Folk or Pre-industrial	Modern or Industrial	Comprehensive or Post-Industrial
Relationship with Nature	In harmony	Control over	Harmonised control
Capital Investment	Low	High	High and low
Flexibility	High	Low	Medium
Geographic Variation	High	Low	Medium
Social Organisation	Individual or small groups	Interlocking and interdependent social structures	Combination

that pre-industrial societies are relatively harmonised with nature, do not invest heavily in defence against natural hazards, are highly flexible and spatially variable in their strategies and do not organise themselves far beyond the level of small groups. When applied to the medieval evidence, however, there are some notable discrepancies between this model and the evidence presented throughout this monograph. It could certainly be argued, for instance, that flood defence infrastructure sought to control nature while the investment needed to construct and maintain these defences was often, by contemporary standards, high. Many of these tangible responses, as well as the cultural milieu in which disaster was conceptualised during the medieval period, were also widely applicable across Christian Europe – geographic variation was, therefore, low.[95] The organisation of responses to disaster during the medieval period, also frequently transcended the individual and local spheres – as this section explores. Based on White's classification, therefore, medieval European responses to disaster display more of the facets of industrial, rather than pre-industrial, responses to disaster.

In the case of the 1287/88 floods, authorities only seem to have involved themselves in quite minor ways. The most significant case of state involvement seems to have focussed on one of the worst hit towns, Winchelsea, which appears to have been all but erased by the storms. To remedy this situation, the state organised the re-location of the town, which had been suffering degradation from the sea at least since an earlier series of floods in the 1250s, to a new hilltop site that would be safe from coastal flooding.[96] This re-foundation of Winchelsea was an impressive undertaking – land had to be

95 Although, of course there were important regional differences – compared to England and the Low Countries, for example, where authorities managed the maintenance of flood defence infrastructure, flood defence on the French Atlantic coast was managed by individual landowners or communities without overarching institutional power structures. Athimon, Maanan 2018: 1493.

96 Martin, Martin 2004: 4–6.

purchased, streets laid out and plots for housing organised. However, as this relocation was planned from at least 1280 it can hardly be viewed as a state reaction to the storms and floods of 1287/88 – but a protracted response to the earlier storms of the 1250s and ongoing problems ever since their occurrence. The 1287/88 storm series, however, likely expedited the relocation plans as the transfer of land to the townspeople took place the following summer in July 1288.[97] Although the planned relocation of Old Winchelsea speaks to the medieval English state's ability to carry out such large scale projects, it must be viewed in the wider context of the reign of Edward I which saw the foundation of many other new towns, where no natural disasters had occurred, as part of a concerted policy.[98] The king's involvement in such matters was likely, therefore, key. The evidence in the aftermath of the 1287/88 floods of the imposition of bridge tolls at Bastwick, Norfolk, therefore is particularly interesting as it suggests his absence from England during the time of the disaster[99] might have allowed local landowners and religious orders greater latitude to manage their own affairs when recovering from the floods, with little state intervention or regulation.

The only other instance of state involvement in the recovery process following the 1287/88 storms was the instigation of a royal commission *de wallis et fossatis* to investigate the extent of flooding and how best to remedy the damage caused.[100] Such commissions, however, usually empowered local landowners and officials to investigate the state of the flood defence system and order works to be carried out where necessary.[101] In co-ordinating a practical response to the problems created by the storm, therefore, the primary responsibility fell to local elites and communities. The meeting at Snargate in 1287, for example, to agree on how best to repair the flood defences at Holewest,[102] is probably illustrative of how such issues were typically resolved by local landowners and officials, although this was especially true in areas such as Romney Marsh and Marshland where specific regulations governed the flood defence and drainage infrastructure. The initiatives instigated by Canterbury Cathedral Priory[103] and Christ Church Priory[104] in the aftermath of the storms reveals the lengths to which landowners with sufficient capital were willing to go to make their lands profitable and secure in the aftermath of these disasters.

Following the occurrence of the floods of 1338 in the Fens, the administration of Edward III responded to petitions from the Marshland dwellers.[105] Since 1337, a new

97 Martin, Martin 2004: 5.
98 Lilley 2014.
99 From 1286 to 1289 Edward I was in Gascony.
100 Anon 1726: 155–163; Maxwell Lyte 1893: 309.
101 Bankoff 2013: 30.
102 Tatton-Brown 1988.
103 Gross, Butcher 1995: 108–109.
104 Rippon 2001: 22–26.
105 As described in Chapter 4.

tax based on the value of movable goods had been imposed across England to help fund the war with France. As the flood had damaged or destroyed many of the goods on which this tax was assessed, it followed that the tax burden for the area affected by the flood required re-assessment. The royal administration assented to the request to re-assess the tax of those who had been afflicted by flooding in Marshland although there appear to have been administrative errors which meant that, some years later, the tax rate reverted to the pre-flood assessment – provoking renewed petitioning and a second re-assessment.

To the central administration of the medieval state, these two flood episodes may have only registered as local problems which did not warrant any specific response from the central administration. In the aftermath of the 1362 storm, however, which, at least initially, was perceived as a universal hazard,[106] the medieval state took a more active role in the management of the disarray created by the disaster. While, in common with the floods in the 1280s and 1330s, repairs seem to have been left entirely to individual landowners and tenants to organise at their own expense – as is made clear by the administration of the Black Prince's estates and the court case of the London burler Henry Maynburgh[107] – royal administration seems to have regarded the regulation of markets as an arena which did warrant state intervention. This is evident in the legislation to fix the prices of labour and roofing materials in the aftermath of the storm issued throughout England.[108] Such a regulation perhaps stemmed from the medieval theological concept of the 'just price' championed by Aquinas which posited that "if … one man [may] derive a great advantage by becoming possessed of the other man's property, and the seller be not at a loss through being without that thing, the latter ought not to raise the price".[109] That the regulation was not wholly successful, however, can be judged by the numbers of people falling foul of the regulations[110] as well as the fact that, rather than suppressing wages or at least keeping them steady, wages of craftsmen in the building trade in London rose to unprecedented levels – and this only increased in subsequent decades.[111]

Although not entirely effective, such provisions in the wake of crises had slowly developed over the course of the 14th century. In attempting to bring an end to the Great Famine, the administration of Edward II introduced price controls on livestock, attempted to limit meat consumption among the peasantry and tried to stimulate the grain market – coaxing those with a surplus to bring it to market and offering

106 This is revealed by the fact that Edward III's regulation of the prices and wages that tilers and roofers could demand was issued to Sheriffs throughout England – not just those who administered areas affected by the storm. Maxwell Lyte 1909a: 238–239.

107 Thomas 1929: 61–62; see Chapter 4.1.

108 Maxwell Lyte 1909a: 238.

109 Aquinas 1947: 1513–1514.

110 As discussed in Chapter 4.

111 Keene 2011: 49, Fig. 3.1.

favourable circumstances to merchants who might bring extra grain to England.[112] In the aftermath of the Black Death, Edward III went one step further. The plunge in the number of workers had allowed the survivors to demand excessively high wages – at least by pre-plague standards. To remedy this, Edward III introduced the *Ordinance of Labourers* and subsequently the *Statute of Labourers* which sought to restore prices and wages to pre-plague levels and limit the ability of workers to refuse work.[113] Set in the context of these earlier pieces of legislation to manage the economic shocks of disasters, the ordinances imposed after the storm of 1362 clearly followed an established formula which had gradually evolved from Edward II's response to the Great Famine, but drawing particularly heavily upon the ordinances that had been introduced after the Black Death a decade earlier.

There is little evidence for state involvement in the aftermath of smaller, more local disasters such as the mid-15th century landslide at Cleeve, Somerset, considered in Chapter 3. Typically, such events, at least compared to severe floods and storms, generally impacted quite a limited geographic area – often affecting only a single landowner or tenant – and, as such, were probably not usually viewed as events warranting intervention at the state level. In the case of Cleeve Abbey, following the destruction of the chapel of St Mary the Virgin,[114] by a slope failure at some point before 1455, the grant of rights to hold markets and fairs does demonstrate a recognition that such events exerted an economic strain on the affected communities. That this grant seems to have come more than 10 years after the cliff collapse, though, illustrates that such assistance was often not immediately forthcoming.

From the available evidence it appears that, at the state level, medieval authorities seldom played an active role in the practical management of the conditions created by disasters. Royal authority did, however, exert control over financial matters, such as the taxation of the inhabitants of Marshland in 1338, and the regulation of market forces, as occurred with respect to tiles and labourer's wages after the 1362 storm. Landowners and local communities, on the other hand, were the most important actors when it came to responding practically to the challenges imposed by the occurrence of a natural hazard – repairing damage and redesigning or rebuilding structures and defences to withstand such events in the future. Damage to property was invariably the responsibility of landowners or their tenants to repair although occasionally indirect aid came through remission of rent or tax commitments or, in some cases, charity.[115] One exception, where royal authority did become involved at the local scale were royal commissions *de wallis et fossatis* which, although imbued with royal authority, were carried out at the local level by landowners, officials and royal agents.[116] In some

112 Braid 2010: 348–349.
113 Braid 2010: 359–360.
114 Maxwell Lyte 1897: 527.
115 See Chapter 4.6.
116 Galloway 2009: 178.

communities, where these hazards occurred repeatedly and threatened the existing economic system, such measures were overseen by local power structures such as Marshland's dike reeves or the 'Lords of the Level' of Romney Marsh.[117] This picture is mirrored along the French Atlantic coast where the royal administration rarely became involved in the management of disasters beyond financial concerns prior to the post-medieval period, while local landowners and communities were typically the ones who managed the practical measures necessitated by the occurrence of natural hazards.[118]

7.2.3 The intangible impacts of disasters

The local actors who frequently organised relief measures were also often the ones who bore the brunt of the impact of disasters. These were events with extremely complex repercussions which touched many aspects of life – many of which are rarely preserved through documentary or material sources of evidence. As we have seen in Chapter 5, glimpses of some people's personal responses may be observed through various categories of artefact which had associations linked to the risks posed by disasters or the medieval beliefs which surrounded them. While these offer some insight into medieval thoughts and prayers relating to how people sought to protect themselves, disasters must have affected the people who lived through them in ways which leave little or no trace in the surviving archaeological or documentary records. Contemporary documentary sources are virtually silent on the effects of specific disasters on the mental health and psychology of the populations who experienced them over the longer term – although we can certainly speculate as to the mental state of particular individuals, such as the anonymous inscriber of the bewailing graffito at Ashwell St Mary.[119] It is inconceivable, however, that these events did not have a significant impact on the victims who were exposed to nature's extremes during the 'moment of disaster'.

Archaeological evidence is unlikely to be able to add a great deal to this picture. Though non-specific stressors, including psychological factors, affect individuals during pre-adult development – often resulting in growth abnormalities such as reduced stature, vertebrae development and reduced life expectancy[120] – it is rarely possible to identify the root cause of such stress. Furthermore, it would be nearly impossible to weigh what contribution a specific stressor, such as a natural disaster, made toward any developmental abnormalities observable in a given individual. The apparent silence on the issue of mental health and psychology in the historical sources, meanwhile, is largely explained by the fact that the concept of post-traumatic stress disorder (PTSD) was not recognised as such in the historical past. PTSD was first defined in the 1950s al-

117 Bankoff 2013: 30–31.
118 Athimon, Maanan 2018: 1493–1494.
119 See Fig. 6.3.
120 Watts 2011.

though with 19th century origins,[121] and, instead, medieval Europeans usually classed such conditions as 'mania'. The 13th century English physician Gilbertus Anglicus ascribed the condition to an imbalance of the bodily humours which led sufferers to experience "much sorrow, and dread much of thing[s] that is not to dread, and think on thing[s] that is not to think on".[122] Certainly, not all those with symptoms matching Gilbertus' description were PTSD sufferers – nor were they all victims of natural disasters – but his description would certainly encompass these cases. Although it is not possible to make definitive 'palaeo-diagnoses', likely PTSD sufferers from the medieval period might have included crusaders, who experienced violent and traumatic episodes,[123] as well as those who survived plague epidemics such as the Black Death – who may have lost loved ones in desperate circumstances.[124] Another category must have been survivors of traumatic natural disasters – though, for the reasons outlined above, no specific examples are known.

Since the medieval sources do not explicitly deal with the mental health and psychology of those who experienced disasters, modern studies of individual and community responses and reactions to natural disasters offer the best hope to understand the impact of these events in the past. Modern studies analysing victims of severe flood events, for example, demonstrate that affected individuals often experience pronounced psychological effects following a flood event. These results appear to be common – studies have compared different cultural and demographic populations, in some cases even incorporating control populations unaffected by flooding in order to isolate the contribution of the flood event.[125] In relation to flood victims, the conditions most frequently reported are PTSD, depression and anxiety[126] though the extent to which individuals in any particular population develop such mental health conditions appears to be highly variable – ranging from 8.6% to 53% in different studies.[127] Such a large discrepancy should not be surprising since no two floods are alike and many variables affect how human communities are impacted.

Although the specifics vary, modern studies do consistently suggest that different demographics – defined by age, gender or socio-economic status – are affected by post-event conditions to different degrees. For example, some studies suggest that women who have been affected by disasters face a greater risk of PTSD than men[128] while, in other studies, no appreciable differences have been identified.[129] Similarly, in victims of

121 Andreasen 2010: 67–68.
122 Getz 1991: 13–14.
123 Heebøll-Holm 2016.
124 Boccaccio 1825: 15–16.
125 See for example Canino et al. 1990.
126 Alderman et al. 2012; Stanke et al. 2012.
127 Lamond et al. 2015.
128 Bonanno et al. 2007: 672, 678.
129 Fernandez et al. 2015.

the 2009 L'Aquilla earthquake, behaviours exhibited by those who were affected by the disaster appear to have been associated with gender. Females, for example, were more likely to experience sleeping disorders, including nightmares, the avoidance of external reminders of the event and a generally negative emotional state. Males, on the other hand, adopted 'mal-adaptive behaviour' such as smoking or alcohol consumption at a comparatively high rate.[130] This latter behaviour has also been observed in studies of flood victims.[131]

It is clear from modern disasters that individuals can experience events in very different ways. For example, being in a slightly different location – a different room in a house, for example – might drastically alter the way in which one person perceives a disaster compared to another. This kind of variance in how different individuals experience disasters also appears to be linked to the extent to which they experience mental health conditions in the aftermath. Quite understandably, one study has demonstrated that loss of loved ones and/or possessions was associated with heightened occurrence of post-event trauma[132] while another study indicates that individuals exposed to flooding of a higher severity during a flood event faced an increased likelihood of developing PTSD in the aftermath.[133] Due to the long periods of time involved, few modern studies have considered the length of time over which flood victims experience mental health conditions. One study suggests that mental health conditions related to flood exposure were still apparent 5 years after the occurrence of the flood[134] while in another study, after a gradual downward trend, emotional distress seemed to increase in severity in tandem with the anniversary of the flood event.[135] At a psychological level, the different ways in which populations that have lived through flood events and other disasters have been affected over longer – decadal – timescales, however, is a topic which requires more detailed study.

Although the experiences of modern populations offer an important insight into understanding how these events affect people at a deep psychological level, the extent to which they provide an appropriate analogue for how medieval populations experienced disasters is debatable. A variety of fundamental differences exist between contemporary and medieval societies which colour the extent to which mental health issues are likely to have affected those who lived through natural disasters. Some of these differences probably increased the resilience of members of medieval society to any deterioration in mental health. Kin bonds, for example, were likely stronger during the medieval period compared to many modern settings – although economic migration could still cause family groups to become fragmented. Furthermore, throughout

130 Carmassi et al. 2014: 59.
131 Fernandez et al. 2015.
132 Assanangkornchai et al. 2004: 87.
133 Feng et al. 2007.
134 Fernandez et al. 2015.
135 Assanangkornchai et al. 2007.

medieval Europe, the extent to which the pervasive and institutionalised nature of the Church influenced the way in which people interpreted disasters – including why specific disasters occurred as well as the correct course of action to be followed in their aftermath – cannot be underestimated. In the form of family groups and the local congregation of the parish church, both of these factors provided support networks, which, compared to many modern populations, probably went some way to combating the development of mental health issues. Certainly, studies analysing modern populations frequently highlight the importance of community support and assistance in lowering the prevalence of mental health issues.[136]

The specific realities of the medieval world, on the other hand, likely exerted a negative impact on the mental health of medieval populations affected by disasters. Most obviously, the socio-economic balance of medieval society probably exposed certain groups to a higher risk of poor mental health outcomes. High levels of poverty, for example, would likely have compounded the mental health problems arising from disasters and, indeed, modern communities reveal that lower income groups face a greater risk of deterioration in mental health in the aftermath of flooding.[137] Medieval beliefs, on the other hand, might have negatively impacted victims' mental health in the aftermath of a disaster. For example, the widespread belief that extreme weather events were signs of divine displeasure, rather than explainable natural phenomena, must have frequently ignited panic and worry in a population that was strongly influenced by teachings which stressed the importance of pleasing God in order to secure salvation in the hereafter. Beyond the tragedy of loss, this worldview logically leads to a fearful interpretation of disasters related to their spiritual connotations. It is easy to imagine, therefore, how the local occurrence of a disaster could quickly trigger anxiety and fear for the future.

Relatedly, the extent to which populations were familiar with a hazard and/or prepared for its occurrence are also likely to have been important factors. No strangers to flooding, medieval lowland populations were well adapted to dealing with the risks posed by flooding and, accordingly, have been characterised as a 'risk culture'.[138] For example, in Romney Marsh the so-called 'Marsh Law', which had been in place since at least 1250, provided a legal framework to ensure co-operation between landowners in order to guarantee the maintenance of flood defences.[139] Whether such familiarity would have contributed to an increased or reduced chance of developing mental health issues in the aftermath of a flood, though, is unknown. Modern studies which have looked at repeat flood victims point to varying conclusions – in some cases they exhibited improved resilience from conditions such as PTSD, while in others they

136 Feng et al. 2007; Fernandez et al. 2015.
137 Lamond et al. 2015.
138 Bankoff 2013.
139 Bankoff 2013: 30.

displayed increased susceptibility.[140] The extent to which a disaster is expected is another important consideration as preparedness, both mentally and physically, guard against any subsequent mental health deterioration.[141] In the absence of any accurate weather forecasting, there was no way that specific meteorological events could be pre-empted – though traditional knowledge allowed some level of estimating the conditions to be expected over the short-term. Hazards which were routine therefore, such as seasonal flooding, were much less likely to cause mental health issues than disasters for which no communal memory existed – be it the type of hazard, its magnitude or temporal occurrence.

In the absence of documentation, the thoughts and mental processes that cycled through people's minds in the aftermath of specific historical disasters can never be reconstructed with certainty. By drawing on modern analogues, though, it is possible to identify a few broad points that it must be assumed applied to affected medieval populations. In all probability, medieval disaster victims would have experienced a spike in mental health conditions such as PTSD, depression and anxiety after the occurrence of the types of hazards considered in Chapter 3. These conditions would have particularly acutely affected those who were poor and/or those who directly experienced a disaster. Through the worldview of medieval Christianity, the unexpected occurrence of these events would have provoked fear and anxiety although such spiritual concerns may have been quelled to some degree by Church ceremonies and festivals such as communal processions which were believed to provide spiritual protection. More regular natural hazards, on the other hand, which were well understood and more or less expected, are comparatively unlikely to have caused any significant mental health impacts.

7.2.4 Summary

The impact of specific disasters has been enumerated at length – both above and in Chapter 3. Typically when high-magnitude natural hazards affected medieval populations the emergency phase of the disaster cycle[142] was characterised by widespread and severe damage and destruction and often, if contemporary chroniclers are to be believed, accompanied by high numbers of casualties. During this phase, effective responses were quite limited – where possible medieval people could flee from hazards or try to endure through them in locations thought to be safe, such as strong buildings or high ground beyond the reach of floodwaters. In the aftermath of the event, however, the available options expanded significantly and practical solutions to the problems created by hazards became viable. It was at this stage that local authorities, and in some

---

140 Lamond et al. 2015: 192.

141 Assanangkornchai et al. 2004: 87.

142 Alexander 2002: 6; see Fig. 1.1.

circumstances the Crown, became involved to mediate a return to normality. During this recovery phase, debris could be cleared, broken dikes could be restored and repairs to damaged structures could be organised. Doubtless, such activities provided a positive focus, which brought people together, for communities and individuals who, as well as coming to terms with their newly altered physical circumstances, may have been pondering the troubling eschatological implications that the Church often attached to the occurrence of natural disasters.

Over the longer-term, continued memory of such events – maintained in a wide variety of forms both material and intangible (as discussed in Chapter 6) – drove some level of adaptation. During the mitigation phase, therefore, structures were occasionally reinforced against the recurrence of future hazards. The structural interventions after 1362, for example, including the repairs to the spire at Salisbury Cathedral,[143] the buttresses added to the church tower at Hitchin[144] and the secondary arch inserted beneath the spire at Austin Friars, London,[145] may be examples of such adaptive strategies within structures. In reclaimed wetlands, the continuity of dike maintenance was doubtless driven by the memory of what happened when the flood protection infrastructure had failed and this, at times, led to high investment in improving such systems – as can be seen in the approach adopted by ecclesiastical landowners after the storms of 1287/88.[146] Such precautions and mitigation strategies provided a degree of protection from recurrent hazards – which was both tangible and also provided visible reassurance that necessary precautions had been taken to protect against the hazard(s).[147] This was important because, during the pre-impact phase, once again, there was little that medieval populations could practicably do to prevent the catastrophic impact of hazards. Where storms were foreshadowed by the weather, there may have been an opportunity to herd livestock to safety and move goods out of the way of floodwaters but even this was often not possible. A 1524 municipal regulation from the town of Krems, Austria, for example, gave permission for citizens to move their wine to a safe place, away from the banks of the Danube when threatened by flooding, as many of the town's wine cellars were at risk when the river flooded its banks. In the case of the disastrous flood of 1573, however, the fact that the inundation was sudden and unexpected, and that it occurred at night, meant that there was insufficient time to get the wine out of harm's way, resulting in devastating losses throughout the town.[148]

143 Miles et al. 2004: 20–22.
144 Parker 1904: 38.
145 I'Anson 1866: 71.
146 Gross, Butcher 1995: 108–109; Rippon 2001: 22–26.
147 Although note that such reassurance could be misleading. The 'levee effect' – a complacent outlook common among residents in flood-prone locales that assumes, once a flood defence is in place, that security is guaranteed whether or not defences are sufficient to protect against future flooding is well documented. Reser 2007: 374.
148 Rohr 2013: 138.

While, in most cases, storms and floods gave some form of warning of their impending occurrence, other hazards were highly unpredictable. Landslides, for example, can occur after a significant period of time has passed since the triggering mechanism, either meteorological or seismic, has occurred.[149] For medieval populations, with no way of monitoring the physical properties of landslide-prone slopes, such unexpected and inexplicable occurrences would have made any preventative measures impossible and must have stoked interpretations of these disasters as the result of divine intervention.

This overview of the disaster cycle in reference to the case studies considered in Chapter 3 is conceptualised in Figure 7.3. The medieval capacity to react to the different challenges in each phase, reveals some interesting conclusions. Clearly, the practical responses adopted by medieval communities were most applicable to the restoration, reconstruction and quiescence phases, during which meaningful steps could be taken to restore conditions to their pre-event norms. In some cases, it was also possible to provide a degree of increased security in the event of a future recurrence of the hazard through structural responses. During the pre-impact and emergency phases, on the other hand, medieval responses were incredibly limited and this, combined with a poor understanding of the physical mechanisms behind most natural hazards, created a gap in medieval populations' technical understanding and physical ability to respond to natural hazards during these phases. According to RESER: "the search for causes and explanation ... is a very common ... psychological phenomenon typical of natural disasters".[150] Where medieval populations were unable to explain disasters through physical means, therefore, it was natural, for those of the Christian faith, to turn to an explanation that was consistently reinforced by the Church through a wide variety of media – divine intervention. It naturally follows that, against divine causes, disasters could be mitigated, or prevented from occurring, using measures that were believed to interact with the divine sphere. This explains the widespread recourse to religious and superstitious practices, discussed in Chapter 5. While practical measures proved insufficient in providing protection in the moment of disaster, religious practices, most notably prayer but also practices manifest through material culture, such as folded coins or *ampullae*, promised immediate deliverance at the utterance of specific words, the bending of a coin or at the dispersal of sanctified material. Far from being "blinded with profound stupidity that they believe[d] these things could happen",[151] medieval peoples' reliance on spiritual and superstitious practices – the efficaciousness of which was no doubt vouched for and reinforced by members of the community – was, in fact, a logical response which 'plugged the gaps' in the available physical responses

149 It has been suggested that the occurrence of landslides increased in the 5 years following the Taiwan Chi-Chi earthquake in 1999, while weather-induced landslides frequently occur up to 5 days from the event thought to be the primary trigger. TATARD et al. 2010: 16.

150 RESER 2007: 382.

151 As the 9th century bishop, Agobard of Lyon, described those who believed in the magical genesis of hail and thunder. LEWIS 2001.

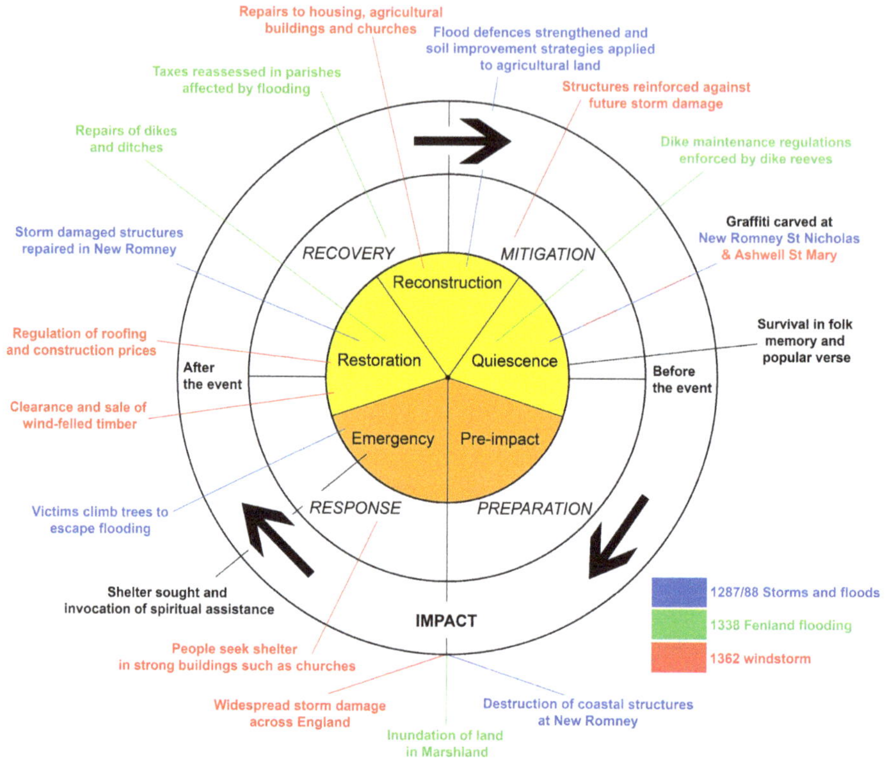

Fig. 7.3: The disaster cycle applied to each of the three most significant events considered in Chapter 3: the storms and floods of 1287/88 (blue), the Fenland flood of 1338 (green) and the 1362 windstorm (red). Annotations in black can be assumed to have taken place in all disasters. Notably, while the impacts of disasters were often severe, during the restoration, reconstruction and quiescence phases (yellow), there were generally practical steps which could be taken to improve the situation. Effective responses were extremely limited during the pre-impact and emergency phases (orange). Redrawn and annotated by the author after Alexander 2002: 6.

constrained by the limited scientific understanding and technological capabilities of the age.

7.3 Physical signatures and material culture

The disasters considered in this book can be traced through numerous strands of evidence linked to their occurrence. Certain types of hazard, left a physical signature in the moment of disaster itself – as the storm deposits linked to the 1287/88 storms detected across New Romney attest. Where this did not occur – in the case of more ephemeral hazards or post-depositional processes that erase material evidence – the pattern of destruction and the subsequent repairs these engendered are the most obvious trace that can be linked to these historical disasters – as in the case of the breaches in dikes in Marshland and the widespread structural damage in buildings damaged by the storm of 1362. These repairs not only sought to repair the damage but, in some cases, also to improve resilience against future hazards. Other strands of evidence relate to responses taken as a direct or indirect result of the occurrence of disasters. As discussed in Chapter 4, for instance, the burial of humans and livestock in unusual ways, such as mass graves, may relate to disastrous circumstances while inscriptions carved in the aftermath may attest to the magnitude and occurrence of hazards – in the case of inscribed flood marks – as well as the fears of the lay population and their desires to gain protection against any possible recurrence – as in the case of apotropaic graffiti. Such fears are also reflected in the artefactual evidence discussed in Chapter 5, the beliefs associated with which – although aimed at accruing protection from a broad gamut of threats – formed an integral component of medieval responses to natural hazards.

7.3.1 Disasters as 'watersheds'

The acute impacts of hazards, across a wide area of effect, have the potential to spur material and social change. In the case of low magnitude events, such as a fierce but routine storm which threatens structures but does not actually cause any material damage, this may be at such a low level that, from an archaeological perspective, it is imperceptible. As discussed above, however, more extreme events – those which contemporary populations were most likely to perceive as disasters – may be expected to leave more recognisable signatures. In his assessment of the adaptations to seismic risk in medieval Cypriot ecclesiastical architecture, for example, O'NEILL posits that the devastating earthquake of 1350 led to a swarm of repair and redesign in the churches across Cyprus that can be traced in the surviving standing building evidence.[152] This wave of activity following at the heels of the earthquake, comprising the clearance of debris, demolition and reconstruction, the organisation and execution of structural repairs as well as more ephemeral social interactions and indirect repercussions, could, in one sense, be conceptualized as a 'watershed'.

Running with the concept of high magnitude disasters as 'watersheds', some of the case studies considered above may accord with this concept. The storm series of 1287/88, for example, could be viewed as a watershed along the affected coastlines; covering much of eastern England as well as the Low Countries, with an area of effect roughly delineated by the evidence presented in Figure 3.1. While the archaeological evidence from New Romney demonstrates the preservation of *in situ* remains connected

152 O'NEILL 2020.

to the storm series, these were the exception rather than the rule. Therefore, in the affected areas, while evidence for late 13th century coastal flooding is likely to be connected to these events, less direct evidence is more likely to be encountered. This might include settlement contraction, altered settlement layouts, reorganisation of seafront and harbour facilities and/or renewed sea defences. Notably the dislocation in pottery manufacture in the aftermath of the storms, manifest by the disappearance of shell and sand tempered wares,[153] is another kind of watershed. On a regional scale this may assist in identifying pre- and post-storm contexts and elucidating human activity impacted by the occurrence of the storms.

In a similar vein, the storm of 1362 in England provides another example. Due to the uncertainty in identifying storm damage where it is not explicitly listed in the historical record, it is not possible to make a definite statement on this issue but the 1362 windstorm may serve as a watershed in architectural terms across the area of effect – especially in eastern England. The vast number of churches and other ecclesiastical buildings damaged by the storm, attested by both contemporary chronicles and the standing evidence recounted above (see Chapter 3.3), with a high number displaying evidence for post-storm repair, makes it worth considering how far reaching the impact of the storm may have been. Certainly, tower repairs and/or the modification of buttresses supporting tall structures, as at Hitchin, Hertfordshire, which date to the late 14th century in southern, and especially eastern, England should be re-evaluated in light of the 1362 storm. While no all-encompassing claims are possible at this stage, it is certainly conceivable that a large number of such cases owe either to direct damage from the storm, and the subsequent phases of repair, or to adaptation in the design of such structures motivated by the damage experienced in 1362. Even if no damage occurred at a particular location, the storm's occurrence may have affected architectural decisions in new structures – especially churches and tall buildings – which were, themselves, unaffected in 1362.

In common with the picture gleaned from considering the economic impact of disasters, viewing disasters as 'watersheds' emphasises their role in bringing about change in human societies over the short-term – during their occurrence and in the immediate aftermath. This finds an analogy with a theoretical framework from the biological sciences, the concept of 'punctuated equilibrium'.[154] This posits that evolution does not occur at a constant, fixed, rate but suddenly, as long periods of stasis are broken by rapid, bursts of change. Although, disasters are certainly not the sole instigators of cultural change, they do ignite a level of activity in certain spheres of human activity not seen in preceding periods of quiescence which validates this comparison. Immediately, after the occurrence of a sudden-onset hazard, repairs must be organised, along with relief efforts and necessary provisions for circumstances created

153 Barber 2008: 172.
154 Eldredge, Gould 1972.

or altered by the calamity. Whether this frenzy of activity could be conceptualized as 'evolution', however, is a different question. Societal 'learning' is a potential product of this post-event phase, as explored in Chapters 4 and 6, but this may be more appropriately termed adaptation rather than evolution. Certainly, the ability of medieval populations to adapt to new conditions created by disasters was key to coping with the challenges they engendered. Indeed, in an investigation of the varying trajectories of rural settlements in medieval Norway and Sweden, for those settlements where occupation went uninterrupted throughout the various crises of the 14th century, the unifying explanation for the continuity of settlement, based on the archaeological evidence, was adaptation – a population's ability to alter the way of doing things in light of the new reality in which they found themselves.[155] Even though adaptation helped such societies deal with disasters after their occurrence, however, there was no guarantee that these same adaptations would help them deal with future disasters – as every event is unique and may affect different geographic areas or spheres of life. The ability to adapt, therefore, to new and unforeseen challenges – rather than any specific adaptation – would always have allowed communities a good chance of overcoming the unique challenges created by different disasters.

7.3.2 Signatures of disaster

As the changes which disasters stimulate within society appear to manifest themselves most obviously in the immediate aftermath of 'the moment of disaster', it is rare to be able to trace far-reaching changes in material culture back to any specific disaster. Devastating calamities at the population level, however, may have more readily sparked shifts in society and its material culture which can be detected. The social upheaval created by the Agrarian Crisis of 1315–22, for example, particularly the heightened need for security, has been interpreted as the driving force behind the construction of moated homesteads across England during the early 14th century.[156] Equally, in the aftermath of the Black Death analysis of tree-ring dates from timber structures across much of England points to a surge in the construction of non-aristocratic timber framed buildings.[157] A key factor in the proliferation of these structures was the improved living standards in which the survivors found themselves as a result of the demographic collapse precipitated by plague while another possible consideration is that these structures were better suited to the altered climatic conditions of the Little Ice Age.[158] Many other changes, manifest in late medieval material culture, took place at approximately the same time. According to STANDLEY, the human suffering

155 SVENSSON et al. 2012: 101.
156 PLATT 2012.
157 DYER 2013.
158 CAMPBELL 2016b: 383.

which had been patently visible to every European during the 14th century brought on the emotional involvement and affective piety which characterised lay devotion during the period.[159] Funerary practices became less ostentatious affairs,[160] though in some cases grave goods which possibly held apotropaic significance become more common[161] – perhaps a reflection of peoples' heightened concerns for the hereafter. An evolution also occurred in the culture of gift giving, affecting provision for feeding the poor and a general replacement of foodstuffs as gifts by their cash value.[162]

While it is easy to view these types of developments solely through the lens of plague, broader changes took place in the 14th century. The 'great transition'[163] – which hinged on the shift in climate from the generally favourable conditions of the Medieval Climate Anomaly to the colder, more hostile climatic regime of the Little Ice Age. These climatic oscillations created an environment in which natural hazards occurred with greater frequency and ferocity, as discussed above in Section 7.1.1, a fact which is well illustrated by the litany of events spanning this period; the storms of 1287/88 and the other sea floods which continued throughout the period, the destruction of crops by unprecedented rains in 1315, 1316 and 1317 (sparking harvest failure and bringing on the Great Famine), the highest magnitude flood to have affected central Europe in 1342,[164] the most deadly landslide in Norwegian history in 1345[165] and the unprecedented high-magnitude windstorm of 1362. As well as an age of plague, therefore, the transition from the Medieval Climate Anomaly to the Little Ice Age was also an age of disasters.

Some of the changes in the material record which occurred during this pivotal period, therefore, may be interpreted as 'teleconnections' between the dramatic fluctuations in the inter-related spheres of climate, disease, and society (see Fig. 7.4 for a conceptualisation of this conjunction inspired by CAMPBELL 2016b: 22; Fig. 1.2.). Material culture is the most visible way in which human society mediated the challenges imposed by deteriorating climatic conditions and the increased incidence of meteorological hazards this brought. In addition to the structural responses considered throughout this volume, this may also be reflected in the types of protective artefact considered in Chapter 5 and, in some cases, a relationship between their circulation and wider affairs may be posited. The chronologies of *ampullae*, pilgrim badges, saintly relics and evidence relating to the cults of St Barbara, St Christopher and the Magi, when compared, display a pronounced divergence from the previous trend hinging on the mid 14th century and continuing throughout the 15th century. This is displayed in Figure 7.5, which uses 25-point moving averages in order, to some degree, to correct

159 STANDLEY 2020.
160 WOOLGAR 2011: 18.
161 GILCHRIST 2008: 130–131.
162 WOOLGAR 2011: 15, 18.
163 CAMPBELL 2016b.
164 DOTTERWEICH et al. 2003.
165 ROKOENGEN et al. 2001.

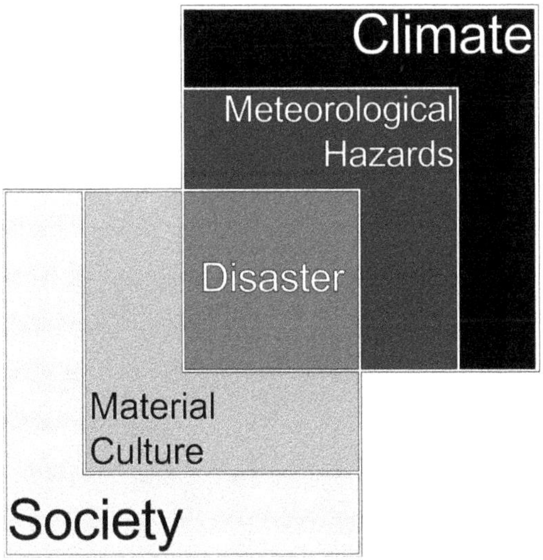

Fig. 7.4: Disasters as the intersection of climate and society. Climatic processes cause meteorological hazards which spawn disasters when they interact with human society. Disasters both affect and are mediated by a society's material culture. Created by the author.

for the artificial jumps in the graphs which result from the use of broad periodisations to assign a date to each artefact as part of the PAS recording process. Alongside these data, Figure 7.5 presents estimates of medieval English population.[166] The demographic collapse, for which the Black Death was the catalyst, of the mid 14th century registers as a major disjuncture between the period of sustained ascension, which characterised the preceding three centuries, followed by stagnation until the early 16th century. Although the precise linkages are highly debatable, this 'before and after' pattern can similarly be discerned in the chronological distribution of artefacts related to ritual protection considered in Chapter 5. A possible interpretation, therefore, is that, at a population level, the trauma of the 14th century – spearheaded by the Black Death but with a key component related to the numerous and recurrent disasters of the period – bolstered peoples' desire to protect themselves from the environmental dangers that defined the age. In reference to the calamities of the 14th century, Jost argues that "society underwent a psychological need for … protection in the face of such adversity"[167] and this, combined with lay populations' increased participation in religious practices,[168] offers a compelling explanation for the rise in material culture linked to protection in the aftermath of a protracted period of crisis punctuated by high-magnitude disasters

166 Broadberry et al. 2015: 20.
167 Jost 2016: 199.
168 Jost 2016: 236.

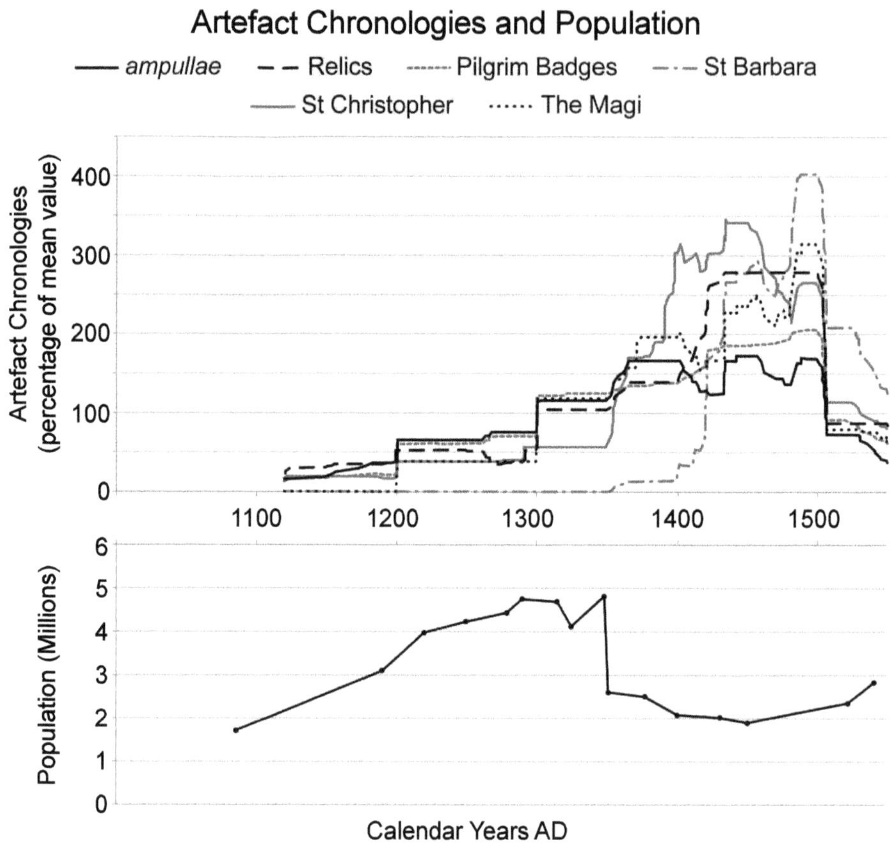

Fig. 7.5: Artefact chronologies from the categories examined in Chapter 5 and population. Values for the cult of St Barbara have been divided by 2 to allow easier comparison. Created by the author. Population estimates after Broadberry et al. 2015: 20.

and a sudden drop in population. Adding a degree of credence to this interpretation is the fact that, in systematic test-pitting across a wide area of eastern England, the prevalence of medieval ceramics dropped on average by c.45% in the post-Black Death era.[169] If this represents a reliable proxy for material culture more generally, then clearly the protective artefacts considered in Chapter 5 significantly bucked the trend for material culture at this time. Certainly, the manifold disasters of the period created "the need for the utmost power to protect against misfortune and disaster".[170]

169 Lewis 2016.
170 Standley 2020: 248.

Folded coins and *bullae* were not included in these comparisons as the circulation of both categories of artefact were dependant on other factors – the wider availability of coinage, in the case of folded coins, and documents from the Papal curia, in the case of *bullae*. As such, although the evidence certainly suggests that they were relied upon for spiritual protection during the Middle Ages, it follows that their chronologies closely track that of their 'raw materials' – i.e. coin folding as a practice was more prevalent when more coins were in circulation and ritualised re-use of *bullae* increased when more *bullae* were available.[171] This being the case, one of the greatest divergences evident between the chronology of folded coins and the availability of coinage more generally occurred during the mid-14th century – when we might assume that individuals, and society as a whole, were particularly focussed on gaining some relief from temporal realities through spiritual intervention. *Bullae* with evidence for ritualised re-use, on the other hand, make up only a small percentage of *bullae* total finds so, in this case, such divergences are unlikely to be meaningful.

7.4 Summary

This book has sought to reconstruct the occurrence of various meteorological disasters using archaeological and historical approaches and to evaluate and contextualise their impact and, specifically, what the occurrence of these types of event meant for contemporary society. A number of conclusions stand out. Rather than entirely anomalous events, disastrous circumstances arose as a result of a complex interplay between a changing environment and human society. As such, climatic factors which increased the incidence or intensity of these hazards played a role in the onset of many of the disasters which characterised the transition from the Medieval Climate Anomaly to the Little Ice Age. At the same time, however, endogenous economic and social conditions exacerbated the vulnerability of contemporary society and, correspondingly, the impact of disasters. In general, however, medieval society was relatively well adapted to deal with these events in the aftermath through the organisation of repairs, the construction or repair of structural responses – such as flood defences – and their continued maintenance. These practical strategies were largely organised and undertaken at the local level although, typically, the Crown did become involved when disasters occurred at a national scale and to regulate markets and financial considerations. Especially where disasters were unprecedented and/or arose without warning, medieval society was relatively unprepared to cope during the pre-impact and emergency phases. Modern psychological studies suggest that unexpected disasters are especially troubling and this must have particularly been the case when the natural mechanisms which caused disasters were poorly understood. At these times, however,

171 This is supported by the finds from the PAS database (see Figs. 5.22 and 5.16).

spiritual protection invoked through prayer and specific types of artefact thought to provide protection provided a response which, to medieval people, seemed a logical recourse. The evidence obtained from the finds reported to the PAS, suggests that the uptake and reliance on these spiritual routes to protection experienced a pronounced expansion in the era that followed the Black Death and the climatic fluctuations of the 14th century. This suggests that, at a population level, the numerous disasters of the 14th century, combined with contemporary populations' psychological requirement for security, contributed to the evolution of belief that took place within medieval society during this period.

8 An endless chain of catastrophes?

The filmmaker and documentarian Werner Herzog describes existence as "an endless chain of catastrophes".[1] Indeed, as this book has explored, the existence of medieval communities in northern Europe was punctuated by the occurrence of natural hazards which, under the right circumstances, generated human disasters. The jump from hazard to disaster is ill-defined and varies between cultures[2] but relates to both the characteristics of the hazard(s) – such as the recurrence interval[3] and magnitude – as well as factors endogenous to the affected societies – including the vulnerability, resilience and capacity for adaptation of the communities in the path of the hazard. These considerations differentiated the 'everyday risks' from the 'disasters'[4] and are evident in all of the disasters considered in Chapter 3. The storms of 1287/88 had an acute impact along the coasts of south-eastern England exacerbated by their high magnitude and close temporal clustering. The archaeological evidence for structural damage and the deposition of sediments by storm action at New Romney demonstrates the severe implications that these events had for the contemporary inhabitants. The worst floods in Marshland over the 13th and 14th centuries, although a part of life in the Fens, came at a time when society was particularly vulnerable as a result of increased taxation, high demographic pressure and perhaps the delayed impact of earlier disasters such as the Agrarian Crisis of 1315–1322. The St Maur's Day storm of 1362, on the other hand – which damaged churches, other structures and blew down trees across much of England – came amidst a renewed outbreak of plague and was likely an unprecedented hazard for the contemporary population.

The evidence and discussion presented in the preceding chapters has refines the existing understanding of the occurrence of these case study events. This permits reconstructions of the areas of effect and types of damage unleashed by natural hazards on medieval society (as in 1287/88 at New Romney, see Fig. 3.2; in Marshland between 1287–1349, see Fig. 3.8; and across England in the storm of 1362, see Fig. 3.15). The historical and archaeological data, when used in concert, often allow a relatively clear picture of the short-term impact of these events to be established. Where archaeological evidence relating to the 'moment of disaster' survives, this often reveals tantalising evidence; crystallising a window of time in which the interplay between natural hazards and human societies created disastrous consequences. In the case of the 1287/88 storms,

1 Encounters at the End of the World (2007) Directed by Werner Herzog. Discovery Films, USA.

2 Bankoff 2004.

3 Events perceived as disasters were most likely to have either very long recurrence intervals, beyond the period of time that memory of the disaster could easily be transmitted, or shorter yet variable intervals which could result in the clustering of hazards over a short span of time – as occurred in the 1287/88 floods.

4 Rohr 2013: 144–145.

https://doi.org/10.1515/9783110719628-008

this is reflected in the structures damaged by storm flooding in New Romney – mainly in the plots along the town's medieval coastline – while the damage pattern traceable through structural evidence across the British Isles attests to the damage wrought by the storm on 15th January 1362.[5] Such material evidence is, however, the exception rather than the rule and it is, therefore, complemented by the surviving historical record which allows a more comprehensive picture of the occurrence of disasters to be established. Unifying data from the two disciplines provides the most holistic understanding of disasters but this frequently proves problematic due to the, sometimes significant, disparities in the precision of dating evidence between different categories of evidence. This problem will hopefully be mediated in the future by the refinement and increased adoption of scientific dating techniques, such as luminescence dating, which have the potential to provide substantially greater chronological resolution than dating based on ceramic typologies – upon which many of the contexts considered in this book rely.

During the tantalising 'moment of disaster', medieval populations were relatively helpless – finding themselves at the mercy of nature's extremes. During the aftermath, however, a broad suite of options and possible pathways to recovery were available. First and foremost, the costs of the catastrophe had to be counted, debris had to be cleared and the worst of the damage had to be repaired. This phase is widely reflected in the historical record – with the repairs to roofing by tilers and thatchers after the 1362 windstorm an obvious example – but may also be seen in the clearance of building debris in sites damaged by the 1287/88 storms in New Romney. The repairs to breaches visible in the Marshland sea wall were probably also fairly rapid repairs in the short-medium term after a disaster, as a subsequent flood could have arisen at any time – and existing breaches would have let the water come through unhindered, perhaps causing further damage to the sea wall and inundating an even larger area. As an agrarian society, the success of the harvest, agricultural yields and the survival of livestock were of foremost significance and where disasters had destroyed crops or decimated herds a period of re-adjustment followed in which new strategies were introduced in an attempt to mitigate any losses – flooding, for example, frequently promoted the cultivation of legumes, which were more resilient to saline soils and helped restore soil fertility, or a switch to pastoralism. In order to prevent disasters pushing medieval populations to a 'tipping point', landowners, higher authorities or the Crown often responded sympathetically to difficulties faced by individuals and communities in meeting their financial commitments. This can be seen in the aftermath of the damaging floods of 1338, when Edward III consented to a reassessment of the tax owed by the affected communities in Marshland, Norfolk, as well as in the aftermath of the 1362 storm. In such cases, authorities seem to have judged pragmatically that

5 Coincidentally, another weather pattern from this year is fossilised through the tephra cloud generated by the Icelandic Öræfajökull eruption in June 1362, which has been detected in Greenland, Scandinavia and Ireland. See Wastegård, Davies 2009: 505–506.

by overlooking a portion of rent or tax in the short-term the recovery process could be accelerated allowing normal payments to be resumed as quickly as possible. Charity too played a role during this phase although, as this was largely motivated by spiritual and eschatological concerns rather than outright philanthropy, the Church was one of the primary beneficiaries of such charitable giving and the extent to which donations and alms were dispersed to genuine disaster victims is unclear.

Once the most pressing issues created by the occurrence of disaster had been addressed, society had a chance to reflect on what had brought about the disaster and how protection from a similar event in the future might be secured. This was often approached in a, perhaps surprisingly, pragmatic manner – the high investment in sea defences by the monastic landowners on Romney Marsh in the wake of the 1287/88 storms, for example, demonstrates an acceptance that storms and floods were an expected reality against which practical precautions were necessary. The archaeological evidence demonstrates that in the aftermath of floods, dikes were sometimes heightened beyond their pre-flood dimensions – clearly recognising that what had stood before had been insufficient – and, in structures exposed to flooding, floor levels were sometimes raised above the height reached by floodwaters. As a class of structure, bridges, which were frequently damaged by torrents of water in times of flood, saw particular experimentation in methods and materials of construction as well as design, to some extent becoming stronger and more resilient over the course of the Middle Ages with the development of innovations such as cutwaters and enlarged flood arches as well as the general switch from timber to more resilient masonry or brick construction. Similarly, in ecclesiastical structures, the occurrence of storms provoked material interventions – including the insertion or strengthening of buttresses, the use of stronger roofing materials or supporting internal scaffolds – which aimed to strengthen the structure against the forces of a future storm. Evidently, the benefits of such structural adaptations, in light of the perceived risk from natural hazards, outweighed the cost of carrying out the work. Although it is most obvious that these types of interventions occurred relatively soon after a disaster had struck, as discussed in Chapter 6, the transmittance and continued survival of memories relating to disastrous occurrences allowed knowledge relating to these hazards – and the risk they posed to buildings, infrastructure and settlement – to be passed on to subsequent generations. In most cases this amounted only to relatively low resolution information – as in the case of flood levels – but even this was important in ensuring society grasped the importance of, for example, ensuring flood defences were properly maintained. This was particularly pertinent in reclaimed wetlands where – as well as being especially vulnerable should the flood defences fail – such works were usually mandated by local regulations and overseen by specially appointed officials. These areas were amongst those exposed to the most recurrent hazards and as such, the need for some level of institutional oversight to oversee the flood defence system is perhaps unsurprising. From an entirely practical standpoint, therefore, such a system demonstrates a high acceptance of the risk posed by flooding which was met with a correspondingly high investment to try and mitigate its occurrence. In this regard, a parallel can be drawn with traditional societies in the present day, which, according to BANKOFF, "historically developed sophisticated strategies and complex institutions to reduce the constant insecurity of their lives".[6] This was also very much the case in medieval northern Europe. During the restoration, reconstruction and quiescence phases of the disaster cycle, therefore, there were a wide variety of practical steps which could be taken to improve the situation and bring about a return to normality while, over longer timespans, medieval communities both recognised the requirement to improve the ability of their buildings, infrastructure and settlements to withstand the forces of natural hazards and usually acted, as far as possible, to ensure this was the case.

The suite of options available to medieval communities in the aftermath of disaster must have had a positive psychological effect as they could clearly see that, through their own agency, they were usually able to restore conditions to pre-disaster norms and take pro-active steps to prevent such conditions recurring. This contrasts starkly with the medieval experience during the pre-impact and emergency phases, however, when, especially at the 'moment of disaster', available options were extremely limited. Medieval disaster victims could choose to flee but even this was not always a valid option – floodwaters might make escape impossible, as occurred when the unfortunates described by John of Oxnead were forced to climb trees in 1287.[7] Equally, there was no escape against a high magnitude windstorm – contemporaries must have been forced to 'hunker down' and wait out the 1362 storm in structures which creaked from the force of the wind outside. In such circumstances, it is easy to see why religious and superstitious practices – which offered some hope of guaranteeing the safety of the participants and brining such frightening circumstances to a close – were popular throughout pre-modern Europe. It was impossible to disprove the efficacy of such strategies – storms and floods always subsided eventually and this was invariably seen as confirmation that such practices were effective. The need for additional spiritual protection also permeated into some of the, otherwise practical, physical responses taken against hazards. For instance, the deposition of cauldrons within the fabric of dikes in the Netherlands demonstrates that the dikes, on their own, were often not perceived to provide sufficient protection, or that an extra 'insurance policy' was desirable. This may relate to the lack of technological understanding of both the causative mechanisms behind hazards and the physical response of structures to the forces unleashed by these hazards. As a result, medieval builders were left somewhat in the dark, unable to know whether what they had built could withstand the forces of the natural hazards it was likely to face. Over time, therefore, the impression could have developed that strong buildings alone were insufficient and only in tandem with spiritual protection could they be expected to survive against nature's extremes.

6 BANKOFF 2004: 32.
7 ELLIS 1859: 270–271.

The evidence obtained from the material culture considered in Chapter 5 suggests that, during the medieval period, and hinging on the 14th century, the extent to which these spiritual practices were relied upon underwent a significant change. While one might conclude this was solely a product of the Black Death, in fact, the 14th century was plagued by other significant disasters – the Agrarian Crisis of 1315–1322, central Europe's worst flood in 1342, Norway's most damaging landslide in 1345, the recurrence of plague in the early 1360s and the ferocious windstorm of 1362, together with the flooding it caused along the Frisian coast.[8] It is a valuable exercise to consider the unfortunate life histories of individuals born in the late 13th and early 14th centuries who would have lived through the near universal trauma of the 1315–1322 Agrarian Crisis followed, some 25 years later, by the Black Death – and, if they were especially unlucky, they may also have witnessed one or two of the disasters considered in Chapter 3.[9] It is hard to imagine that witnessing such disastrous circumstances could not have had a profound psychological impact on the psyche of this generation and their descendants. Indeed, this pronounced period of disasters, appears to have triggered a response at the population level – causing people to seek protection from the dangers posed by natural hazards to a greater extent than had been the case in the past. Based on the medieval Christian worldview, and the commonly accepted explanation of the causation of natural disasters, such a response was logical – without a scientific understanding of the factors behind hazards and the response of structures to their effects, more practical solutions were not necessarily obvious. While this volume has provided evidence for one way in which people responded to the challenges of the 14th century, there is much about this time that remains uncertain. Our understanding of the impact of this tumultuous time on contemporary belief could be refined in the future by the consideration of additional categories of material culture to those considered in Chapter 5, artefacts which are not included within the PAS database – such as museum collections and excavation archives as well as evidence from neighbouring geographic regions.

Returning to the overarching aims stated in Chapter 1, it has largely been possible to address these – though further work in the future will allow more detailed knowledge and clarification in certain areas. The first of these, to reconstruct a number of detailed case studies, is the subject of Chapter 3, which investigates a range of different categories of hazard through different sources of evidence, permitting a particularly wide-ranging analysis of disasters as a broad category. The second aim was to characterise medieval responses to these events. Chapters 4 and 5 tackled this topic identifying many areas in which responses to disaster during the medieval period can be traced. Although beyond the scope of the present study, many of these responses would benefit

8 Where it became known as the "*Grote Mandrenke*" or Great Drowning of Men.

9 It is perhaps also worth considering whether living through such traumatic times would leave any signatures in such an individuals bones that scientific analysis might reveal.

from further consideration – such as the specific agricultural strategies adopted in the aftermath of disasters and the structural adaptations necessitated by disasters in bridges and ecclesiastical architecture. Importantly, a seemingly incongruous relationship between physical and structural responses and superstitious and religious responses was identified which are reconciled when interpreted in light of the level of understanding and technological abilities of medieval society in reference to the different stages of the disaster cycle. The third aim was to assess the extent to which exposure to disaster affected the resilience of medieval society in the face of disasters. This was mainly considered by Chapter 6 which found that, although only relatively low resolution information concerning disasters was transmitted to subsequent generations, this could have played an important role in improving preparedness against hazards – particularly those which recurred relatively frequently. For those hazards which recurred over timescales too long to be meaningful for humans, however, it was unlikely that memory of the event would survive from one occurrence of the disaster to the next – meaning that resilience would be unlikely to be affected. Finally, the fourth aim was to examine the role of disasters as drivers of cultural change. This has been accomplished in a variety of ways including a consideration of the long-term trajectories of settlements affected by disasters. This topic, however, would benefit from a more detailed consideration as the myriad of factors which affect the pathways taken by settlements struck by disasters, and the many factors which play into what form this takes, require a large dataset to provide meaningful answers. The material culture considered in Chapter 5 provides another avenue to investigate cultural change influenced by disaster and it is suggested here that the experience of multiple disasters over the 14th century, in aggregate, contributed to the sea-change in belief patterns visible at the population level.

Despite the fatalistic interpretation of natural disasters as events orchestrated from heaven conveyed in many of the documentary sources, medieval populations clearly had a sophisticated understanding of natural hazards, the threats they posed and what steps might be taken to protect and mitigate against their occurrence. In a number of ways, however, they were constrained in what could practically be accomplished by the limits of scientific and technological understanding of the age. These gaps in understanding created a psychological requirement for alternative explanations which, to medieval Christians, were logically filled by the agency of God – ushering in the power of prayers, protective amulets and ritual deposits as logical and efficacious responses against disaster. Although, therefore, in many cases medieval peoples' first response to disaster was prayer for divine intercession, in the aftermath they almost always responded to the new reality in which they found themselves practically – adapting to changed circumstances by repairing what had been damaged, and, in some cases, improving what had stood before so that it might better withstand hazards in the future. In a variety of ways, therefore, medieval populations adapted to the challenges of disaster and were thus able to, at least temporarily, uncouple Herzog's 'endless chain'.

Bibliography

AKASOY, Anna: Islamic attitudes to disasters in the Middle Ages: a comparison of earthquakes and plagues. In: Medieval History Journal 10, 1–2 (2010), 387–410.
ALCOCK, Nat W.: Leverhulme cruck project. Results 1988. In: Vernacular Architecture 20 (1989), 43–45.
ALDERMAN, Katarzyna; TURNER, Lyle R.; TONG, Shilu: Floods and human health: a systematic review. In: Environment International 47 (2012), 37–47.
ALDRETE, Gregory S.: Floods of the Tiber in ancient Rome. Baltimore 2007.
ALEXANDER, David E.: Principles of emergency planning and management. Harpenden 2002.
ALLEN, John R. L.: Trees and their response to wind: mid Flandrian strong winds, Severn Estuary and inner Bristol Channel, Southwest Britain. In: Philisophical Transactions of the Royal Society B 338 (1992), 335–364.
ALLEN, John R. L.: The geoarchaeology of land-claim in coastal wetlands: a sketch from Britain and the north-west European Atlantic—North Sea coasts. In: Archaeological Journal 154, 1 (1997), 1–54.
ALLEN, Martin: The volume of the English currency, 1158-1470. In: The Economic History Review 54, 4 (2001), 595–611.
ALLEN BROWN, Reginald: Rochester Castle, Ministry of Public Works official guidebook. London 1969.
ALSTON, Leigh: Slough Farmhouse, Stackyard Green, Monks Eleigh, Suffolk: heritage asset assessment. Unpublished Report 2013.
ANDERSON, Robert W.; JOHNSON, Noel D.; KOYAMA, Mark: Jewish persecutions and weather shocks: 1100-1800. In: The Economic Journal 127, 602 (2016), 924–958.
ANDERSON, Sue: A medieval moated site at Cedars Field, Stowmarket, Suffolk. East Anglian Archaeology Occasional Paper Number 15. Ipswich 2004.
ANDERSON, William: Blessing the fields? A study of late-medieval ampullae from England and Wales. In: Medieval Archaeology 54 (2010), 182–203.
ANDREASEN, Nancy C.: Posttraumatic stress disorder: a history and a critique. In: Annals of the New York Academy of Sciences 1208 (2010), 67–71.
ANGELER, David G.; ALLEN, Craig R.; GARMESTANI, Ahjond S.; GUNDERSON, Lance H.; HJERNE, Olle; WINDER, Monika: Quantifying the adaptive cycle. In: PLoS ONE 10, 12. DOI: 10.1371/journal.pone.0146053.
ANNIS, Richard; CAFFELL, Anwen; GERRARD, Christopher M.; GRAVES, Pam; MILLARD, Andrew: Lost lives, new voices: unlocking the stories of the Scottish soldiers at the battle of Dunbar 1650. Oxford and Philadelphia 2018.
ANON: The laws of sewers or the office and authority of commissioners of sewers. London 1726.
ANON: Walpole St Peter Church: an illustrated guide n. d.
AQUINAS, Thomas: Summa theologica, containing second part of the second part, QQ. 1-189 and third part, QQ. 1-90. Volume 2. New York 1947.
ARAÚJO, Ana C.: The Lisbon earthquake of 1755—public distress and political propaganda. In: e-Journal of Portuguese History 4, 1 (2006), 1–25.
ARMITAGE ROBINSON, Joseph (Ed.): The history of Westminster Abbey by John Flete. Cambridge 1909.
ARNOLD, John H.; GOODSON, Caroline: Resounding community: the history and meaning of medieval church bells. In: Viator 43, 1 (2012), 99–130.
ARNOLD, Thomas (Ed.): Memorials of St. Edmund's Abbey. Volume III. London 1896.
ARNOLD, Werner: DI 19, Stadt Göttingen, Nr. 5. 1980. www.inschriften.net/goettingen/inschrift/nr/di019-0005.html. Accessed 16 August 2018.

https://doi.org/10.1515/9783110719628-009

ASSANANGKORNCHAI, Sawitri; TANGBOONNGAM, Sa-nguansri; EDWARDS, J. Guy: The flooding of Hat Yai: predictors of adverse emotional responses to a natural disaster. In: Stress and Health 20 (2004), 81–89.

ASSANANGKORNCHAI, Sawitri; TANGBOONNGAM, Sa-nguansri; SAM-ANGSRI, Nisan; EDWARDS, J. Guy: A Thai community's anniversary reaction to a major catastrophe. In: Stress and Health 23 (2007), 43–50.

ASTON, Mick: The earthworks of Bordesley Abbey, Redditch, Worcestershire. In: Medieval Archaeology 16 (1972), 133–136.

ATHIMON, Emmanuelle; MAANAN, Mohamed: Vulnerability, resilience and adaptation of societies during major extreme storms during the Little Ice Age. In: Climate of the Past 14 (2018), 1487–1497.

AUNGIER, George J. (Ed.): Croniques de London depuis l'an 44 Hen. III. Jusqu' à l'an 17 Edw. III. London 1844.

AVERY, Harold: Plague churches, monuments and memorials. In: Proceedings of the Royal Society of Medicine 59, 2 (1966), 110–116.

BAGGS, Anthony P.; JURICA, A. R.: A history of the county of Gloucester: Bledisloe Hundred, St. Briavels Hundred, the Forest of Dean. Volume 5. London 1996.

BAILEY, Mark: Per impetum maris: natural disaster and economic decline in eastern England, 1275-1350. In: CAMPBELL, Bruce M. S. (Ed.): Before the Black Death: Studies in the "crisis" of the Early Fourteenth Century. Manchester 1991, 184–208.

BAILEY, Mark: Medieval Suffolk: an economic and social history 1200-1500. Woodbridge 2007a.

BAILEY, Michael D.: From sorcery to witchcraft: clerical conceptions of magic in the later Middle Ages. In: Speculum 76, 4 (2001), 960–990.

BAILEY, Michael D.: Magic and superstition in Europe. Lanham 2007b.

BAILEY, Michael D.: A late-medieval crisis of superstition? In: Speculum 84 (2009), 633–661.

BAILEY, Thomas: Annals of Nottinghamshire. History of the county of Nottingham, including the borough. Volume II. London 1853.

BAILIFF, Ian K.; FRENCH, Charles A.; SCARRE, Chris J.: Application of luminescence dating and geomorphological analysis to the study of landscape evolution, settlement and climate change on the Channel Island of Herm. In: Journal of Archaeological Science 41 (2014), 184–208.

BAKER, Chris J.: Wind engineering—past, present and future. In: Journal of wind engineering and industrial aerodynamics 95 (2007), 843–870.

BALEN, Matthys: Beschryvinge der stad Dordrecht, vervatende haar begin, opkomst, toeneming, en verdere stant ... Volume 2. Dordrecht 1677.

BANERJEE, Debabrata; MURRAY, Andrew S.; FOSTER, Ian D. L.: Scilly Isles, UK: optical dating of a possible tsunami deposit from the 1755 Lisbon earthquake. In: Quaternary Science Reviews 20 (2001), 715–718.

BANKOFF, Greg: Cultures of disaster: society and natural hazard in the Philippines. London 2003.

BANKOFF, Greg: The historical geography of disaster: 'vulnerability' and 'local knowledge' in western discourse. In: BANKOFF, Greg; FRERKS, Georg; HILHORST, Dorothea (Eds.): Mapping vulnerability: disasters, development & people. Abingdon 2004, 25–36.

BANKOFF, Greg: The 'English Lowlands' and the North Sea basin system: a history of shared risk. In: Environment and History 35 (2013), 3–37.

BARBER, Luke: The pottery. In: BARBER, Luke; PRIESTLEY-BELL, Greg (Eds.): Medieval adaptation, settlement and economy of a coastal wetland. The evidence from around Lydd, Romney Marsh, Kent. Oxford 2008, 113–173.

BARKER, Juliet R. V.: The tournament in England 1100-1400. Woodbridge 1986.

BARNEY, Stephen A.; LEWIS, W. J.; BEACH, Jennifer A.; BERGHOF, Oliver (Eds.): The etymologies of Isidore of Seville. Cambridge 2006.

Barredo, José I.: Major flood disasters in Europe: 1950-2005. In: Natural Hazards 42 (2007), 125–148.
Barrowclough, David A.; Hallam, John: The Devil's footprints and other folklore: local legend and archaeological evidence in Lancashire. In: Folklore 119, 1 (2008), 93–102.
Bartholy, Judit; Pongrácz, Rita; Pattantyús-Ábrahám, Margit: European cyclone track analysis based on ECMWF ERA-40 data sets. In: International Journal of Climatology 26 (2006), 1517–1527.
Bartlett, Robert: The natural and the supernatural in the Middle Ages. Cambridge 2008.
Bartlett, Robert: Why can the dead do such great things? Saints and worshippers from the Martyrs to the Reformation. Princeton and Oxford 2013.
Batho, Edith C; Husbands, H Winifred (Eds.): The Chronicles of Scotland compiled by Hector Boece. Volume II. Edinburgh 1941.
Baxter, Peter J.: The east coast Big Flood, 31 January–1 February 1953: a summary of the human disaster. In: Philosophical Transactions of the Royal Society A 363 (2005), 1293–1312.
Beaumont James, Tom; Gerrard, Christopher M.: Clarendon: Landscape of Kings. Macclesfield 2007.
Beaumont James, Tom; Robinson, Anne M.: Clarendon Palace: the history and archaeology of a medieval palace and hunting lodge near Salisbury, Wiltshire. London 1988.
Behringer, Wolfgang: Climatic change and witch-hunting: the impact of the Little Ice Age on mentalities. In: Climatic Change 43, 1 (1999), 335–351.
Bekar, Cliff T.; Reed, Clyde G.: Open fields, risk, and land divisibility. In: Explorations in Economic History 40 (2003), 308–325.
Bellenden, John (Ed.): The History and Chronicles of Scotland: Written in Latin by Hector Boece. Volume II. Edinburgh 1821.
Bellovacensi, Vincentino: Speculum quadruplex, naturale, doctrinale, morale, historale. Douai 1624.
Beloe, Edward M.: Freebridge Marshland Hundred and the making of Lynn. In: Norfolk Archaeology 12 (1895), 311–334.
Benson, Don G.; Evans, John G.; Williams, G. H.; Darvill, Timothy; David, A.: Excavations at Stackpole Warren, Dyfed. In: Proceedingsof the Prehistoric Society 56 (1990), 179–245.
Beresford, Guy: The medieval clay-land village: excavations at Goltho and Barton Blount. SMA Monograph No. 6. London 1975.
Beresford, Guy: Climatic change and its effect upon the settlement and desertion of medieval villages in Britain. In: Delano Smith, Catherine; Parry, Martin (Eds.): Consequences of climatic change. Nottingham 1981, 30–39.
Beresford, Maurice; Hurst, John G.: Deserted Medieval Villages. London 1971.
BGS: Cwmllynfell, South Wales. 2018. www.earthquakes.bgs.ac.uk/earthquakes/recent_events/20180217142554.html. Accessed 26 September 2018.
Birdshall, Jean; Newhall, Richard A. (Eds.): The chronicle of Jean de Venette. New York 1953.
Blake, Hugo; Egan, Geoff; Hurst, John; New, Elizabeth: From popular devotion to resistance and revival in England: the cult of the holy name of Jesus and the Reformation. In: Gaimster, David; Gilchrist, Roberta (Eds.): The Archaeology of Reformation 1480-1580. Leeds 2003, 175–203.
Blanchard-Boehm, R. Denise; Cook, M. Jeffrey: Risk communication and public education in Edmonton, Alberta, Canada on the 10th anniversary of the 'Black Friday' tornado. In: International Research in Geographical and Environmental Education 13, 1 (2004), 38–54.
Bliss, William H. (Ed.): Calendar of entries in the Papal registers relating to Great Britain and Ireland. Petitions to the Pope. Volume I. London 1896.
Bliss, William H.; Twemlow, Jessie A. (Eds.): Calendar of entries in the Papal registers relating to Great Britain and Ireland. Papal letters. A.D. 1362-1404, Volume IV. London 1902.

Bliss, William H.; Twemlow, Jessie A. (Eds.): Calendar of entries in the Papal registers relating to Great Britain and Ireland. Papal letters. A.D. 1404-1415, Volume VI. London 1904a.

Bliss, William H.; Twemlow, Jessie A. (Eds.): Calendar of entries in the Papal registers relating to Great Britain and Ireland. Papal letters. A.D. 1398-1404, Volume V. London 1904b.

Blockley, Kevin: The Anglo-Saxon churches of Canterbury archaeologically reconsidered. Unpublished MPhil Thesis, Durham University 2000.

Bloomfield, Morton W.: Piers Plowman as a fourteenth-century apocalypse. New Brunswick 1962.

Boccaccio, Giovanni: The Decameron, or ten days' entertainment. London 1825.

Bonanno, George A.; Galea, Sandro; Bucciarelli, Angela; Vlahov, David: What predicts psychological resilience after disaster? The role of demographics, resources, and life stress. In: Journal of Consulting and Clinical Psychology 75, 5 (2007), 671–682.

Bond, Edward A. (Ed.): Chronica monasterii de Melsa. Volume III. London 1868.

Bond, James; Weller, John B.: The Somerset barns of Glastonbury Abbey. In: Abrams, Lesley; Carley, James P. (Eds.): The archaeology and history of Glastonbury Abbey. Essays in honour of the ninetieth birthday of C. A. Ralegh Radford. Woodbridge 1991, 57–87.

Borg, Alan: Some medieval war memorials. In: Fernie, Eric; Crossley, Paul (Eds.): Medieval architecture and its context: studies in honour of Peter Kidson. London and Ronceverte 1990, 1–8.

Boulter, Stuart: Leiston Old Abbey Chapel (LCS 002; SAM 21404); building recording & excavation. Unpublished Suffolk County Council Archaeological Service Report No. 2008/90 2008.

Boxer, Charles R. (Ed.): South China in the sixteenth century. London 1953.

Boyer, Marjorie N.: Medieval French bridges: a history. Cambridge 1976.

Boyer, Marjorie N.: Moving ahead with the fifteenth century: new ideas in bridge construction at Orleans. In: Rupert Hall, Alfred; Smith, Norman (Eds.): History of Technology: Volume 6. London 1981, 1–22.

Bradley, Richard: The past in prehistoric societies. London and New York 2002.

Brady, Niall: The Gothic barn of England: icon of prestige and authority. In: Bradford Smith, Elizabeth; Wolfe, Michael (Eds.): Technology and resource use in medieval Europe: cathedrals, mills and mines. Aldershot 1997, 76–105.

Brady, Niall: Agricultural buildings. In: Gerrard, Christopher M.; Gutiérrez, Alejandra (Eds.): The Oxford handbook of later medieval archaeology in Britain. Oxford 2018, 259–272.

Braid, Robert: Economic behaviour, markets and crises: the English economy in the wake of plague and famine in the 14th century. In: Cavaciocchi, Simonetta (Ed.): Le Interazioni fra Economia e Ambiente Biologico Nell'Europa Preindustriale Secc. XIII-XVIII. Firenze 2010, 335–372.

Brandon, Peter F.: Late-medieval weather in Sussex and its agricultural significance. In: Transactions of the Institute of British Geographers 54 (1971), 1–17.

Braudel, Fernand: The Mediterranean and the Mediterranean World in the Age of Philip II. New York 1972.

Brázdil, Rudolf, Pfister, Christian; Wanner, Heinz; von Storch, Hans; Luterbacher, Jürg: Historical climatology in Europe – the state of the art. In: Climate Change 70 (2005), 363–430.

Brennan, Michael: Condolence books: language and meaning for Hillsborough and Diana. In: Death Studies 32, 4 (2008), 326–351.

Brian, Anthea: The allocation of strips in Lammas Meadows by the casting of lots. In: Landscape History 21, 1 (1999), 43–58.

Brie, Friedrich W. D. (Ed.): The brut or the chronicles of England. Part I. London 1906.

Brill, Willem G. (Ed.): Rijmkroniek van Melis Stoke, eerste deel. Utrecht 1885.

Britten, John (Ed.): Remaines of Gentilisme and Judaisme, 1686–87. London 1881.

Broadberry, Stephen N.; Campbell, Bruce M. S.; Klein, Alexander; Overton, Mark; van Leeuwen, Bas: British economic growth 1270-1870. Cambridge 2015.

Brook, Timothy: The troubled empire: China in the Yuan and Ming dynasties. Cambridge and London 2010.

Brooks, Rob: Land east of The Granary, Stoke Road Clare. Unpublished 2015.

Brown, Antony G.; Cooper, Lynden; Salisbury, Chris R.; N., Smith David: Late Holocene channel changes of the Middle Trent: channel response to a thousand-year flood record. In: Geomorphology 39 (2001), 69–82.

Brown, David M.: Irish and English dendrochronology. In: Vernacular Architecture 41 (2010), 119–122.

Brown, Peter J.: Coasts of catastrophe? The incidence and impact of aeolian sand on British medieval coastal communities. In: European Journal of Post Classical Archaeologies 5 (2015), 127–148.

Brown, Peter J.: The contribution of archaeology to the study of historical disasters. In: O'Sullivan, Rebecca; Marini, Christina; Binnberg, Julia (Eds.): Archaeological approaches to breaking boundaries: interaction, integration and division: proceedings of the Graduate Archaeology at Oxford conferences 2015–2016. Oxford 2017, 239–245.

Brundage, James A.: Sex, law and marriage in the Middle Ages. Aldershot 1993.

Buckland, Paul C.; Amorosi, Thomas; Barlow, Lisa K.; Dugmore, Andrew J.; Mayewski, Paul A.; McGovern, Thomas H.; Ogilvie, Astrid E. J.; Sadler, Jonathan P.; Skidmore, Peter: Bioarchaeological and climatological evidence for the fate of Norse farmers in medieval Greenland. In: Antiquity 70 (1996), 88–96.

Bullock, James H. (Ed.): The Norfolk poriton of the chartulary of the priory of St. Pancras of Lewes. Norfolk Record Society Volume XII. Norwich 1939.

Büntgen, Ulf; Frank, David C.; Nievergelt, Daniel; Esper, Jan: Summer temperature variations in the European Alps, A.D. 755-2004. In: Journal of Climate 19 (2006), 5606–5623.

Büntgen, Ulf; Kyncl, Tomá; Ginzler, Christian; Jacks, David S.; Esper, Jan; Tegel, Willy; Heussner, Karl-Uwe; Kyncl, Josef: Filling the Eastern European gap in millennium-long temperature reconstructions. In: Proceedings of the National Academy of Sciences 110, 5 (2013), 1773–1778. DOI: 10.1073/pnas.1211485110.

Büntgen, Ulf; Tegel, Willy; Nicolussi, Kurt; McCormick, Michael; Frank, David; Trouet, Valerie; Kaplan, Jed O; Herzig, Franz; Karl-Uwe, Heussner; Wanner, Heinz; Luterbacher, Jürg; Esper, Jan: 2500 years of European climate variability and human susceptibility. In: Science 331, 6017 (2011), 578–582.

Burt, Caroline: Edward I and the governance of England 1272-1307. Cambridge 2013.

Burt, Stephen D.; Mansfield, D. A.: The Great Storm of 15-16 October 1987. In: Weather 43 (1988), 90–110.

Cambridge, Eric: The masons and building works of Durham priory, 1339-1539. Ph.D. thesis, Durham University 1992.

Camden, William: Britain, or a chorographicall description of the most flourishing kingdomes, England, Scotland, and Ireland, and the ilands adjoyning. London 1610.

Campbell, Bruce M. S.: The agrarian problem in the early fourteenth century. In: Past and Present 188 (2005), 3–70.

Campbell, Bruce M. S.: Three centuries of English crops yields, 1211-1491. 2007. www.cropyields.ac.uk. Accessed 09 December 2016.

Campbell, Bruce M. S.: Agriculture in Kent in the High Middle Ages. In: Sweetinburgh, Sheila (Ed.): Later Medieval Kent 1220-1540. Woodbridge 2010a, 25–53.

Campbell, Bruce M. S.: Nature as historical protagonist: environment and society in pre-industrial England. In: Economic History Review 63, 2 (2010b), 281–314.

CAMPBELL, Bruce M. S.: Grain yields on English demesnes after the Black Death. In: BAILEY, Mark; RIGBY, Stephen H. (Eds.): Town and countryside in the age of the Black Death: essays in honour of John Hatcher. Turnhout 2012, 121–174.

CAMPBELL, Bruce M. S.: Global climates, the 1257 mega-eruption of Samalas Volcano, Indonesia, and the English food crisis of 1258. In: Transactions of the Royal Historical Society 27 (2016a), 87–121.

CAMPBELL, Bruce M. S.: The great transition. climate, disease and society in the late-medieval world. Cambridge 2016b.

CAMPBELL, Bruce M. S.; BARRY, Lorraine: The population geography of Great Britain c.1290: a provisional reconstruction. In: BRIGGS, Chris; KITSON, Peter M.; THOMPSON, Stephen J. (Eds.): Population, welfare and economic change in Britain 1290-1834. Woodbridge 2014, 43–78.

CAMPBELL, Bruce M. S.; OVERTON, Mark: A new perspective on medieval and early modern agriculture: six centuries of Norfolk farming c.1250-c.1850. In: Past & Present 14, 1 (1993), 38–105.

CAMPBELL, Greg: Ampullae, re-imbursed: a formal analysis of medieval "shell-shaped" lead-alloy pilgrim ampullae. In: Peregrinations 5, 2 (2015), 97–134.

CANINO, Glorisa; BRAVO, Milagros; RUBIO-STIPEC, Maritza; WOODBURY, Michael: The impact of disaster on mental health: prospective and retrospective analyses. In: International Journal of Mental Health 19, 1 (1990), 51–69.

CANTERBURY ARCHAEOLOGICAL TRUST: The New Romney and Greatstone First Time Sewerage Scheme pipeline watching brief, New Romney, Kent, archaeological assessment report. Unpublished Report 2010.

CAPLE, Chris: The apotropaic symbolled threshold to Nevern Castle - Castell Nanhyfer. In: The Archaeological Journal 169, 1 (2012), 422–452.

CARION, Johannes: Prognosticatio vnd erklerung der grossen wesserung, Auch anderer erschrockenlichenn wuerckungen. Leipzig 1521.

CARMASSI, Claudia; AKISKAL, Hagop S; BESSONOV, Daniel; MASSIMETTI, Gabriele; CALDERANI, Enrico; STRATTA, Paolo; ROSSI, Alessandro; DELL'OSSO, Liliana: Gender differences in DSM-5 versus DSM-IV-TR PTSD prevalence and criteria comparison among 512 survivors to the L'Aquila earthquake. In: Journal of Affective Disorders 160 (2014), 55–61.

CATLOS, Brian A.: Muslims of medieval Latin Christendom, c. 1050-1614. Cambridge 2014.

CHAMPION, Matthew: The graffiti inscriptions of St Mary's Church, Troston. In: Proceedings of the Suffolk Institute of Archaeology and History 43, 2 (2014), 235–258.

CHAMPION, Matthew: Medieval graffiti: the lost voices of England's churches. London 2015a.

CHAMPION, Matthew: Medieval ship graffiti in English churches: interpretation and function. In: The Mariner's Mirror 101, 3 (2015b), 343–350.

CHARALAMBOUS, Katerina; BRUGGEMAN, Adriana; BAKIRTZIS, Nikolas; LANGE, Manfred A.: Historical flooding of the Pedieos River in Nicosia, Cyprus. In: Water History 8 (2016), 191–207.

CHARDONNENS, László S.; HEBING, Rosanne: Two charms in a late medieval English manuscript at Nijmegen University Library. In: The Review of English Studies 62, 254 (2010), 181–192.

CHENEY, Christopher R.; JONES, Michael: A handbook of dates. 2nd edition. Cambridge 2000.

CHESTER, David K.; DUNCAN, Angus M.; SANGSTER, Heather: Human responses to eruptions of Etna (Sicily) during the late-pre-industrial era and their implications for present-day disaster planning. In: Journal of Volcanology and Geothermal Research 225–226 (2012), 65–80.

CHRISTIE, Richard C (Ed.): Annales Cestrienses: or chronicle of the Abbey of S. Werburg, at Chester. London 1887.

CLARIDGE, Jordan; LANGDON, John: Storage in medieval England: the evidence from purveyance accounts, 1295-1349. In: The Economic History Review 64, 4 (2011), 1242–1265.

CLARKE, Helen; CARTER, Alan: Excavations in King's Lynn, 1963-1970. London 1977.

CLARKE, Michèle L.; RENDELL, Helen M.: The impact of North Atlantic storminess on western European coasts: a review. In: Quaternary International 195, 1–2 (2009), 31–41.

COCKAYNE, Thomas O.: Leechdoms, wortcunning and starcraft of early England. London 1864.

COLEMAN, Russel: The archaeology of burgage plots in Scottish medieval towns: a review. In: Proceedings of the Society of Antiquaries of Scotland 134 (2004), 281–323.

COMFORT, Louise K.: Cities at risk: Hurricane Katrina and the drowning of New Orleans. In: Urban Affairs Review 41, 4 (2006), 501–516.

CONNELL, Brian; JONES, Amy G.; REDFERN, Rebecca; WALKER, Don: A bioarchaeological study of medieval burials on the site of St Mary Spital. MOLA Monograph 60. London 2012.

CONYBEARE, Edward: Highways and byways in Cambridge and Ely. London 1910.

COOPER, Jago; PEROS, Matthew: The archaeology of climate change in the Caribbean. In: Journal of Archaeological Science 37 (2010), 1226–1232.

COOPER, Lynden; RIPPER, Susan: The medieval Trent bridges at Hemington Fields, Castle Donington (SK 459 302). In: Transactions of the Leicestershire Archaeological and Historical Society 68 (1994), 153–161.

CORNER, David J.; SCOTT, A. Brian; SCOTT, William W.; WATT, Donald E. R. (Eds.): Scotichronicon by Walter Bower. Volume IV. Edinburgh 1998.

COSGEL, Metin M.: Scattering and contracts in medieval agriculture: challenges ahead. In: The Journal of Economic History 50, 3 (1990), 663–668.

COX, J. Charles: Churchwardens' accounts from the fourteenth century to the close of the seventeenth century. London 1913.

CROWSON, Andy; LANE, Tom; PENN, Kenneth; TRIMBLE, Dale: Anglo-Saxon Settlement on the Siltland of Eastern England. Lincolnshire Archaeology and Heritage Reports Series 7. Sleaford 2005.

CULL, Robert J.; HOFFMAN, Philip T.; HUGHSON, Eric: New evidence for an old controversy: scattered landholdings and open fields. In: California Institute of Technology Social Science Working Paper 788 (1992), 1–28.

CUNLIFFE, Barry; MUNBY, Julian: Excavations at Portchester Castle: medieval, the inner bailey. Volume IV. London 1985.

CURTIS, Daniel R.; SOENS, Tim; VAN BAVEL, Bas: History and the social sciences: shock therapy with medieval economic history as the patient. In: Social Science History 40, 4 (2016), 751–774.

DALY, Aoife: The dendrochronological dating of timber crossings in West Jutland, Denmark. In: Journal of Wetland Archaeology 6, 1 (2006), 19–48.

DANIELL, Christopher: Death and burial in Medieval England 1066-1550. London 1997.

DAVIES, Althea L.: Upland agriculture and environmental risk: a new model of upland land-use based on high spatial-resolution palynological data from West Affric, NW Scotland. In: Journal of Archaeological Science 34 (2007), 2053–2063.

DAVIS, Alfred H. (Ed.): William Thorne's Chronicle of Saint Augustine's Abbey Canterbury. Oxford 1934.

DAWES, Michael C. B. (Ed.): Register of Edward the Black Prince, Part II, Duchy of Cornwall, A. D. 1351-1365. London 1931.

DAWES, Michael C. B. (Ed.): Register of Edward the Black Prince, Part IV, England, A. D. 1351-1365. London 1933.

DAWES, Michael C. B.; CHAPMAN, J. B. W. (Eds.): Calendar of inquisitions post mortem and other analogous documents preserved in the Public Record Office. Edward III. Volume XII. London 1938.

DAWSON, Alastair G.; HICKEY, Kieran; MAYEWSKI, P. A.; NESJE, Atle: Greenland (GISP2) ice core and historical indicators of complex North Atlantic climate changes during the fourteenth century. In: The Holocene 17, 4 (2007), 427–434.

De Blasi, Giovanna: Wells Cathedral: the crossing and the central tower. A new hypothesis on the constructional sequence and dating of the central tower. In: D'Alya, Dina; Fodde, Enrico (Eds.): Structural analysis of historic construction: preserving safety and significance: Volume 1. London 2008, 199–206.

de Boer, G.: Spurn Head: Its history and evolution. In: Transactions and papers (Institute of British Geographers) 34 (1964), 71–89.

de Kraker, Adriaan M. J.: Flooding in river mouths: human caused or natural events? five centuries of flooding events in the SW Netherlands, 1500–2000. In: Hydrology and Earth System Sciences 19 (2015), 2673–2684.

Dean, John; Hill, Nick: Burn marks on buildings: accidental or deliberate? In: Vernacular Architecture 45 (2014), 1–15.

Defoe, Daniel: The storm: or, a collection of the most remarkable casualties and disasters, which happen'd in the late dreadful tempest, both by sea and land, on Friday the twenty-sixth of November, seventeen hundred and three. 2nd edition. London 1704.

Delaygue, Gilles; Bard, Edouard: An Antarctic view of Beryllium-10 and solar activity for the past millennium. In: Climate Dynamics 36, 11 (2010), 2201–2218.

Denton, Jeffrey: From the foundation of Vale Royal Abbey to the Statute of Carlisle: Edward I and ecclesiastical patronage. In: Cross, Peter R.; Lloyd, Simon D. (Eds.): Proceedings of the Newcastle Upon Tyne Conference 1991, Thirteenth Century England IV. Woodbridge 1992, 123–137.

Dhoop, Thomas: Shaped by ships and storms: a maritime archaeology of medieval Winchelsea. Ph.D. thesis, University of Southampton 2016.

Dickinson, Tania M.: Symbols of protection: the significance of animal ornamented shields in early Anglo-Saxon England. In: Medieval Archaeology 49 (2005), 109–163.

Diemar, Hermann: Chroniken von Hessen und Waldeck, Erster Band. Marburg 1909.

Donkin, R. A.: The site changes of mediaeval Cistercian monasteries. In: Geography 44, 4 (1959), 251–258.

Dotterweich, Markus; Schmitt, Anne; Schmidtchen, Gabriele; Bork, Hans-Rudolf: Quantifying historical gully erosion in northern Bavaria. In: Cantena 50 (2003), 135–150.

Draper, Gillian; Meddens, Frank: The sea and the marsh: The medieval Cinque Port of New Romney. Pre-Construct Archaoelogy Monograph No. 10. London 2009.

Draper, Gillian M.: Writing English, French and Latin in the fifteenth century: a regional perspective. In: Clark, Linda (Ed.): The fifteenth century VII: conflicts, consequences and the Crown in the late Middle Ages. Woodbridge 2007, 213–235.

Duffy, Eamon: The stripping of the altars: traditional religion in England c. 1400 - c. 1580. New Haven and London 1992.

Dugdale, William: The history of imbanking and drayning of divers fenns and marshes, both in foreign parts and in this kingdom and of the improvements thereby. London 1662.

Dugmore, Andrew J.; Church, Mike J.; Mairs, Kerry-Anne; McGovern, Thomas H.; Perdikaris, Sophia; Vésteinsson, Orri: Abandoned farms, volcanic impacts, and woodland management: revisiting Þjórsárdalur, the "Pompeii of Iceland". In: Arctic Anthropology 44, 1 (2007), 1–11.

Durkan, John: The Great Fire at Glasgow Cathedral. In: The Innes Review 26 (1975), 89–92.

Dyer, Christopher: Dispersed settlements in medieval England. A case study of Pendock, Worcestershire. In: Medieval Archaeology 34 (1990), 97–121.

Dyer, Christopher: Standards of living in the later Middle Ages: Social change in England c. 1200-1520. Revised edition edition. Cambridge 1998.

Dyer, Christopher: Gardens and garden produce in the later Middle Ages. In: Woolgar, Christopher M.; Serjeantson, Dale; Waldron, Tony (Eds.): Food in medieval England: diet and nutrition. Oxford 2006, 27–40.

DYER, Christopher: Poverty and its relief in late medieval England. In: Past & Present 216 (2012), 41–78.
DYER, Christopher: Medieval peasant buildings 1250-1550: documents and historical significance. In: ALCOCK, Nat; MILES, Dan (Eds.): The Medieval Peasant House in Midland England. Oxford 2013, 105–118.
DYER, Christopher: Rural living 1100-1540. In: GERRARD, Christopher M.; GUTIÉRREZ, Alejandra (Eds.): The Oxford handbook of later medieval archaeology in Britain. Oxford 2018, 193–209.
DYER, Christopher: Recovering from catastrophe: How medieval society in England coped with disasters. In: GERRARD, Christopher M.; FORLIN, Paolo; BROWN, Peter J. (Eds.): Waiting for the End of the World? New Perspectives on Natural Disasters in Medieval Europe. London 2020, 218–238.
EGAN, Geoff; PRITCHARD, Frances: Dress Accessories 1150-1450. 2nd edition. Woodbridge 2002.
ELDREDGE, Niles; GOULD, Stephen J.: Punctuated equilibria: an alternative to phyletic gradualism. In: SCHOPF, Thomas J. M. (Ed.): Models in Paleobiology. San Francisco 1972, 82–115.
ELLIS, Henry (Ed.): The chronicle of John Harding. London 1812.
ELLIS, Henry (Ed.): Chronica Johanis de Oxenedes. London 1859.
ENTICOTT, Delyth: Medieval pendant from Gleaston. In: Contrebis 21 (1996), 10.
EVERETT, Linzi: Managed retreat, Trimley Marshes, Trimley St. Martin archaeological evaluation, monitoring and excavation archive report TYN 073. Unpublished Report 2000.
EVERETT, Linzi; BOULTER, Stuart: An assessment on the Aldham Mill Hill, Hadleigh Excavations (HAD 059). Unpublished Report 2010.
FABYAN, Robert: The new chronicles of England and France. London 1811.
FALK, Ann-Britt: My home is my castle: protection against evil in medieval times. In: ANDRÉN, Anders; JENNBERT, Kristina; RAUDVERE, Catharina (Eds.): Old Norse religion in long-term perspectives: origins, changes and interactions. Lund 2006, 200–205.
FAVIER, René; GRANET-ABISSET, Anne-Marie: Society and natural risks in France, 1500-2000: changing historical perspectives. In: JANKU, Andrea; SCHENK, Gerrit Jasper; MAUELSHAGEN, Franz (Eds.): Historical disasters in context: science, religion and politics. London 2012, 101–136.
FEILDEN, Bernard M.: Restorations and repairs after World War II. In: ATHERTON, Ian; FERNIE, Eric; HARPER-BILL, Christopher; SMITH, Hassell (Eds.): Norwich Cathedral: church, city and diocese, 1096-1996. London 1996, 728–744.
FENG, Shuidong; TAN, Hongzhuan; BENJAMIN, Abuaku; WEN, Shiwu; LIU, Aizhong; ZHOU, Jia; LI, Shuoqi; YANG, Tubao; ZHANG, Yanbo; LI, Xinhua; LI, Gangqian: Social support and posttraumatic stress disorder among flood victims in Hunan, China. In: Annals of Epidemiology 17, 10 (2007), 827–833.
FENOALTEA, Stefano: Risk, transaction costs, and the organization of medieval agriculture. In: Explorations in Economic History 13 (1976), 129–151.
FENWICK, Valerie: The Graveney boat: a tenth-century find from Kent. Oxford 1978.
FERNANDEZ, Ana; BLACK, John; JONES, Mairwen; WILSON, Leigh; SALVADOR-CARULLA, Luis; ASTELL-BURT, Thomas; BLACK, Deborah: Flooding and mental health: a systematic mapping review. In: PLoS ONE 10, 4 (2015), e0119929.
FERNÁNDEZ, Jesús; MOSHENSKA, Gabriel; IRIARTE, Eneko: Archaeology and climate change: evidence of a flash-flood during the LIA in Asturias (NW Spain) and its social consequences. In: Environmental Archaeology 24, 1 (2019), 38–48. DOI: 10.1080/14614103.2017.1407469.
FERRER, Vincent: Sermones aestivales. Antwerp 1572.
FIGLIUOLO, Bruno: Medieval earthquakes in Italy: Perceptions and reactions. In: GERRARD, Christopher M.; FORLIN, Paolo; BROWN, Peter J. (Eds.): Waiting for the End of the World? New Perspectives on Natural Disasters in Medieval Europe. London 2020, 43–61.

FILOTAS, Bernadette: Pagan survivals, superstitions and popular cultures in early medieval pastoral literature. Toronto 2005.

FINUCANE, Ronald C.: Miracles and pilgrims: popular beliefs in medieval England. Basingstoke 1995.

FLETCHER, Taleyna: Archaeological excavations at Wisbech Castle - a community archaeology project. Unpublished Report 2010.

FORLIN, Paolo; GERRARD, Christopher M.: The archaeology of earthquakes: the application of adaptive cycles to seismically-affected communities in late medieval Europe. In: Quaternary International 446 (2017), 95–108.

FOSTER, Megan H.: Pilgrimage through the pages: pilgrim's badges in late medieval devotional manuscripts. Ph.D. thesis, University of Illinois at Urbana-Champaign 2011.

FOX, Harold S. A.: Outfield cultivation in Devon and Cornwall: a reinterpretation. In: HAVINDEN, Michael (Ed.): Husbandry and marketing in the south-west 1500-1800: Exeter Papers in Economic History 8. Exeter 1973, 19–38.

FOX, Harold S. A.: The alleged transformation from two-field to three-field systems in medieval England. In: Economic History Review 4 (1986), 526–548.

FRENCH, Katherine L.: The good women of the parish. Philadelphia 2008.

FUGLESANG, Signe Horn: Viking and medieval amulets in Scandinavia. In: Fornvännen - Journal of Swedish Antiquarian Research 84 (1989), 15–27.

FYFE, Ralph M.; BROWN, Anthony G.; RIPPON, Stephen J.: Characterising the late prehistoric, 'Romano-British' and medieval landscape, and dating the emergence of a regionally distinct agricultural system in South West Britain. In: Journal of Archaeological Science 31 (2004), 1699–1714.

GADE, Martin; KOHLUS, Jörn; KOST, Cornelia: SAR imaging of archaeological sites on intertidal flats in the German Wadden Sea. In: Geosciences 7 (2017), 1–14.

GALADINI, Fabrizio; HINZEN, Klaus-G.; STIROS, Stathis: Archaeoseismology: methodological issues and procedure. In: Journal of Seismology 10 (2006), 395–414.

GALBRAITH, Vivian H. (Ed.): The anonimalle chronicle 1333 to 1381. Manchester 1927.

GALLOWAY, James A.: Storm flooding, coastal defence and land use around the Thames estuary and tidal river c.1250–1450. In: Journal of Medieval History 35, 2 (2009), 171–188.

GALLOWAY, James A.: 'Tempests of weather and great abundance of water': the flooding of the Barking marshes in the later middle ages. In: DAVIES, Matthew; GALLOWAY, James A. (Eds.): London and beyond. Essays in honour of Derek Keene. London 2012, 67–83.

GALLOWAY, James A.; POTTS, Jonathan S.: Marine flooding in the Thames Estuary and tidal river c.1250 –1450: impact and response. In: Area 39, 3 (2007), 370–379.

GAO, Chaochao; ROBOCK, Alan; AMMANN, Caspar: Volcanic forcing of climate over the past 1500 years: an improved ice core-based index for climate models. In: Journal of Geophysical Research: Atmospheres 113. DOI: 10.1029/2008JD010239.

GARCIA, Michael: Medieval medicine, magic and water. In: Peregrinations 1, 3 (2005), 1–13.

GARDINER, Mark: Medieval settlement and society in the Broomhill area and excavations at Broomhill Church. In: EDDISON, Jill; GREEN, Christopher (Eds.): Romney Marsh: evolution, occupation and reclamation 1988, 112–127.

GARDINER, Mark: Old Romney: An examination of the evidence for a lost Saxo-Norman port. In: Archaeologia Cantiana 114 (1994), 329–345.

GARDINER, Mark: Medieval farming and flooding in the Brede Valley. In: EDDISON, Jill (Ed.): Romney Marsh: The debatable ground: Monograph 41. Oxford 1995, 127–137.

GARDINER, Mark: Shipping and trade between England and the continent during the eleventh century. In: HARPER-BILL, Christopher (Ed.): Anglo-Norman Studies, Proceedings of the Battle Conference 1999: Volume XXII. Woodbridge 2000, 71–93.

Gardiner, Mark: The late medieval 'antediluvian' landscape of Walland Marsh. In: Long, Antony; Hipkin, Stephen; Clarke, Helen (Eds.): Romney Marsh: coastal and landscape change through the ages: Oxford University School of Archaeology Monograph 56. Oxford 2002, 101–120.

Gardiner, Mark: Stacks, barns and granaries in early and high medieval England: crop storage and its implications. In: Guirado, Alfonso V.; Bianchi, Giovanna; Quirós, Juan A. (Eds.): Horrea, barns and silos: storage and incomes in early medieval Europe: Documentos de Arqueología Medieval 5. Vitoria-Gasteiz 2013, 23–38.

Gardiner, Mark; Hartwell, Barrie: Landscapes of failure: the archaeology of flooded wetlands at Titchell and Thornham (Norfolk), and Broomhill (East Sussex). In: Journal of Wetland Archaeology 6, 1 (2006), 137–160.

Gardner, John: Sir Gawain and the Green Knight. Chicago and London 2011.

Gardner, Rhodri: Former Smyth Works, Church Street, Peasenhall, PSH 012, a report on the archaeological excavations, 2004. Unpublished Report 2005.

Garrioch, David: 1666 and London's fire history: a re-evaluation. In: The Historical Journal 59, 2 (2016), 319–338.

Gerrard, Christopher M.: Paisaje y señorío: la casa conventual de Ambel (Zaragoza). Arqueología, arquitectura e historia de las Órdenes militares del Temple y del Hospital. Zaragoza 2003.

Gerrard, Christopher M.: Not all archaeology is rubbish: the elusive life histories of three artefacts from Shapwick, Somerset. In: Costen, Michael (Ed.): People and places: essays in honour of Mick Aston. Oxford 2007a, 166–180.

Gerrard, Christopher M.: The Shapwick project, Somerset. a rural landscape explored. Leeds 2007b.

Gerrard, Christopher M.: Contest and co-operation: strategies for medieval and later irrigation along the upper Huecha valley, Aragón, north-east Spain. In: Water History 3 (2011), 3–28.

Gerrard, Christopher M.; Forlin, Paolo; Brown, Peter J. (Eds.): Waiting for the End of the World? New Perspectives on Natural Disasters in Medieval Europe. London 2020.

Gerrard, Christopher M.; Forlin, Paolo; Froude, Melanie; Petley, David; Gutiérrez, Alejandra; Treasure, Edward; Milek, Karen; Oliveira, N'zinga: The Archaeology of a Landslide: Unravelling the Azores Earthquake Disaster of 1522 and its Consequences. In: European Journal of Archaeology 24, 3 (2021), 388–411.

Gerrard, Christopher M.; Petley, David N.: A risk society? Environmental hazards, risk and resilience in the later Middle Ages in Europe. In: Natural Hazards 69, 1 (2013), 1051–1079.

Gerritsen, Herman: What happened in 1953? The Big Flood in the Netherlands in retrospect. In: Philosophical Transactions of the Royal Society A 363 (2005), 1271–1291.

Getz, Faye M. (Ed.): Healing and society in medieval England: a middle English translation of the pharmaceutical writings of Gilbertus Anglicus. Madison 1991.

Giardini, Domenico; Grünthal, Gottfried; Shedlock, Kaye M.; Zhang, Peizhen: The GSHAP global seismic hazard map. In: Lee, William H. K.; Kanamori, Hiroo; Jennings, Paul; Kisslinger, Carl (Eds.): International Handbook of Earthquake & Engineering Seismology. Amsterdam 2003, 1233–1239.

Giardini, Domenico; Woessner, Jochen; Danciu, Laurentiu; Crowley, Helen; Cotton, Fabrice; Grünthal, Gottfried; Pinho, Rui; Valensise, Gianluca: SHARE European seismic hazard map for peak ground acceleration, 10% exceedance probabilities in 50 years. 2013. www.share-eu.org/node/90. Accessed 13 October 2017.

Gibbons, Alfred (Ed.): Ely episcopal records. Lincoln 1891.

Gidney, Louisa J.: Recognising catastrophic cattle-mortality events in England and their repercussions. In: Gerrard, Christopher M.; Forlin, Paolo; Brown, Peter J. (Eds.): Waiting for the End of the World? New Perspectives on Natural Disasters in Medieval Europe. London 2020, 328–344.

GILBERT, John T. (Ed.): Chartularies of St. Mary's Abbey, Dublin: with the register of its house at Dunbrody, and Annals of Ireland. Volume II. London 1884.

GILCHRIST, Roberta: Norwich Cathedral: a biography of the north transept. In: Journal of the British Archaeological Association 151, 1 (1998), 107–136.

GILCHRIST, Roberta: Magic for the dead? The archaeology of magic in later medieval burials. In: Medieval Archaeology 52 (2008), 119–159.

GILCHRIST, Roberta: The intimacy of death: interpreting gender and the life course in medieval and early modern burials. In: BEAUDRY, M. C.; SYMONDS, J. (Eds.): Interpreting the Early Modern World: Transatlantic Perspectives. New York 2011, 159–173.

GILCHRIST, Roberta: Medieval life: archaeology and the life course. Woodbridge 2012.

GILES, John Allen (Ed.): Chronicon Angliae Petriburgense. London 1845.

GILES, John Allen (Ed.): The Venerable Bede's Ecclesiastical history of England also the Anglo-Saxon chronicle. London 1894.

GILES, John Allen (Ed.): Matthew Paris's English history. From the year 1235 to 1273. Volume II. London 1953.

GILES, John Allen (Ed.): Matthew Paris's English history. From the year 1235 to 1273. Volume III. London 1954.

GILL, David: Land adjacent to the Greyhound Inn, Flempton FMP 020. Unpublished Report 1997.

GILL, David: Chevington Hall, CHV 021, archaeological evaluation report. Unpublished Report 2004.

GILL, David: Ballingdon Bridge, Sudbury. BCB 012, A report on the archaeological monitoring, 2002. Unpublished Archaeological Monitoring Report, Suffolk County Council Archaeological Service 2007.

GILL, David: Clare Priory Church new extension Clare Priory CLA 037. Unpublished Archaeological Monitoring Report, Suffolk County Council Archaeological Service 2013.

GILL, David; CASS, S.; CRAVEN, J.: Greater Gabbard Wind Farm onshore works, Sizewell Wents, Leiston LCS 148. Unpublished Report 2013.

GIMBEL, Lena M.: Bawdy badges and the Black Death: late medieval apotropaic devices against the spread of the plague. Master's thesis, University of Louisville 2012.

GLASER, Rüdiger; BRÁZDIL, Rudolf; PFISTER, Christian; DOBROVOLNÝ, Petr; VALLVÉ, Mariano Barriendos; BOKWA, Aanita; CAMUFFO, Dario; KOTYZA, Oldrich; LIMANÓWKA, Danuta; RÁCZ, Lajos; RODRIGO, Fernando S: Seasonal temperature and precipitation fluctuations in selected parts of Europe during the sixteenth century. In: Climatic Change 43, 1 (1999), 169–200.

GLASSCOCK, Robin E. (Ed.): The Lay Subsidy of 1334. Records of Social and Economic History New Series II. Oxford 1975.

GOLB, Norman: Nature et destination du monument Hebraique Decouvert a Rouen. In: Proceedings of the American Academy for Jewish Research 48 (1981), 101–182.

GOLLOP, Adrian G.: Archaeological excavation at St Martin's Field, New Romney, Kent: draft assessment report. Unpublished Report 2008.

GOOD, Jonathan: The cult of Saint George in medieval England. Woodbridge 2009

GOSDEN, Chris; MARSHALL, Yvonne: The cultural biography of objects. In: World Archaeology 31, 2 (1999), 169–178.

GOTTSCHALK, Maria K. E.: Stormvloeden en rivieroverstromingen in Nederland (I de periode vóór 1400). Assen 1971.

GRAHAM-CAMPBELL, James; BATEY, Colleen E.: Vikings in Scotland: an archaeological survey. Edinburgh 1998.

GREEN, Charles: East Anglian coast-line levels since Roman times. In: Antiquity 35, 137 (1961), 21–28.

GREEN, Royston D.: Soils of Romney Marsh. Soil Survey of Great Britain England and Wales Bulletin No. 4. Harpenden 1968.

Greene, J. Patrick: Norton Priory: the archaeology of a medieval religious house. Cambridge 1989.

Grieve, Hilda: The great tide: the story of the 1953 flood disaster in Essex. Chelmsford 1959.

Griffiths, David; Philpott, Robert A.; Egan, Geoff: Meols: the archaeology of the north Wirral coast. Oxford University School of Archaeology Monographs, Volume 68. Oxford 2007.

Griffiths, Hywel; Salisbury, Eurig; Tooth, Stephen: 'May God place a bridge over the River Tywi': interrogating flood perceptions and memories in Welsh medieval poetry. In: Endfield, Georgina H.; Veale, Lucy (Eds.): Cultural histories, memories and extreme weather: a historical geography perspective. London and New York 2017, 93–111.

Grimson, John: The Channel coasts of England. London 1978.

Gross, Anthony; Butcher, Andrew: Adaptation and investment in the age of the great storms: agricultural policy on the manors of the principal lords of the Romney Marshes and the marshland fringe, c. 1250–1320. In: Eddison, Jill (Ed.): Romney Marsh: The debatable ground: Monograph 41. Oxford 1995, 107–117.

Grove, L. R. A.; Warhurst, A.: A thirteenth-century kiln site at Ashford. In: Archaeologia Cantiana 65 (1952), 183–187.

Grundmann, Herbert; Heimpel, Hermann (Eds.): Die Schriften des Alexander von Roes. Weimar 1949.

Gunderson, Lance: Ecological and human community resilience in response to natural disasters. In: Ecology and Society 15, 2 (2010), 18.

Gurney, David: Archaeological finds in Norfolk in 2004. In: Norfolk Archaeology 44, 4 (2005), 736–750.

Guttormsen, Torgrim S.: Natural catastrophes and landscape archaeology: landslide history in the Romerike landscape in south-eastern Norway. In: Childis, Konstantinos; Lund, Julie; Prescott, Christopher (Eds.): Facets of archaeology: essays in honour of Lotte Hedeager on her 60th birthday: number 10 InOslo Arkeologiske serie. Oslo 2008, 451–464.

Guy, S.: Heigham Bridge historic building report, Norfolk County Council. 2011. web.archive.org/web/20170401125733/hbsmrgateway2.esdm.co.uk/norfolk/DataFiles/Docs/AssocDoc35921.pdf. Accessed 01 June 2023.

Hadler, H.; Vött, A.; Newig, J.; Emde, K.; Finkler, C.; Fischer, P.; Willershäuser, T.: Geoarchaeological evidence of marshland destruction in the area of Rungholt, present-day Wadden Sea around Hallig Südfall (North Frisia, Germany), by the Grote Mandrenke in 1362 AD. In: Quaternary International 473, A (2018), 37–54.

Hall, Derek: 'Unto yone hospitall at the tounis end': the Scottish medieval hospital. In: Tay and Fife Archaeological Journal 12 (2006), 89–105.

Hall, Edward: Hall's chronicle. London 1809.

Hall, Mark; Price, Neil: Medieval Scotland: a future for its past, ScARF summary medieval document. 2012. www.scottishheritagehub.com/sites/default/files/. Accessed 25 July 2016.

Hall, Mark A.: The cult of saints in medieval Perth: everyday ritual and the materiality of belief. In: Journal of Material Culture 16, 1 (2011), 80–104.

Hall, Mark A.: Money isn't everything: the cultural life of coins in the medieval burgh of Perth, Scotland. In: Journal of Social Archaeology 12, 1 (2012), 72–91.

Hallam, Herbert E.: Drainage techniques. In: Hallam, Herbert E. (Ed.): The agrarian history of England and Wales: Volume II. Cambridge 1988a, 497–507.

Hallam, Herbert E.: Eastern England. In: Hallam, Herbert E. (Ed.): The agrarian history of England and Wales: Volume II. Cambridge 1988b, 139–173.

Hamerow, Helena: 'Special deposits' in Anglo-Saxon settlements. In: Medieval Archaeology 50 (2006), 1–30.

Hanks, Patrick; Coates, Richard; McClure, Peter: The Oxford dictionary of family names in Britain and Ireland. Volume 1. Oxford 2016.

Hansen, Gitte: After the town burned! Use and reuse of iron and building timber in a medieval town. In: Baug, Irene; Larsen, Janicke; Mygland, Sigrid Samset (Eds.): Nordic Middle Ages—Artefacts, Landscapes and Society. Essays in Honour of Ingvild Øye on her 70th Birthday. Bergen 2015, 159–174.

Hanska, Jussi: Strategies of survival: religious responses to natural disasters in the Middle Ages. Helsinki 2002.

Hansom, Jim D.; Hall, Adrian M.: Magnitude and frequency of extra-tropical North Atlantic cyclones: a chronology from cliff-top storm deposits. In: Quaternary International 195 (2009), 42–52.

Hantemirov, Rashit M.; Gorlanova, Ludmila A.; Shiyatov, Stephen G.: Extreme temperature events in summer in northwest Siberia since AD 742 inferred from tree rings. In: Palaeogeography, Palaeoclimatology, Palaeoecology 209, 1 (2004), 155–164.

Hardy, Thomas D. (Ed.): Rymer's Foedera, 1066-1377. Volume I. London 1869.

Harper-Bill, Christopher (Ed.): The cartulary of the Augustinian Friars of Clare. Woodbridge 1991.

Harris, Sylvia: The "historia trium regum" and the medieval legend of the Magi in Germany. In: Medieval Ævum 28, 1 (1959), 23–30.

Harrison, David: The bridges of medieval England: transport and society 400-1800. Oxford 2004.

Hasegawa, Koichi: Facing nuclear risks: lessons from the Fukushima nuclear disaster. In: International Journal of Japanese Sociology 21 (2012), 84–91.

Haslett, Simon; Bryant, Edward A.: Evidence for historic coastal high-energy wave impact (tsunami?) in North Wales, United Kingdom. In: Atlantic Geology 43 (2007a), 137–147.

Haslett, Simon; Bryant, Edward A.: Reconnaissance of historic (post-AD 1000) high-energy deposits along the Atlantic coasts of southwest Britain, Ireland and Brittany, France. In: Marine Geology 242, 1-3 (2007b), 207–220.

Hass, Christian H.: Northern Europe climate variations during late Holocene: evidence from marine Skagerrak. In: Palaeogeography, Palaeoclimatology, Palaeoecology 123 (1996), 121–145.

Hawkes, Christopher: Archaeological theory and method: some suggestions from the Old World. In: American Anthropologist 56, 2 (1954), 155–168.

Hayatsu, Kenji: The activity and age of Niigata-Yakeyama Volcano. Central Japan. In: Journal of Geography 103, 2 (1994), 149–165.

Haydon, Frank S. (Ed.): Eulogium (historiarum sive temporis): chronicon ab orbe condito usque ad annum domini M.CCC.LXVI. Volume III. London 1863.

Heale, Martin: Training in superstition? Monasteries and popular religion in late medieval and Reformation England. In: Journal of Ecclesiastical History 58, 3 (2007), 417–439.

Heanley, Robert M.: Lincolnshire superstitions. In: Folklore 9, 2 (1898), 186–187.

Hearne, Carrie M.; Perkins, D. R. J.; Phil, Andrews: The Sandwich Bay wastewater treatment scheme archaeological project, 1992-1994. In: Archaeologia Cantiana 115 (1995), 239–354.

Hearne, Thomas (Ed.): Walteri Hemingford, canonici de Gisseburne, historia de rebus gestis Edvardi I. Edvardi II. & Edvardi III. Volume II. Oxford 1731.

Heebøll-Holm, Thomas K.: Apocalypse then? The first crusade, traumas of war and Thomas de Marle. In: Hundahl, Kerstin; Kjæ, Lars; Lund, Niels (Eds.): Denmark and Europe in the Middle Ages, c. 1000-1525: essays in honour of Professor Michael H. Gelting. London 2016, 237–254.

Heege, Andreas: Brandstiftung! Der Einbecker Stadtbrand vom 26. Juli 1540—Archäologischer Befund und politische Hintergründe. Einbeck 2005.

Helland, Amund; Steen, Helge: Lerfaldet i Guldalen i 1345. In: Archiv for mathematik og naturvidenskab 17, 6 (1895), 1–48.

Helmus, Liesbeth M.: Het altaarstuk met de Sint Elisabethsvloed uit de Grote Kerk van Dordrecht. De oorspronkelijke plaats en de opdrachtgevers. In: Oud Holland 105, 2 (1991), 127–139.

Herdman, M. L.; Jarman, C.: An archaeological watching brief on a new gas main, New Romney, Kent 1995. Unpublished Report 1996.

HERRGOT, Marquard: Vetus disciplina monastica. Paris 1726.
HILDBURGH, Walter L.: Notes on some amulets of the three magi kings. In: Folklore 19, 1 (1908), 83–87.
HINDMARCH, Erlend; ORAM, Richard: Eldbotle; the archaeology and environmental history of a medieval rural settlement in East Lothian. In: Proceedings of the Society of Antiquaries of Scotland 142 (2012), 245–299.
HINGESTON, Francis C. (Ed.): The chronicle of England by John Capgrave. London 1858.
HINMAN, Mark; POPESCU, Elizabeth: Extraordinary inundations of the sea: excavations at Market Mews, Wisbech, Cambridgeshire. East Anglian Archaeology Monograph, Volume 142. Bar Hill 2012.
HINTON, David A.: A medieval cistern from Churchill. In: Oxoniensia 33 (1968), 66–70.
HINTON, David A.: Gold and gilt, pots and pins: possessions and people in medieval Britain. Oxford 2005.
HINZEN, Klaus-G.; SCHREIBER, Stephan; FLEISCHER, Claus; REAMER, Sharon K.; WIOSNA, Isabel: Archaeoseismic study of damage in Roman and medieval structures in the center of Cologne, Germany. In: Journal of Seismology 17, 2 (2013), 399–424.
HIRST, Sue M.; WALSH, David A.; WRIGHT, Susan M.: Bordesley Abbey II: second report on the excavations at Bordesley Abbey, Redditch, Hereford - Worcestershire. BAR British Series 111. Oxford 1983.
HOBHOUSE, Edmund (Ed.): Calendar of the register of John de Drokensford, Bishop of Bath and Wells (A.D. 1309-1329). London 1887.
HOFFMAN, Richard C.: An environmental history of medieval Europe. Cambridge 2014.
HOG, Thomas (Ed.): Adami Murimuthensis chronica. London 1846.
HOLLAND, John: Cruciana. Liverpool 1835.
HORDEN, Peregrine; PURCELL, Nicholas: The corrupting sea: a study of Mediterranean history. Oxford 2000.
HORROX, Rosemary (Ed.): The Black Death. Manchester and New York 1994.
HORSTMANN, Carl (Ed.): The three kings of Cologne: an early English translation of the "historia trium regum" by John of Hildesheim. London 1886.
HOUSE, Jonathan: Archaeological excavation at Ashend, East Barton Road, Great Barton, Suffolk. Post-excavation assessment. Unpublished Report 2017.
HOWARD, Robert E.: Nottingham University tree-ring dating laboratory results. general list. In: Vernacular Architecture 25 (1994), 36–40.
HUANG, Cheng; LI, Zhu; WANG, Meng; MARTORELL, Reynaldo: Early life exposure to the 1959-1961 Chinese famine has long term consequences. In: The Journal of Nutrition 140, 10 (2010), 1874–1878.
HUDSON, Anne: The legacy of *Piers Plowman*. In: ALFORD, John A. (Ed.): A companion to *Piers Plowman*. Berkeley and Los Angeles 1988, 251–266.
HUGGON, Martin: Medieval medicine, public health, and the medieval hospital. In: GERRARD, Christopher M.; GUTIÉRREZ, Alejandra (Eds.): The Oxford handbook of later medieval archaeology in Britain. Oxford 2018, 836–855.
HYBEL, Nils; POULSEN, Bjørn: The Danish resources c. 1000-1550. growth and recession. Leiden 2007.
I'ANSON, Edward: Account of the restoration of the Dutch Church, Austin Friars. In: Papers Read at the Royal Institute of British Architects Session 1865-66 (1866), 67–75.
IVERSON, Richard M.: Landslide triggering by rain infiltration. In: Water Resources Research 36, 7 (2000), 1897–1910.
JAEDICKE, Christian; LIED, Karstein; KRONHOLM, Kalle: Integrated database for rapid mass movements in Norway. In: Natural Hazards and Earth Systems Science 9 (2009), 469–479.

JAEDICKE, Christian; VAN DEN EECKHAUT, Miet; NADIM, Farrokh; HERVÁS, Javier; KALSNES, Bjørn; VANGELSTEN, Bjørn Vidar; SMITH, Jessica T.; TOFANI, Veronica; CIUREAN, Roxana; WINTER, Mike G.; SVERDRUP-THYGESON, Kjetil; SYRE, Egil; SMEBYE, Helge: Identification of landslide hazard and risk 'hotspots' in Europe. In: Bulletin of Engineering Geology and the Environment 73 (2014), 325–339.

JEAKE, Samuel: Charters of the Cinque Ports, two ancient towns, and their members. London 1728.

JERVIS, Ben: Assessing urban fortunes in six late medieval ports: an archaeological application of assemblage theory. In: Urban History 44, 1 (2017), 2–26.

JONES, Annette: Portals to the past: distribution patterns in stave church inscriptions. 2010. Preprint from: Seventh International Symposium on Runes and Runic Inscriptions, Oslo 2010. www.khm.uio.no/english/research/publications/7th-symposium-preprints/Jones.pdf. Accessed 26 May 2015.

JONES, Richard: Signatures in the soil: the use of pottery in manure scatters in the identification of medieval arable farming regimes. In: Archaeological Journal 161, 1 (2004), 159–188.

JONES, Richard: The medieval natural world. London 2013.

JONES, William R.: The heavenly letter in medieval England. In: Medievalia et Humanistica 6 (1975), 163–178.

JORDAN, William Chester: The Great Famine: northern Europe in the early fourteenth century. Princeton 1996.

JØRGENSEN, Jon G. (Ed.): Alsak Bolts jordebok. Oslo 1997.

JOST, Jean E.: The effects of the Black Death: the plague in fourteenth-century religion, literature and art. In: CLASSEN, Albrecht (Ed.): Death in the Middle Ages and early modern time: the material and spiritual conditions of the culture of death. Berlin 2016, 193–238.

JUNEJA, Monica; MAUELSHAGEN, Franz: Disasters and pre-industrial societies: historiographic trends and comparative perspectives. In: The Medieval History Journal 10, 1 & 2 (2007), 1–31.

JUNIUS, Hadrianus: Batavia. Leiden 1588.

KACKI, Sacha: Digging up the victims of the Black Death: A bioarchaeological perspective on the second plague pandemic. In: GERRARD, Christopher M.; FORLIN, Paolo; BROWN, Peter J. (Eds.): Waiting for the End of the World? New Perspectives on Natural Disasters in Medieval Europe. London 2020, 259–279.

KEATS-ROHAN, Katherine S. B.: 'Most securely fortified': Wallingford Castle 1071-1540. In: KEATS-ROHAN, Katherine S. B.; CHRISTIE, Neil; ROFFE, David (Eds.): Wallingford: The Castle and the Town in Context. Oxford 2015, 34–115.

KEENE, Derek: Crisis management in London's food supply, 1250–1500. In: DODDS, Ben; LIDDY, Christian D. (Eds.): Commercial activity, markets and entrepreneurs in the Middle Ages: Essays in honour of Richard Britnell. Woodbridge 2011, 45–62.

KELLEHER, Richard: Interpreting single finds in medieval England - the secondary lives of coins. In: HOLMES, Nicholas (Ed.): Proceedings of the XIV International Numismatic Congress, Glasgow: Volume II. Glasgow 2011, 1492–1499.

KELLEHER, Richard: Coins, monetisation and re-use in medieval England and Wales: new interpretations made possible by the Portable Antiquities Scheme. Ph.D. thesis, Durham University 2013.

KENDALL, Calvin B.; WALLIS, Faith (Eds.): Isidore of Seville, on the nature of things. Liverpool 2016.

KILBY, Susan: Mapping peasant discontent: trespassing on manorial land in fourteenth-century Walsham-le Willows. In: Landscape History 36, 2 (2015), 69–88.

KINTIGH, Keith W.; ALTSCHUL, Jeffrey H.; BEAUDRY, Mary C.; DRENNAN, Robert D.; KINZIG, Ann P.; KOHLER, Timothy; LIMP, W. Fredrick; MASCHNER, Herbert D. G.; MICHENER, William K.; PAUKETAT, Timothy R.; PEREGRINE, Peter; SABLOFF, Jeremy A.; WILKINSON, Tony J.; WRIGHT, Henry T.; ZEDER, Melinda: Grand challenges for archaeology. In: American Antiquity 79, 1 (2014), 5–24.

KIRBY, John L.; STEVENSON, Janet H. (Eds.): Calendar of Inquisitions Post Mortem and other analogous documents preserved in the Public Record Office. 6 to 10 Henry V (1418-1422). Volume XXI. Woodbridge 2002.

KIRK, William: Sands of Forvie. In: Discovery and Excavation Scotland 4.

KISS, Andrea: Weather and weather-related environmental hazards in medieval Hungary III: The fourteenth century. In: Medium Aevum Quotidianum 73 (2016), 5–55.

KISS, Andrea; LASZLOVSZKY, József: 14th - 16th Century Danube floods and long-term water-level changes in archaeological and sedimentary evidence in the western and central Carpathian Basin: An overview with documentary comparison. In: Journal of Environmental Geography 6, 3–4 (2013), 1–11.

KISS, Andrea; NIKOLIĆ, Zrinka: Droughts, dry spells and low water levels in medieval Hungary (and Croatia) I: the great droughts of 1362, 1474, 1479, 1494 and 1507. In: Journal of Environmental Geography 8, 1–2 (2015), 11–22.

KLEIN, Naomi: The shock doctrine: the rise of disaster capitalism. New York 2007.

KLEINHANS, Maarten G.; WEERTS, Henk J. T.; COHEN, Kim M.: Avulsion in action: reconstruction and modelling sedimentation pace and upstream flood water levels following a medieval tidal-river diversion catastrophe (Biesbosch, The Netherlands, 1421–1750 AD). In: Geomorphology 118 (2010), 65–79.

KNOX, Ronald; LESLIE, Shane (Eds.): The miracles of King Henry VI. Cambridge 1923.

KOOPMANS, Rachel: 'Water mixed with the blood of Thomas': contact relic manufacture pictured in Canterbury Cathedral's stained glass. In: Journal of Medieval History 42, 5 (2016), 535–558.

KRISTJÁNSDÓTTIR, Steinunn: The tip of the iceberg: the material of Skriðuklaustur monastery and hospital. In: Norwegian Archaeological Review 43, 1.

LAGERÅS, Per; BROSTRÖM, Anna; FREDH, Daniel; LINDERSON, Hans; BERG, Anna; BJÖRKMAN, Leif; HULTBERG, Tove; KARLSSON, Sven; LINDBLADH, Matts; MAZIER, Florence; SEGERSTRÖM, Ulf; SKÖLD, Eva: Abandonment, agricultural change and ecology. In: LAGERÅS, Per (Ed.): Environment, society and the Black Death: an interdisciplinary approach to the late-medieval crisis in Sweden. Oxford 2016, 30–68.

LAMB, Hubert; FRYDENDAHL, Knud: Historic storms of the North Sea, British Isles and northwest Europe. Cambridge 1991.

LAMOND, Jessica Elizabeth; JOSEPH, Rotimi D.; PROVERBS, David G.: An exploration of the factors affecting the long term psychological impact and deterioration of mental health in flooded households. In: Environmental Research 140 (2015), 325–334.

LATHAM, Jacob A.: Inventing Gregory "the Great": memory, authority, and the afterlives of the *Letania Septiformis*. In: Church History 84, 1 (2015), 1–31.

LATHAM, Ronald (Ed.): The travels of Marco Polo. Harmondsworth 1958.

LAVIGNE, Franck; DEGEAI, Jean-Philippe; KOMOROWSKI, Jean-Christophe; GUILLET, Sébastien; ROBERT, Vincent; LAHITTE, Pierre; OPPENHEIMER, Clive; STOFFEL, Markus; VIDAL, Céline M.; SURONO; PRATOMO, Indyo; WASSMER, Patrick; HAJDAS, Irka; HADMOKO, Danang Sri; DE BELIZAL, Edouard: Source of the great A.D. 1257 mystery eruption unveiled, Samalas volcano, Rinjani Volcanic Complex, Indonesia. In: Proceedings of the National Academy of Sciences 110, 42 (2013), 16742–16747. DOI: 10.1073/pnas.1307520110.

LEE, Jennifer: Medieval pilgrims' badges in rivers: the curious history of a non-theory. In: Open Access Journal of Art Historiography 11 (2014), 1–11.

LEE, John S.: Decline and growth in the late medieval fenland: the examples of Outwell and Upwell. In: Proceedings of the Cambridge Antiquarian Society 104 (2015), 137–147.

LEE, Sanguen; VINK, Karina: Assessing the vulnerability of different age groups regarding flood fatalities: case study in the Philippines. In: Water Policy 17 (2015), 1045–1061.

LEES, D. J.; SELL, S.: Excavation of a Medieval dwelling at Pennard. In: Journal of the Gower Society 34 (1983), 44–52.

LESLIE, John; LESLIE, William: Parish of Fintray. In: ANON (Ed.): The New Statistical Account of Scotland: Volume XII. Edinburgh and London 1845, 166–174.

LEWIS, Carenza: Disaster recovery: new archaeological evidence for the long-term impact of the 'calamitous' fourteenth century. In: Antiquity 90, 351 (2016), 777–797.

LEWIS, Carenza: Reconstructing the impact of 14th-century demographic disasters on late medieval rural communities in England. In: GERRARD, Christopher M.; FORLIN, Paolo; BROWN, Peter J. (Eds.): Waiting for the End of the World? New Perspectives on Natural Disasters in Medieval Europe. London 2020, 298–327.

LEWIS, W. J.: Agobard of Lyons (9th century): on hail and thunder. 2001. https://sourcebooks.fordham.edu/source/Agobard-OnHailandThunder.asp. Accessed 11 October 2018.

LILLEY, Keith D.: Royal authority and urban formation: King Edward I and the making of his "new towns". In: LARKHAM, Peter J.; CONZEN, Michael P. (Eds.): Shapers of urban form: Explorations in morphological agency. London 2014, 27–45.

LINKLATER, Andrew: An archaeological assessment during the removal of petrol and oil tanks, during the demolition of a garage at Sussex Road, New Romney, Kent. Unpublished Report 2003.

LINKLATER, Andrew: New Romney sewer scheme, Shepway Kent. Unpublished Report 2004.

LLOYD, Virginia; DEAN, John; WESTWOOD, Jennifer: Burn marks as evidence of apotropaic practices in houses, farm buildings and churches in post-medieval East Anglia. In: WALLIS, Robert J.; LYMER, Kenneth (Eds.): A Permeability of Boundaries? New Approaches to the Archaeology of Art, Religion and Folklore: BAR International Series, Volume 936 2001, 57–70.

LLOYD-HUGHES, Benjamin; SAUNDERS, Mark A.: A drought climatology for Europe. In: International Journal of Climatology 22 (2002), 1571–1592.

LOCKWOOD, William B.: A manuscript in the British Library and Flemish-Dutch and Low German accounts of the life and miracles of Saint Barbara. In: Bulletin of the John Rylands Library 36, 1 (1953), 23–37.

LUARD, Henry R. (Ed.): Bartholomaei de Cotton Monachi Norwicensis Historia Anglicana; (A.D. 449-1298). London 1859.

LUARD, Henry R. (Ed.): Annales Monastici. Volume III. London 1866.

LUARD, Henry R. (Ed.): Annales Monastici. Volume IV. London 1869.

LUARD, Henry R. (Ed.): Matthæi Parisiensis Monachi Sancti Albani, Chronica Majora A.D. 1216 - A.D. 1239. Volume III. London 1876.

LUARD, Henry R. (Ed.): Matthæi Parisiensis Monachi Sancti Albani, Chronica Majora, A.D. 1248 - A.D. 1259. Volume V. London 1880.

LUARD, Henry R. (Ed.): Flores Historiarum. Volume II. London 1890.

LÜBKEN, Uwe: Chasing a ghost? environmental change and migration in history. In: Global Environment 9 (2012), 5–24.

LUCAS, Henry S.: The Great European famine of 1315, 1316, and 1317. In: Speculum 5, 4 (1930), 343–377.

LUMBY, J. Rawson (Ed.): Polychronicon Ranulph Higden Monachi Cestrensis; English Translations of John Trevisa and of an unknown writer of the fifteenth century. Volume VIII. London 1882.

LUND, Julie: Banks, borders and bodies of water in a Viking Age mentality. In: Journal of Wetland Archaeology 8 (2008), 53–72.

LUTERBACHER, Jürg; WERNER, Johannes P.; SMERDON, J. E.; FERNÁNDEZ-DONADO, L.; GONZÁLEZ-ROUCO, F. J.; BARRIOPEDRO, D.; LJUNGQVIST, F. C.; BÜNTGEN, Ulf; ZORITA, E; WAGNER, S.; ESPER, J.; MC-CARROLL, D.; TORETI, A.; FRANK, D.; JUNGCLAUS, J. H.; BARRIENDOS, M.; BERTOLIN, C.; BOTHE, O.; BRÁZDIL, R.; CAMUFFO, D.; DOBROVOLNÝ, P.; GAGEN, M.; GARCÍA-BUSTAMANTE, E.; GE, Q.; GÓMEZ-NAVARRO, J. J.; GUIOT, J.; HAO, Z.; HEGERL, G. C.; HOLMGREN, K.; KLIMENKO, V. V.; MARTÍN-

CHIVELET, J.; PFISTER, Christian; ROBERTS, N.; SCHINDLER, A.; SCHURER, A.; SOLOMINA, O.; VON GUNTEN, L.; WAHL, E.; WANNER, H.; WETTER, Oliver; XOPLAKI, E.; YUAN, N.; ZANCHETTIN, D.; ZHANG, H.; ZEREFOS, C.: European summer temperatures since Roman times. In: Environmental Research Letters 11, 2. DOI: 10.1088/1748-9326/11/2/024001.

LUTHER, Martin: D. Martin Luthers Werke: Kritische Gesamtausgabe. Weimar 1912.

MACDONALD, Neil: On epigraphic records: a valuable resource in reassessing flood risk and long-term climate variability. In: Environmental History 12 (2007), 136–140.

MACLEOD, Mindy; MEES, Bernard: Runic Amulets and Magic Objects. Woodbridge 2006.

MAGUIRE, Henry: Magic and Money in the Early Middle Ages. In: Speculum 72, 4 (1997), 1037–1054.

MALDEN, Henry E. (Ed.): The Victoria County History of the County of Surrey. Volume III. London 1911.

MANN, Michael E.; ZHANG, Zhihua; RUTHERFORD, Scott; BRADLEY, Raymond S.; HUGHES, Malcolm K.; SHINDELL, Drew; AMMANN, Caspar; FALUVEGI, Greg; NI, Fenbiao: Global signatures and dynamical origins of the Little Ice Age and Medieval Climate Anomaly. In: Science 326, 5957 (2009), 1256–1260.

MARGETTS, Andrew: An archaeological evaluation of land at Rolfe Lane, New Romney, Kent. Unpublished Report 2010.

MARSTON, John M.: Archaeological markers of agricultural risk management. In: Journal of Anthropological Archaeology 30 (2011), 190–205.

MARTIN, David; MARTIN, Barbara: New Winchelsea, Sussex: A Medieval Port Town. Field Archaeology Unit Monograph No. 2, Institute of Archaeology, University College London. Great Dunham 2004.

MARTIN, Geoffrey H. (Ed.): Knighton's Chronicle 1337-1396. Oxford 1995.

MARVIN, Julia: Cannibalism as an aspect of famine in two English chronicles. In: CARLIN, Martha; ROSENTHAL, Joel T. (Eds.): Food and eating in medieval Europe. London 1998, 73–86.

MASSÉ, Guillaume; ROWLAND, Steven J.; SICRE, Marie-Alexandrine; JACOB, Jeremy; JANSEN, Eystein; BELT, Simon T.: Abrupt climate changes for Iceland during the last millennium: Evidence from high resolution sea ice reconstructions. In: Earth and Planetary Science Letters 269, 3-4 (2008), 565–569.

MAUELSHAGEN, Franz: Disaster and political culture in Germany since 1500. In: JANKU, Andrea; SCHENK, Gerrit Jasper; MAUELSHAGEN, Franz (Eds.): Historical disasters in context: science, religion and politics. London 2012, 41–75.

MAXWELL, Herbert (Ed.): The Chronicle of Lanercost 1272-1346. Glasgow 1913.

MAXWELL LYTE, Henry C. (Ed.): Calendar of the Patent Rolls, Edward I, A.D. 1281-1292. Volume II. London 1893.

MAXWELL LYTE, Henry C. (Ed.): Calendar of the Close Rolls, Edward III, A.D. 1334-1338. London 1895a.

MAXWELL LYTE, Henry C. (Ed.): Calendar of the Patent Rolls, Edward I, A.D. 1292-1301. Volume III. London 1895b.

MAXWELL LYTE, Henry C. (Ed.): Calendar of the Patent Rolls, Edward IV. A.D. 1461-1467. Volume XII. London 1897.

MAXWELL LYTE, Henry C. (Ed.): Calendar of the Close Rolls, Edward III, A.D. 1337-1339. London 1900a.

MAXWELL LYTE, Henry C. (Ed.): Calendar of the Patent Rolls, Edward III, A.D. 1340-1343. Volume V. London 1900b.

MAXWELL LYTE, Henry C. (Ed.): Calendar of the Patent Rolls, Edward III, A.D. 1345-1348. Volume VII. London 1903.

MAXWELL LYTE, Henry C. (Ed.): Calendar of the Close Rolls, Edward III, A.D. 1360-1364. Volume XI. London 1909a.

MAXWELL LYTE, Henry C. (Ed.): Calendar of the Patent Rolls, Henry VI, A.D. 1446-1452. Volume V. London 1909b.

Maxwell Lyte, Henry C. (Ed.): Calendar of the Patent Rolls, Edward III, A.D. 1361-1364. Volume XII. London 1912.

Maxwell Lyte, Henry C. (Ed.): Calendar of the Fine Rolls, Edward III, A.D. 1337-1347. Volume V. London 1915.

Maynor, Ashley R.: Response to the unthinkable: collecting and archiving condolence and temporary memorial materials following public tragedies. In: Decker, Emy N.; Townes, Jennifer A. (Eds.): Handbook of research on disaster management and contingency planning in modern libraries. Hershey 2016, 582–624.

McCloskey, Donald N.: The open fields of England: rent, risk, and the rate of interest, 1300-1815. In: Galenson, David W. (Ed.): Markets in history: Economic studies of the past. Cambridge 1989, 5–51.

McCloskey, Donald N.: The prudent peasant: new findings on open fields. In: The Journal of Economic History 51, 2 (1991), 343–355.

McCloskey, Donald N.; Nash, John: Corn at interest: the extent and cost of grain storage in medieval England. In: The American Economic Review 74, 1 (1984), 174–187.

McIlvenna, Una: Ballads of Death and Disaster: The Role of Song in Early Modern News Transmission. In: Spinks, Jennifer; Zika, Charles (Eds.): Disaster, Death and the Emotions in the Shadow of the Apocalypse, 1400–1700: Palgrave Studies in the History of Emotions. London 2016, 275–294.

McIntosh, Marjorie K.: Local responses to the poor in late medieval and Tudor England. In: Continuity and Change 3, 2 (1988), 209–245.

McRoberts, David: The Medieval Church of St Andrews. Glasgow 1976.

Meager, Richard: Lynn PM7 Kings Lynn Norfolk, Archaeological Desk Based Assessment. Unpublished Report 2007.

Meddens, Frank; Wooldridge, Kevin: An assessment of archaeological excavations on the site of Southlands School, Fairfield Road, New Romney, Kent. Unpublished Report 2002.

Meeson, Bob: Ritual marks and graffiti: curiosities or meaningful symbols? In: Vernacular Architecture 36 (2005), 41–48.

Megaw, Arthur H. S.: Archaeology in Cyprus, 1957. In: Archaeological Reports (The Society for the Promotion of Hellenic Studies) 43–50.

Menotti, Francesco; Jennings, Benjamin; Gollnisch-Moos, Hartmut: 'Gifts for the gods': lake-dwellers' macabre remedies against floods in the Central European Bronze Age. In: Antiquity 88, 340 (2014), 456–469.

Merrifield, Ralph: The Archaeology of Ritual and Magic. London 1987.

Michell, Robert; Forbes, Nevill (Eds.): The chronicle of Novgorod 1016-1471. Camden Third Series XXV. London 1914.

Middleton, Guy D.: Bang or whimper? In: Science 361, 6408 (2018), 1204–1205.

Miles, Dan; Howard, Robert E.; Simpson, W. G.: The tree-ring dating of the tower and spire at Salisbury Cathedral, Wiltshire. Centre for Archaeology Report 44/2004. Portsmouth 2004.

Miles, Dan; Worthington, Michael; Bridge, Martin: General List, Oxford Dendrochronology Laboratory. In: Vernacular Architecture 34 (2003), 109–111.

Mitchell, Stephen A.: Witchcraft and Magic in the Nordic Middle Ages. Philadelphia and Oxford 2011.

Moorhouse, Stephen: Documentary evidence for the uses of medieval pottery: an interim statement. In: Medieval Ceramics 2 (1978), 3–21.

Moorhouse, Stephen: The ceramic contents of a 13th century timber building destroyed by fire at Pennard, Gower. In: Medieval and Later Pottery in Wales 8 (1985), 1–9.

Morgan, John Emrys: Understanding flooding in early modern England. In: Journal of Historical Geography 50 (2015), 37–50.

MORRIS, A.: Glamorgan: being an outline of its geography, history and antiquities. Newport 1907.
MUNDIN, Andy; WOOLHOUSE, Thomas: Phase 5B excavation Cedars Park, Stowmarket, Suffolk interim report. Unpublished Report 2006.
MURPHY, Peter: The English coast: a history and a prospect. London 2009.
NATIONAL ACADEMIES OF SCIENCES, ENGINEERING, AND MEDICINE: Attribution of Extreme Weather Events in the Context of Climate Change. Washington 2016.
NEWMAN, J.; BOULTER, Stuart: Moat Farmhouse, Dennington SAM 21309/DNNO 14 Archaeological Evaluation Report No 97/14. Unpublished Report 1997.
NEWTON, Kenneth C.: Thaxted in the Fourteenth Century. Chelmsford 1960.
NIBLETT, Rosalind; THOMPSON, Isobel: Alban's Buried Towns: An Assessment of St Albans' Archaeology up to AD 1600. Oxford 2005.
NILSON, Ben: Cathedral Shrines of Medieval England. Woodbridge 1998.
NOBLE, Gordon; TURNER, Joe; HAMILTON, Derek; HASTIE, Lee; KNECHT, Rick; STIRLING, Lindsey; SVEINBJARNARSON, Oskar; UPEX, Beth R; MILEK, Karen B.; STIRLING, L.: Early medieval shellfish gathering at the Sands of Forvie, Aberdeenshire: feast or famine? In: Proceedings of the Society of Antiquaries of Scotland 146 (2017), 121–152.
NORTH, Jeffrey J.: English Hammered Coinage, Early Anglo-Saxon to Henry III c. 600-1272. Volume I. London 1994.
O'DONOVAN, John (Ed.): Annals of the Four Masters. Volume III. Dublin 1856.
OGILVIE, Astrid; FARMER, Graham: Documenting the Medieval Climate. In: HULME, Mike; BARROW, Elaine (Eds.): Climates of the British Isles: present, past and future. London and New York 1997, 112–133.
OLIVER-SMITH, Anthony: Anthropological research on hazards and disasters. In: Annual Review of Anthropology 25 (1996), 303–328.
O'NEILL, Rory: Seismic adaptation in the Latin churches of Cyprus. In: GERRARD, Christopher M.; FORLIN, Paolo; BROWN, Peter J. (Eds.): Waiting for the End of the World? New Perspectives on Natural Disasters in Medieval Europe. London 2020, 62–81.
ORAM, Richard: 'The worst disaster suffered by the people of Scotland in recorded history': climate change, dearth and pathogens in the long 14th century. In: Proceedings of the Society of Antiquaries of Scotland 144 (2014), 223–244.
ORMROD, W. Mark: The crown and the English economy, 1290-1348. In: CAMPBELL, Bruce M. S. (Ed.): Before the Black Death: Studies in the "crisis" of the Early Fourteenth Century. Manchester 1991, 149–183.
ORMROD, W. Mark: The politics of pestilence: government in England after the Black Death. In: ORMROD, W. Mark; LINDLEY, Philip G. (Eds.): The Black Death in England. Stamford 1996, 147–181.
ORMROD, W. Mark; KILLICK, Helen; BRADFORD, Phil (Eds.): Early common petitions in the English Parliament, c.1290-c.1420. Camden Fifth Series Volume 52. Cambridge 2017.
OWEN, Arthur E. B. (Ed.): The records of a commission of sewers for Wiggenhall 1319-1324. Norwich 1981.
OWEN, Arthur E. B.: Mablethorpe St. Peter's and the Sea. In: Lincolnshire History and Archaeology 21 (1986), 61–62.
OWEN, Thomas (Ed.): The Fourteen Books of Palladius Rutilius Taurus Aemilianus on Agriculture. London 1807.
OXFORD ARCHAEOLOGY: Northeye Pevensey Levels East Sussex geoarchaeological assessment report. 2009. library.thehumanjourney.net/342. Accessed 12 July 2016.
PAGE, William (Ed.): The Victoria history of the county of Norfolk. Volume II. London 1906.
PAGE, William (Ed.): The Victoria history of the county of Gloucester. Volume II. London 1907.
PAGE, William (Ed.): A History of Hertfordshire. Volume IV. London 1914.
PARKER, Anthony K.; PYE, Denis: The Fenland. London 1976.

PARKER, W. S.: Hitchin, Hertfordshire. In: Harvard Engineering Journal 3, 1 (1904), 34–45.

PARKIN, E. W.: The ancient buildings of New Romney. In: Archaeologia Cantiana 88 (1973), 117–128.

PARRY, M. L.: Secular climatic change and marginal agriculture. In: Transactions of the Institute of British Geographers 64 (1975), 1–13.

PAWLEY, Simon: Lincolnshire Coastal Villages and the Sea c.1300 - c.1600: Economy and Society. Ph.D. thesis, University of Leicester 1984.

PEACOCK, Edward (Ed.): Instructions for Parish Priests by John Mirk. London 1868.

PEDERSEN, Sarah: Piety and Charity in the Painted Glass of Late Medieval York. In: Northern History 36, 1 (2000), 33–42.

PENN, Simon A. C.; DYER, Christopher: Wages and Earnings in Late Medieval England: Evidence from the Enforcement of the Labour Laws. In: The Economic History Review 43, 3 (1990), 356–376.

PERTZ, Georg H. (Ed.): Monumenta Germaniae Historica, Annales, chronica et historiae aevi Carolini et Saxonici, Tomus IV. Hannover 1841.

PERTZ, Georg H. (Ed.): Monumenta Germaniae Historica, Annales aevi Suevici, Tomus XVI. Hannover 1859.

PERTZ, Georg H. (Ed.): Monumenta Germaniae Historica, Annales aevi Suevici, Tomus XVII. Hannover 1861.

PERTZ, Georg H. (Ed.): Monumenta Germaniae Historica, Chronica aevi Suevici, Tomus XXIII. Hannover 1874.

PEVSNER, Nikolaus: The buildings of England: North west and south Norfolk. London 1962.

PEVSNER, Nikolaus: The buildings of England: Suffolk. 2nd edition. Harmondsworth 1974.

PEVSNER, Nikolaus: The buildings of England: Hertfordshire. 2nd edition. New Haven and London 1977.

PEVSNER, Nikolaus; WILLIAMSON, Elizabeth: The buildings of England: Buckinghamshire. 2nd edition. London 1994.

PFISTER, Christian: Winter Severity in Europe: The Fourteenth Century. In: Climatic Change 34 (1996), 91–108.

PFISTER, Christian: "The Monster Swallows You". Disaster Memory and Risk Culture in Western Europe, 1500-2000. In: RCC Perspectives 2011, 1 (2011), 1–22.

PFISTER, Christian: When Europe was burning: the multi-season mega-drought of 1540 and arsonist paranoia. In: SCHENK, Gerrit Jasper (Ed.): Historical disaster experiences: towards a comparative and transcultural history of disasters across Asia and Europe. Cham 2017, 155–185.

PIGGOTT, Stuart: The sources of Geoffrey of Monmouth II. The Stonehenge story. In: Antiquity 15, 60 (1941), 305–319.

PLATT, Colin: The homestead moat: security or status? In: The Archaeological Journal 167, 1 (2010), 115–133.

PLATT, Colin: Archaeology and the Agrarian Crisis in England, 1315–22. In: Medieval Archaeology 56, 1 (2012), 292–295.

POLLMANN, Judith: Memory in early modern Europe, 1500-1800. Oxford 2017.

POWELL, Sue: All Saints' Church, North Street, York: Text and Image in the Pricke of Conscience Window. In: MORGAN, Nigel (Ed.): Prophecy, Apocalypse and the Day of Doom, Proceedings of the 2000 Harlaxton Symposium. Donington 2004, 292–316.

PRIBYL, Kathleen: Farming, Famine and Plague: The impact of climate in late medieval England. Cham 2017.

PRIDGEON, Elizabeth E.: Saint Christopher Wall Paintings in English and Welsh Churches c. 1250-c. 1500. Ph.D. thesis, University of Leicester 2010.

PRIDGEON, Ellie: National and International Trends in Hampshire Churches: A Chronology of St Christopher Wall Painting (c. 1240 to c. 1530). In: Proceedings of the Hampshire Field Club & Archaeological Society 68 (2013), 85–118.

Priestley-Bell, Greg: An Archaeological Evaluation of Land at Church Road, New Romney, Kent. Unpublished Report 1999.
Puhvel, Martin: Snow and mist in Sir Gawain and the Green Knight - portents of the otherworld? In: Folklore 89, 2 (1978), 224–228.
Putnam, Bertha H. (Ed.): Proceedings Before the Justices of the Peace in the Fourteenth and Fifteenth Centuries, Edward III to Richard III. London 1938.
Quarantelli, Enrico L.: Disaster studies: an analysis of the social historical factors affecting the development of research in this area. In: International Journal of Mass Emergencies and Disasters 5, 3 (1987), 285–310.
Quine, Christopher P.: Damage to trees and woodlands in the storm of 15-16 October 1987. In: Weather 43, 3 (1988), 114–118.
Radford, Ursula M.: The Wax Images Found in Exeter Cathedral. In: The Antiquaries Journal 29, 3-4 (1949), 164–168.
Raine, James (Ed.): Reginaldi Monachi Dunelmensis Libellus De Admirandis Beati Cuthberti Virtutibus. London 1835.
Raine, James (Ed.): The historians of the Church of York and and its archbishops. Volume I. London 1879.
Raithby, John (Ed.): The Statutes of the Realm. Volume I. London 1810.
Rajendran, C. P.; Rajendran, Kusala; Jaishri, Sanwal; Sandiford, Mike: Archeological and historical database on the medieval earthquakes of the central Himalaya: ambiguities and inferences. In: Seismological Research Letters 84, 6 (2013), 1–11.
Randall, Henry J.; Rees, William (Eds.): The Storie of the Lower Borowes. Cardiff 1932.
Randolph-Hingeston, Francis Charles (Ed.): The register of Edmund Stafford, (A.D. 1395-1419); an index and abstract of its contents. London 1886.
RCAHMW (The Royal Commission on Ancient and Historic Monuments in Wales): An Inventory of the ancient monuments in Glamorgan, Medieval Non-defensive Secular Monuments. Volume 3 part 2. Cardiff 1982.
RCHM: Second Report of the Royal Commission on Historical Manuscripts. London 1874.
RCHME: A Guide to Saint Albans Cathedral. 2nd edition. London 1982.
Reeves, William (Ed.): The Life of St Columba. Dublin 1857.
Reitsma, H. J.; Ponsen, A.: The Leiden disaster of 1807. In: Icon 13 (2007), 1–18.
Renshaw, Layla; Powers, Natasha: The archaeology of post-medieval death and burial. In: Post-Medieval Archaeology 50, 1 (2016), 159–177.
Reser, Joseph P.: The experience of natural disasters: psychological perspectives and understandings. In: John, Lidstone; M., Dechano Lisa; P., Stoltman Joseph (Eds.): International perspectives on natural disasters: occurrence, mitigation, and consequences: Advances in Natural and Technological Hazards Research, Volume 21. Dordrecht 2007, 369–384.
Rhodes, John: The Severn flood-plain at Gloucester in the medieval and early modern periods. In: Transactions of the Bristol and Gloucestershire Archaeological Society 124 (2006), 9–36.
Richardson, Gary: The prudent village: risk pooling institutions in medieval English agriculture. In: The Journal of Economic History 65, 2 (2005), 386–413.
Riede, Felix: Towards a science of past disasters. In: Natural Hazards 71, 1 (2014), 335–362.
Rigby, Stephen H.: Urban decline in the later Middle Ages: some problems in interpreting the statistical data. In: Urban History 6 (1979), 46–59.
Riley, Henry T. (Ed.): Gesta Abbatum Monasterii Sancti Albani, 1290-1349. Volume II. London 1867.
Riley, Henry T. (Ed.): Gesta Abbatum Monasterii Sancti Albani, A.D. 1349-1411. Volume III. London 1869.

RIPPER, Susan; COOPER, Lynden P.: The Hemington Bridges: the excavation of three medieval bridges at Hemington Quarry near Castle Donington, Leicestershire. Leicester Archaeology Monograph 16. Leicester 2009.

RIPPON, Stephen: The historic landscapes of the Severn Estuary Levels. In: Archaeology in the Severn Estuary 11 (2000a), 119–135.

RIPPON, Stephen: The transformation of coastal wetlands: exploitation and management of marshland landscapes in north west Europe during the Roman and Medieval Periods. Oxford 2000b.

RIPPON, Stephen: Adaptation to a changing environment: the response of marshland communities to the late medieval 'crisis'. In: Journal of Wetland Archaeology 1 (2001), 15–39.

RIPPON, Stephen: Infield and outfield: the early stages of marshland colonisation and the evolution of medieval field systems. In: LANE, T.; COLES, John (Eds.): Through wet and dry: essays in honour of David Hill. Sleaford 2002a, 54–70.

RIPPON, Stephen: Romney Marsh: Evolution of the Historic Landscape and its Wider Significance. In: LONG, Antony; HIPKIN, Stephen; CLARKE, Helen (Eds.): Romney Marsh: Coastal and Landscape Change through the Ages 2002b, 84–100.

RIPPON, Stephen: Making the most of a bad situation? Glastonbury Abbey, Meare, and the medieval exploitation of wetland resources in the Somerset Levels. In: Medieval Archaeology 48 (2004), 91–130.

RIPPON, Stephen: 'Uncommonly rich and fertile' or 'not very salubrious'? The perception and value of wetland landscapes. In: Landscapes 10, 1 (2009), 39–60.

RIPPON, Stephen: Human impact on the coastal wetlands of Britain in the medieval period. In: THOEN, Erik; BORGER, Guus J.; SOENS, Tim; DE KRAKER, Adriaan M J; TYS, Dries; LIES, Vervaet; WEERTS, Henk J. T. (Eds.): Landscapes or seascapes? The history of the coastal environment in the North Sea area reconsidered: number 13 InCORN Publication Series. Turnhout 2013, 333–351.

RIPPON, Stephen; FYFE, Ralph; BROWN, Anthony G.: Beyond villages and open fields: the origins and development of a historic landscape characterised by dispersed settlement in south-west England. In: Medieval Archaeology 50 (2006), 31–70.

RIPPON, Stephen; WAINWRIGHT, Adam; SMART, Chris: Farming regions in medieval England: the archaeobotanical and zooarchaeological evidence. In: Medieval Archaeology 58 (2014), 195–255.

ROBBINS, Katherine J.: Balancing the scales: exploring the variable effects of collection bias on data collected by the Portable Antiquities Scheme. In: Landscapes 14, 1 (2013), 54–72.

ROBERTS, Charlotte; COX, Margaret: Health & disease in Britain: from prehistory to the present day. Stroud 2003.

ROFFEY, Simon: Medieval leper hospitals in England: an archaeological perspective. In: Medieval Archaeology 56, 1 (2012), 203–233.

ROGERSON, Andrew: Excavations on Fuller's Hill, Great Yarmouth. In: WADE-MARTINS, Peter (Ed.): Norfolk, East Anglian Archaeology Report No. 2. Gressenhall 1976, 131–246.

ROHR, Christian: Man and Natural Disaster in the Late Middle Ages: The Earthquake in Carinthia and Northern Italy on 25 January 1348 and its Perception. In: Environment and History 9 (2003), 127–149.

ROHR, Christian: Writing a Catastrophe: Describing and Constructing Disaster Perception in Narrative Sources from the Late Middle Ages. In: Historical Social Research 32, 3 (2007), 88–102.

ROHR, Christian: Floods of the Upper Danube River and Its Tributaries and Their Impact on Urban Economies (c. 1350–1600): The Examples of the Towns of Krems/Stein and Wels (Austria). In: Environment and History 19 (2013), 133–148.

ROKOENGEN, Kåre; JESPERSEN, Martin Nørman; KLEIV, Rolf Arne; SÆTERBØ, Einar: The 1345 slide and flood disaster in the Gauldalen valley, Mid-Norway: A new interpretation. In: Norsk Geografisk Tidsskrift 55, 2 (2001), 57–70.

ROSSER, John: Excavations at Saranda Kolones, Paphos, Cyprus, 1981-1983. In: Dumbarton Oaks Papers 39 (1985), 81–97.

ROSSER, John: Crusader Castles of Cyprus. In: Archaeology 39, 4 (1986), 40–47.

ROSSER, John: Archaeological and literary evidence for the destruction of "Saranda Kolones" in 1222. In: Cyprus Research Centre Annual Review 30 (2004), 39–50.

ROUTLEDGE, C. F.: Roman foundations at St Pancras, Canterbury. In: Archaeologia Cantiana 14 (1882), 103–108.

ROWE, Michael: The Storm of 16th October 1987 and a brief comparison with three other historic gales in southern England (1362, 1662, 1703). In: Journal of Meteorology 13 (1988), 148–155.

RUAS, Marie-Pierre; BOUBY, Laurent; PY, Vanessa; CAZES, Jean-Paul: An 11th century A.D. burnt granary at La Gravette, south-western France: preliminary archaeobotanical results. In: Vegetation History and Archaeobotany 14 (2005), 416–426.

RYAN, William G. (Ed.): Jacobus de Voragine, The Golden Legend. Volume 1. Princeton 1993.

SAUER, Eberhard: Sacred Springs and a pervasive Roman ritual. In: Bulletin of the Association for Roman Archaeology 21 (2012), 51–55.

SAUNDERS, Ann: St Paul's, the story of the cathedral. London 2001.

SAYER, Duncan: Medieval waterways and hydraulic economics: monasteries, towns and the East Anglian fen. In: World Archaeology 41, 1 (2009), 134–150.

SCHENK, Gerrit Jasper: Historical Disaster Research. State of Research, Concepts, Methods and Case Studies. In: Historical Social Research / Historische Sozialforschung 32, 3 (2007a), 9–31.

SCHENK, Gerrit Jasper: '...prima ci fu la cagione de la mala provedenza de' Fiorentini...' disaster and 'life world'—reactions in the Commune of Florence to the flood of November 1333. In: The Medieval History Journal 10, 1 & 2 (2007b), 355–386.

SCHENK, Gerrit Jasper: 'Learning from history'? Chances, problems and limits of learning from historical natural disasters. In: KRÜGER, Fred; BANKOFF, Greg; CANNON, Terry; ORLOWSKI, Benedikt; SCHIPPER, E. Lisa F. (Eds.): Cultures and disasters: understanding cultural framings in disaster risk reduction. Abingdon 2015, 72–87.

SCHENK, Gerrit Jasper: Friend or foe? Negotiating the future on the example of dealing with the rivers Arno and Rhine in the Renaissance (ca. 1300-1600). In: BIANCA, Concetta; SALVESTRINI, Francesco (Eds.): L'Aqua Nemica: Fiumi, inondazioni e città storiche dall'antichità al contemporaneo. Spoleto 2017, 137–156.

SCHMID, Boris V.; BÜNTGEN, Ulf; EASTERDAY, W. Ryan; GINZLER, Christian; WALLØE, Lars; BRAMANTI, Barbara; STENSETH, Nils Chr.: Climate-driven introduction of the Black Death and successive plague reintroductions into Europe. In: Proceedings of the National Academy of Sciences 112, 10 (2015), 3020–3025.

SCHMIDT, Aubrey V. C. (Ed.): Piers Plowman. A Parallel-Text Edition of the A, B, C and Z Versions. Kalamazoo 2011.

SCHOENER, Gustav-Adolf: The flood of 1524: the first mass-media event in European history. In: Esoterica 9 (2007), 166–178.

SCHOFIELD, John: Landscape with Snow. In: Landscapes 2 (2009), 1–18.

SCOTT, G. A. M.: The shingle succession at Dungeness. In: Journal of Ecology 53, 1 (1965), 21–31.

SCOTT, Walter: The Minstrelsy of the Scottish Border. Volume I. Edinburgh 1807.

SCOTT-STOKES, Charity; GIVEN-WILSON, Chris (Eds.): Chronicon Anonymi Cantvariensis, The Chronicle of Anonymous of Canterbury 1346-1365. Oxford 2008.

SEARLE, William G. (Ed.): The Chronicle of John Stone, Monk of Christ Church 1415-1471. Cambridge 1902.

SELART, Anti: Livonia. Rus' and the Baltic Crusades in the Thirteenth Century. Leiden 2015.

SERJEANTSON, Dale: Farming and fishing in the Outer Hebrides AD 600 to 1700: the Udal, North Uist. Oxford 2013.

Sharma, Kirti; Self, Stephen; Blake, Stephen; Thordarson, Thorvaldur; Larsen, Gudrun: The AD 1362 Öræfajökull eruption, S.E. Iceland: Physical volcanology and volatile release. In: Journal of Volcanology and Geothermal Research 178, 4 (2008), 719–739. DOI: 10.1016/j.jvolgeores.2008.08.003.

Sharpe, Reginald R. (Ed.): Calendar of Letter-Books preserved among the archives of the Corporation of the City of London at the Guildhall. Letter Book G. Circa A.D. 1352-1374.. London 1905.

Sharples, Niall (Ed.): A Norse farmstead in the Outer Hebrides: excavations at Mound 3, Bornais, South Uist. Oxford 2005.

Sheets, Payson: Archaeological studies of disaster: their range and value. Institute of Behavioural Science. Boulder 1980.

Shennan, Ian: Impacts of sea-level rise on The Wash, United Kingdom. In: Tooley, Michael J.; Jelgersma, Saskia (Eds.): Impacts of sea-level rise on European coastal lowlands. Oxford 1992, 72–93.

Shiels, Jenny: East Haven (Panbride parish). In: Discovery and Excavation in Scotland 6 (2005), 21.

Shinners, John R.: The Veneration of Saints at Norwich Cathedral in the Fourteenth Century. In: Norfolk Archaeology 40, 2 (1988), 133–144.

Sigl, Michael; Winstrup, M.; McConnell, J. R.; Welten, K. C.; Plunkett, G.; Ludlow, Francis; Büntgen, Ulf; Caffee, M.; Chellman, N.; Dahl-Jensen, D.; Fischer, H.; Kipfstuhl, S.; Kostick, C.; Maselli, O. J.; Mekhaldi, F.; Mulvaney, R.; Muscheler, R.; Pasteris, D. R.; Pilcher, J. R.; Salzer, M.; Schüpbach, S.; Steffensen, J. P.; Vinther, B. M.; Woodruff, T. E.: Timing and climate forcing of volcanic eruptions for the past 2,500 years. In: Nature 523 (2015), 543–549.

Silvester, Robert: Abandoning the uplands: depopulation among dispersed settlements in western Britain. In: Dyer, Christopher; Jones, Richard (Eds.): Deserted villages revisited. Hatfield 2010, 140–161.

Silvester, Robert J.: The Fenland Project Number 3: Marshland and the Nar Valley, Norfolk. East Anglian Archaeology Report no. 45. Dereham 1988.

Sintubin, Manuel: Archaeoseismology: Past, present and future. In: Quaternary International 242 (2011), 4–10.

Skemer, Don C.: Binding Words: Textual Amulets in the Middle Ages. University Park 2006.

Slack, Paul: Responses to pague in early modern Europe: the implications of public health. In: Social Research 55, 3 (1988), 433–453.

Slavin, Philip: Between death and survival: Norfolk cattle, c. 1280-1370. In: Fons Luminis 1 (2009), 14–60.

Smith, Keith; Petley, David N.: Environmental Hazards: assessing risk and reducing disaster. 5th edition. London and New York 2009.

Smith, Reginald A. L.: Canterbury Cathedral Priory, A Study in Monastic Administration. Cambridge 1969.

Snoek, Godefridus J. C.: Medieval Piety from Relics to the Eucharist: A Process of Mutual Interaction. Leiden 1995.

Soens, Tim: Floods and money: funding drainage and flood control in coastal Flanders from the thirteenth to the sixteenth centuries. In: Continuity and Change 26, 3 (2011), 333–365.

Soens, Tim: Flood security in the medieval and early modern North Sea area: a question of entitlement? In: Environment and History 19 (2013), 209–232.

Sognnes, Kalle J.: These rocks were made for walking: rock art at Leirfall, Trøndelag, Norway. In: Oxford Journal of Archaeology 30, 2 (2011), 185–205.

Solanki, Sami K.; Usoskin, Ilya G.; Kromer, Bernd; Schüssler, Manfred; Beer, Juerg: 11,000 Year Sunspot Number Reconstruction. In: IGBP PAGES/World Data Center for Paleoclimatology Data Contribution Series #2005-015, NOAA/NGDC Paleoclimatology Program, Boulder CO, USA ,

Accessed 01June 2023. www.ncei.noaa.gov/pub/data/paleo/climate_forcing/solar_variability/solanki2004-ssn.txt.
SOLOPOVA, Elizabeth: Manuscripts of the Wycliffe Bible in the Bodleian and Oxford College Libraries. Liverpool 2016.
SOMERSET HISTORIC ENVIRONMENT RECORD: 33891: St Mary's Chapel, NE of Chapel Cleeve, Old Cleeve. 1984. www.somersetheritage.org.uk/record/33891. Accessed 03 November 2016.
SOMMERS, Mark: Spexhall Manor, Planning Application No. W/12973/7 SMR Ref. SPX 012. Unpublished Report 2002.
SOMMERS, Mark: Hadleigh Quarry (phase 2), Peyton Hall Farm, Hadleigh, Suffolk. Unpublished Report 2016.
SOMNER, William: A Treatise of the Roman Ports and Forts in Kent. Oxford 1693.
SOUTHALL, Humphrey R.; BURTON, Nick: GIS of the Ancient Parishes of England and Wales, 1500-1850. 2004. UK Data Service. SN: 4828, doi.org/10.5255/UKDA-SN-4828-1. Accessed 08 February 2018.
SPENCER, Brian: The Pewter Brooch. In: JARVIS, Keith S. (Ed.): Excavations in Christchurch 1969-1980. Dorchester 1983, 81–83.
SPENCER, Brian: Pilgrim Souvenirs and Secular Badges. Salisbury 1990.
SPUFFORD, Peter: Continental Coins in Late Medieval England. In: British Numismatic Journal 32 (1963), 127–139.
STAMP, Alfred E.: Calendar of Inquisitions Miscellaneous (Chancery). Volume III. London 1937.
STANDLEY, Eleanor R.: Spinning yarns: the archaeological evidence for hand spinning and its social implications, c AD 1200–1500. In: Medieval Archaeology 60, 2 (2016), 266–299.
STANDLEY, Eleanor R.: Fear, matter and miracles: Personal protection and coping with disasters through material culture c1200–1600. In: GERRARD, Christopher M.; FORLIN, Paolo; BROWN, Peter J. (Eds.): Waiting for the End of the World? New Perspectives on Natural Disasters in Medieval Europe. London 2020, 239–254.
STANKE, Carla; MURRAY, Virginia; AMLÔT, Richard; NURSE, Jo; WILLIAMS, Richard: The effects of flooding on mental health: outcomes and recommendations from a review of the literature. In: PLOS Current Disasters 1 (2012), 1–11.
STENSETH, Nils Chr.; SAMIA, Noelle I.; VILJUGREIN, Hildegunn; KAUSRUD, Kyrre Linné; BEGON, Mike; DAVIS, Stephen; LEIRS, Herwig; DUBYANSKIY, V. M.; ESPER, Jan; AGEYEV, Vladimir S.; KLASSOVSKIY, Nikolay L.; POLE, Sergey B.; CHAN, Kung-Sik: Plague dynamics are driven by climate variation. In: Proceedings of the National Academy of Sciences 103, 35 (2006), 13110–13115.
STEVENSON, Jim: The Elms, (South and East of Sainsbury's), Dymchurch Road, New Romney, Kent: Archaeological Excavations 2005. Unpublished Report 2006.
STEVENSON, Joseph (Ed.): The Church Historians of England. Volume IV. London 1856.
STEVENSON, Joseph (Ed.): Radulphi de Coggeshall Chronicon Anglicanum. London 1875.
STILGOE, John R.: Jack-o'-lanterns to Surveyors: The Secularization of Landscape Boundaries. In: Environmental Review 1, 1 (1976), 14–30.
STOLLE, Amelie; BERNHARDT, Anne; SCHWANGHART, Wolfgang; HOELZMANN, Philipp; ADHIKARI, Basanta R.; FORT, Monique; KORUP, Oliver: Catastrophic valley fills record large Himalayan earthquakes, Pokhara, Nepal. In: Quaternary Science Reviews 177 (2017), 88–103.
STONE, David: Decision making in medieval agriculture. Oxford 2005.
STOPFORD, J.: Some approaches to the archaeology of Christian pilgrimage. In: World Archaeology 26, 1 (1994), 57–72.
STOPP, Frederick John: Henry the Younger of Brunswick-Wolfenbüttel. Wild man and werwolf in religious polemics 1538-1544. In: Journal of the Warburg and Courtauld Institutes 33 (1970), 200–234.
STORM, Gustav (Ed.): Islandske Annaler indtil 1578. Christiania 1888.

STORM, Theodor: Der Schimmelreiter. Boston 1908.
STUBBS, Clare E. P.: Abraham Fleming: writer, cleric and preacher in Elizabethan and Jacobean London. Ph.D. thesis, Royal Holloway, University of London 2011.
STUBBS, William (Ed.): The Historical Works of Gervase of Canterbury. Volume II. London 1880.
SVENSSON, Eva; PETTERSSON, Susanne; NILSSON, Stefan; BOSS, Lotta; JOHANSSON, Annie: Resilience and medieval crises at five rural settlements in Sweden and Norway. In: Lund Archaeological Review 18 (2012), 89–106.
SWANSON, Robert N.: Indulgences in late medieval England: passports to paradise?. Cambridge 2007.
SWEDO, Elizabeth M.: Icelandic Church and Society, 1300-1550. Ph.D. thesis, University of Minnesota 2012.
TAIT, James (Ed.): Chronica Johannis de Reading et Anonymi Cantuariensis, 1346-1367. Manchester 1914.
TATARD, Lucile; GRASSO, Jean-Robert; HELMSTETTER, Agnés; GARAMBOIS, Stéphane: Characterization and comparison of landslide triggering in different tectonic and climatic settings. In: Journal of Geophysical Research 115 (2010), F04040.
TATTON-BROWN, Tim: St Nicholas' Romney, New Romney. In: Archaeologia Cantiana 104 (1987), 344–346.
TATTON-BROWN, Tim: The Topography of the Walland Marsh area between the Eleventh and Thirteenth Centuries. In: EDDISON, Jill; GREEN, Christopher (Eds.): Romney Marsh: Evolution, Occupation, Reclamation. Oxford 1988, 105–111.
TATTON-BROWN, Tim: Archaeology and Chichester Cathedral. In: TATTON-BROWN, Tim; MUNBY, Julian (Eds.): The Archaeology of Cathedrals. Oxford 1996, 47–55.
TAYLOR, Alison: “Roman Bank”: A medieval sea-wall I. a culvert beneath the sea bank at Newton, near Wisbech. In: Proceedings of the Cambridge Antiquarian Society 67 (1977), 63–66.
TESTER, Peter J.: Medieval Pottery found in Bexley Church. In: Archaeologia Cantiana 70 (1956), 260–261.
THOMAS, Arthur H. (Ed.): Calendar of the Plea and Memoranda Rolls of the City of London, 1364-1381. Volume II. London 1929.
THOMAS, Keith: Religion and the Decline of Magic. London 1971.
THOMASON, D.; STAFFORD, Elizabeth: Land to the rear of the Old School House, Church Lane, New Romney, Kent, archeological evaluation report. Unpublished Report 2001.
THOMPSON, Aaron (Ed.): Geoffrey of Monmouth: History of the kings of Britain. Cambridge 1999.
THOMPSON, F. H.: Medieval pottery from Chapel St. Leonards, Lincs. In: The Antiquaries Journal 33 (1953), 3–4.
THORARINSSON, Sigurdur: The Oraefajokull Eruption of 1362. In: Acta Naturalia Islandica 2, 2 (1958), 5–99.
THORPE, Benjamin (Ed.): Florentii Wigorniensis Monachi Chronicon ex Chronicis, Tomus II. London 1849.
TIGHE, Robert R.; DAVIS, James E.: Annals of Windsor. Volume I. London 1858.
TIPPING, Richard: Climatic variability and 'marginal' settlement in upland British landscapes: a re-evaluation. In: Landscapes 3, 2 (2002), 10–29.
TRAVAINI, Lucia: Saints, Sinners and ... a Cow: Offerings, Alms and Tokens of Memory. In: GASPAR, Giles E M; GULLBEKK, Svein H. (Eds.): Money and the Church in Medieval Europe, 1000-1200: Practice, Morality and Thought. London and New York 2015, 209–222.
TRISTRAM, Eileen W.: English Wall Painting of the Fourteenth Century. London 1955.
TROUET, Valerie; SCOURSE, James D.; CORNELIUS, Raible Christoph: North Atlantic storminess and Atlantic Meridional Overturning Circulation during the last Millennium: Reconciling contradictory

proxy records of {NAO} variability. In: Global and Planetary Change 84–85 (2012), 48–55. DOI: 10.1016/j.gloplacha.2011.10.003.

Tyson, Colin: The Battle of Otterburn: when and where it was fought. In: Goodman, Anthony; Tuck, Anthony (Eds.): War and border societies in the middle ages. London and New York 1992, 65–93.

Uekötter, Frank: Review of Groh D, Kempe M and Mauelshagen F (eds.) (2003) Naturkatasrophen. Beitrage zu ihrer Deutung, Wahnehmung und Darstellung in Text und Bild von der Antike bis ins 20. Jahrhundert. Narr Francke Attempto GmbH & Co. KG, Konstanz. 2004. www.hsozkult.de/publicationreview/id/rezbuecher-3081. Accessed 15 June 2016.

Unger, Carl R.; Huitfeldt-Kaas, Henrik J. (Eds.): Diplomatarium Norvegicum. Volume XIII. Christiania 1889.

Vahey, Zoe: Maidenhead Hastings 1987 an excavation report. Hastings 1989.

Vahey, Zoe: Phoenix Brewery, excavation report. Hastings 1991.

van Bavel, Bas; Curtis, Daniel R.; Dijkman, Jessica; Hannaford, Matthew; de Keyzer, Maïka; van Onacker, Eline; Soens, Tim: Disasters and History: The Vulnerability and Resilience of Past Societies. Cambridge 2020.

van Bavel, Bas J. P.; van Zanden, Jan Luiten: The jump-start of the Holland economy during the late-medieval crisis, c.1350–c.1500. In: Economic History Review 57, 3 (2004), 503–532.

Van de Noort, Robert: North Sea archaeologies: a maritime biography, 10,000 BC - AD 1500. Oxford 2011.

Van der Veen, Marijke; Hill, Alistair; Livarda, Alexandra: The archaeobotany of Medieval Britain (c AD 450-1500): identifying research priorities for the 21st century. In: Medieval Archaeology 57 (2013), 151–182.

van Gijn, Simon: Dordracum illustratum. verzameling van kaarten, teekeningen, prenten en portretten, betreffende de stad Dordrecht. Deel I. Dordrecht 1908.

van Vilsteren, Vincent T.: Pay peanuts, get monkeys – on the ritual context of medieval miniature bronze cauldrons. In: Siedlungs- und Küstenforschung im südlichen Nordseegebiet 36 (2013), 171–182.

Venables, Edmund (Ed.): Chronicon Abbatie de Parco Lude. The Chronicle of Louth Park Abbey. Horncastle 1891.

Vésteinsson, Orri: The Christianisation of Iceland. Priests, power and social change 1000-1300. Oxford 2000.

Wadey, Matthew P.; Haigh, Ivan D.; Nicholls, Robert J.; Brown, Jennifer M.; Horsburgh, Kevin; Carroll, Ben; Gallop, Shari L.; Mason, Travis; Bradshaw, Elizabeth: A comparison of the 31 January-1 February 1953 and 5-6 December 2013 coastal flood events around the UK. In: Frontiers in Marine Science 2, 84 (2015), 1–27.

Wallis, T.: The Nic-Nac; or, Literary Cabinet. Volume I. London 1823.

Walters, Henry B.: Church bells of England. London 1912.

Ward, John O.: Disaster and disaster-response in a medieval context: the First Crusade. In: Hinton, Peter (Ed.): Disasters: Image and Context: number 7 InSydney Studies - Society and Culture. Sydney 1992, 105–140.

Warren, James Francis: Philippine typhoons, sources and the historian. In: Water History 7 (2015), 213–231.

Wastegård, Stefan; Davies, Siwan M.: An overview of distal tephrochronology in northern Europe during the last 1000 years. In: Journal of Quaternary Science 24, 5 (2009), 500–512.

Watt, David; Colston, Belinda: Investigating the effects of humidity and salt crystallisation on medieval masonry. In: Building and Environment 8, 1 (2000), 737–749.

Watts, Rebecca: Non-specific indicators of stress and their association with age at death in medieval York: using stature and vertebral neural canal size to examine the effects of stress occur-

ring during different periods of development. In: International Journal of Osteoarchaeology 21 (2011), 568–576.
WEAVER, Frederic William: Cleeve Abbey. In: Proceedings of the Somersetshire Archaeological and Natural History Society 52, 2 (1906), 1–41.
WEBSTER, Paul: The cult of St Thomas Becket. In: WEBSTER, Paul; GELIN, Marie-Pierre (Eds.): The Cult of St Thomas Becket in the Plantagenet World, c.1170-c.1220. Woodbridge 2016, 1–24.
WEIKINN, Curt: Quellentexte zur Witterungsgeschichte Europas von der Zeitwende bis zum Jahre 1850: Hydrographie. Teil 1, Zeitwende – 1500. Berlin 1958.
WENINGER, Bernhard; SCHULTING, Rick; BRADTMÖLLER, Marcel; CLARE, Lee; COLLARD, Mark; EDINBOROUGH, Kevan; HILPERT, Johanna; JÖRIS, Olaf; NIEKUS, Marcel; ROHLING, Eelco J.; WAGNER, Bernd: The catastrophic final flooding of Doggerland by the Storegga Slide tsunami. In: Documenta Praehistorica 35 (2008), 1–24.
WESSEX ARCHAEOLOGY: 16 High Street New Romney Kent, archaeological evaluation report. Unpublished Report 2012a.
WESSEX ARCHAEOLOGY: Kenfig, Bridgend, Glamorgan, Wales: archaeological evaluation and assessment of results. Unpublished Report 2012b.
WESTERDAHL, Christer: Medieval carved ship images found in Nordic churches: the poor man's votive ships? In: The International Journal of Nautical Archaeology 42, 2 (2013), 337–347.
WETTER, Oliver; PFISTER, Christian: An underestimated record breaking event—why summer 1540 was likely warmer than 2003. In: Climate of the Past 9 (2013), 41–56.
WETTER, Oliver; PFISTER, Christian; WERNER, Johannes P.; ZORITA, Eduardo; WAGNER, Sebastian; SENEVIRATNE, Sonia I.; HERGET, Jürgen; GRÜNEWALD, Uwe; LUTERBACHER, Jürg; ALCOFORADO, Maria-Joao; BARRIENDOS, Mariano; BIEBER, Ursula; BRÁZDIL, Rudolf; BURMEISTER, Karl H.; CAMENISCH, Chantal; CONTINO, Antonio; DOBROVOLNÝ, Petr; GLASER, Rüdiger; HIMMELSBACH, Iso; KISS, Andrea; KOTYZA, Oldřich; LABBÉ, Thomas; LIMANÓWKA, Danuta; LITZENBURGER, Laurent; NORDL, Øyvind; PRIBYL, Kathleen; RETSÖ, Dag; RIEMANN, Dirk; ROHR, Christian; SIEGFRIED, Werner; SÖDERBERG, Johan; SPRING, Jean-Laurent: The year-long unprecedented European heat and drought of 1540 - a worst case. In: Climatic Change 125, 3 (2014), 349–363.
WHARTON, Henry: Anglia sacra sive collectio historiarum. Volume I. London 1691.
WHITE, Gilbert F.: Natural hazards research: concepts, methods and policy implications. In: WHITE, Gilbert F. (Ed.): Natural hazards: local, national, global. Oxford 1974, 3–16.
WILLIAMS, Stephen W.: The Cistercian abbey of Strata Florida: its history, and an account of the recent excavations made on its site. London 1889.
WILLIAMSON, Eila: The cult of the three kings of Cologne in Scotland. In: BOARDMAN, Steve; DAVIES, John R.; WILLIAMSON, Eila (Eds.): Saints' Cults in the Celtic World. Woodbridge 2009, 160–179.
WILLIAMSON, Tom: Shaping medieval landscapes: settlement, society, environment. Macclesfield 2003.
WILLIAMSON, Tom; LIDDIARD, Robert; PARTIDA, Tracey: Champion: the making and unmaking of the English Midland landscape. Liverpool 2013.
WILLMOTT, Hugh; TOWNEND, Peter: Thornton Abbey Project, 2nd interim report 2012. 2012. www.thorntonabbeyproject.com. Accessed 13 April 2015.
WILSON, J.; LINKLATER, Andrew: An archaeological watching brief on a plot of land at Church Road, New Romney, Kent. Unpublished Report 2002.
WILSON, John: Medieval Buildings Discovered at New Romney. In: Kent Archaeological Review 89 (1987), 198–212.
WISNER, Ben; BLAIKIE, Piers; CANNON, Terry; DAVIS, Ian: At risk: natural hazards, people's vulnerability and disasters. 2nd edition. Abingdon 1994.
WOLF, Kirsten (Ed.): The Old Norse-Icelandic Legend of Saint Barbara. Toronto 2000.

WOODMAN, Francis: The Gothic Campaigns. In: ATHERTON, Ian; FERNIE, Eric; HARPER-BILL, Christopher; SMITH, Hassell (Eds.): Norwich Cathedral: church, city and diocese, 1096-1996. London and Rio Grande 1996, 158–196.

WOOLGAR, Christopher M.: Gifts of food in late medieval England. In: Journal of Medieval History 37, 1 (2011), 6–18.

WOOLHOUSE, Thomas: Medieval dispersed settlement on the mid Suffolk clay at Cedars Park, Stowmarket. East Anglian Archaeology Report No. 161. Bury St Edmunds 2016.

WRIGHT, Susan M.: Barton Blount: climatic or economic change? In: Medieval Archaeology 20 (1976), 148–152.

YATES, David; BRADLEY, Richard: The siting of metalwork hoards in the Bronze Age of South-East England. In: The Antiquaries Journal 90 (2010a), 41–72.

YATES, David; BRADLEY, Richard: Still water, hidden depths: the deposition of Bronze Age metalwork in the English Fenland. In: Antiquity 84 (2010b), 405–415.

YOUNG, Andrew: Eckweek Medieval Settlement, Peasedown St John, Somerset. Survey and Excavations at a shrunken medieval hamlet 1988-90. The Society for Medieval Archaeology Monograph 40. London 2020.

ZHOU, Liya; LI, Yuansheng; JIHONG, Cole-dal; TAN, Dejun; SUN, Bo; REN, Jiawen; WEI, Lijia; WANG, Henian: A 780-year record of explosive volcanism from DT263 ice core in east Antarctica. In: Chinese Science Bulletin 51, 22 (2006), 2771–2780.

ZIEGLER, Philip: The Black Death. London 1969.

ZIKA, Charles: Hosts, processions and pilgrimages: controlling the sacred in fifteenth-century Germany. In: Past & Present 118 (1988), 25–64.

Index

abandonment, *see also* migration, 99, 130–134, 225
adaptation, 97, 107, 109, 112–114, 123, 232, 234, 235, 237, 239
adaptive cycles, 182, 183, 206–208
Adstone, 41
Aduard, Abbey of, 201
afterlife, 127
Agney, 57
Agobard of Lyon, 33, 38, 235
agricultural strategies, 23, 27, 104–106, 124, 135
alcohol, 231
altars, 142, 203
Alveston, 84
ampullae, 20, 155, 157, 158, 235, 240
amulets, 158–160, 162
ancestors, 202
Ankerwycke Priory, 128
anniversaries, 198, 199, 231
anxiety, 230, 233
apocalypse, 38, 39, 41
Appledore, 58, 104, 105
Aquinas, Thomas, 227
archaeobotanical evidence, 29
architecture
– earthquake resistant, 98, 206
– ecclesiastical, 237
– Gothic, 98, 108
Aristotelian theory, 98, 138, 194
Arrow (river), 117
arson, 43
Arthur, King, 127
Ashby, 41
Ashford, 133, 221
Ashton-under-Hill, 150
Ashwell St Mary, 90, 192–195, 219, 229
Assisi, 165
authorities, *see also* elites, 100
Auvergne, 142
Avignon, 160

Baddesley Clinton, 176
ballads, 190
Ballingdon Bridge, 109, 112
Bangor, Bishop of, 93, 94
baptism, 176
Barkesdale, John, 101
Barton Blount, 212
Basel, 205
Bastwick, 108, 226
Beachamwell St Mary, 36
Beauvais, Vincent of, 34
beliefs, 137–139
bells, 139, 163, 165, 180
Benedict XII, 199
Bergen, 31, 32, 202, 206
Berkhamsted, 126, 127
Berrow, 46
Beverley, John of, 143
Bexley, St Mary's, 160
biases, 18, 85, 91, 140, 158, 173, 180
Bicker, 118
Black Death, 101, 134, 135, 155, 175, 180, 211, 220, 224, 228, 230, 239, 241, 244
Bloemhof, 50
Blythburgh, 35
Boniface IX, 161
Bordesley Abbey, 116, 117
Boston, 55
boundaries, *see also* liminal areas, 117
bridges, 41, 98, 107–109, 112, 226
Brittany, 56
Bromholm, 158
Broomhill, 30, 57, 122, 123, 130, 180
building materials, *see also* timber, lead, 107, 109, 121, 126
bullae, 160–163, 243
Burg, 165
burials, 101, 102, 167, 199, 200, 237, *see also* mass graves, 240
Bury St Edmunds, 82
buttresses, 112, 230
Byfleet, 126

calendar systems, 49, 81
Camden, William, 58
cannibalism, 29
Canterbury, 35, 82, 151, 199
– Cathedral, 155
– Cathedral Priory, 50, 105, 127, 226
– Hall, 127, 222
– St Augustine's Abbey, 142

https://doi.org/10.1515/9783110719628-010

Canterbury, Gervase of, 60
cauldrons, 167
ceramics, 221, 222
cereals, 27, 104, 105, 222, 227
Chabannes, Ademar of, 37
Champeneys, William, 100
Chapel St Leonards, 56
chapels, 82, 93–95, 128, 131, 132, 142, 186, 199, 203, 204
charity, 30, 127, 128, 228
charms, 158–160
Chevington Hall, 27
Chichester Cathedral, 112
Christ Church Priory, 30
Christchurch, 179
Christian worldview, 97, 232, 233, 235
chroniclers, 49, 50
Cinque Ports, 58
civil unrest, 218
Clare, 27, 86, 127, 200
Clare Priory, 116
Clarendon, 84, 126
Cleeve, 93–95, 128, 228
cliff erosion, 93, 95
Cliffsend, 60
climate, 11, 22, 210, 218
climate change, 103, 210–213, 215, 217–220, 240, 244
Cloyne, 129
Cobham, 41
Coggeshall, Ralph of, 35
coins, 63, 137, 138, 171, 172, 243
– folded, 169, 171–173, 175, 235, 243
Colcheter, 129
cold spells, 103
Cologne, 146, 149, 151
commemoration, *see also* memories, 198–203
communities, 120, 184, 232
compass drawn designs, 175, 177
construction, *see also* masons, 107, 112, 114
Cottingham, 126
Cotton, Bartholomæi de, 51
credit, 30, 31
Cressing Temple, 26
Culemborg, 166
cultural change, 135, 238
curation, 185
Cwmllynfell, 7
Dabernoun, John, 85
Dagenham, 122
Danube (river), 234
dating evidence, 48, 55, 63, 135
Day of Judgement, *see* apocalypse
de Sapy, Robert, 86
Deal, 56
decline, 133–135
dedications, 151
demand, *see also* markets, 101
demographic change, 129, 132, 155, 210, 213, 219, 220, 222, 224, 239, 241
demolition, 99, 116, 132, 133, 237
demons, 34–36, 138
dendrochronological evidence, 91, 114, 214
Dennington, 28
depositions
– foundation, 160, 165–169
– in water, 153, 169, 173
– ritual, 153, 173
– special, 168
depression, 230
Deucalion, 38
devil, the, 35, 139, 160, 168, 176, 177
dike reeves, 120, 229
dikes, 57, 117, 118, 166, 167
Dioscorus, 147
disaster capitalism, 223, 224
disaster cycle, 2, 21, 235
disasters, 4, 9, 10, 97, 182, 220, 224, 225, 232, 233, 236, 239
– areas of effect, 237
– economic impacts, 20, 105, 106, 131–133, 135, 222
– psychological impacts, 229, 230, 232, 233, 235
diseases, *see* panzootics, plague
ditches, 16, 28, 69, 79, 80, 117–120, 122
divine intervention, 34–37, 97, 203, 235
DNA, 102
documentary evidence, 2, 14, 18, 184, 185, 188, 197
donations, *see also* charity, 200
Dordrecht, 131, 195, 203
Dover, 8
drainage, 16, 69, 72, 73, 79, 80, 117–120, 122, 218, 219, 226
droughts, 20
– 1540, 43, 201
Dublin, 82, 91

Dunstable, 127
Durham, 47, 175
Durham, Reginald of, 159
Dymchurch, 51

earthquakes, 7–9, 98, 194, 237
– 1117, 194
– 1222, 194
– 1348, 194
– 1350, 237
– 1356, 205
– 1381, 8
– 1522, 222
– 1755, 9
– 2009, 231
– 2018, 7
East Haven, 153
Eaton, 84
Ebony, 99, 104, 105
economic change, 218, 221, 224
Edmund Lacey, 203
Edward I, 130, 203, 226
Edward II, 227, 228
Edward III, 86, 100, 226, 228
Edward IV, 171
Edward, the Black Prince, 84, 85, 125–127, 222, 227
Einbeck, 31, 43
Eldbotle, 29, 46
elites, 223, 226
Ely, 128
– Bishop of, 106, 128
emergency services, 224
errors, 51, 81
Eucharist, 139, 144
ex votos, 202–205, 207, 208
exchequer, 29, 125
Exeter Cathedral, 167, 203
Exeter, Bishop of, 128
Eylard, Abbot of Aduard, 201

families, 231
famines, 29, 38, 103
– 1315–1322, 29, 38, 194, 199–201, 210, 227, 228, 239, 240
fertility, 105
festivals, *see also* processions, 198
field systems, 23
finances, 124–129, 219, 243
Fintray, 142
fires, 31, 32, 47, 125, 127, 166, 202, 206
fishing, 29, 64, 179, 212
Flempton, 117
flood defences, 116–124, 221, 223, 225, 226, 243
flood deposits, 55, 56, 63–65
flood gates, 118, 119
flood marks, 196–198, 207, 237
floods, 7, 41, 63, 64, 106, 107, 116, 117, 147, 219, 233
– 1250s, 63, 225
– 1287/88, 108–110, 120, 123, 135
– 1248, 63
– 1249, 63
– 1330, 198
– 1338, 124, 125, 205, 218, 226
– 1342, 195, 197, 200, 240
– 1362, 87
– 1374, 37
– 1409, 122
– 1421, 131, 195, 202, 203, 207
– 1445, 38
– 1524, 41
– 1573, 197, 234
Florence, 224
folklore, 195
forecasting, 21, 206, 233
Forvie, 29, 46, 131
foundations, 203

Gabbott, Ralph, 170
Gaddesden, 126
gardens, 28
Garyp, 165
Gascony, 125, 197
Gauldalen, 187, 188, 207
gender, 231
gifts, 126–128, 240
Gilbertus Anglicus, 230
Glasgow Cathedral, 47
glass, 39, 147
Glastonbury, 26, 127
Gleaston, 144
Gloucester, 185
Glyn Cothi, Lewys, 41
Goltho, 212
Goltsmet, Hermen, 200
Göttingen, 200
grave goods, 240

Great Barton, 27
Great Henney, 127
Great Meols, 131
Great Yarmouth, 31, 55, 56
Greenland, 211
Gregory I, 198
grey literature, 61
Groningen, 201
Guinevere, 127

Hadleigh Quarry, 27
hail, 20, 137, 138, 235
Hailes Abbey, 203
Hallig Südfall, 87
Hannoversch Münden, 197
harvest failures, 27, 103, 104, 219
Hastings, 56, 135
Hayward, William, 122
health, *see also* mental health, 30
heavenly letters, 160
Heigham, 57, 108
heirlooms, 185, 201, 202
Hemington, 109, 112
Henbury-in-Salt-Marsh, 31
Henry III, 63
Henry VI, 170, 172
Henry VII, 171
Herzog, Werner, 245
Hickling, 51, 197
Hinxworth, St Nicholas, 86
Hitchin, 112, 234, 238
hoards, 169, 172
Holewest, 60, 226
holy water, 139, 152
Horseshoe Corner, 122
Horsey, 108
hospitals, 30
humours, 230
Huntingdon, 151
hurricanes, 5, 39, 184
Hythe, 60

Iceland, 127
Icklingham, 222
illness, 170
indulgences, 128, 129, 161, 199
industries, 221
inequality, 126, 210, 220
inflation, 101
Inglebard, Philip, 32
Innocent VI, 163
inscriptions, 165, 167, 168, 192–196, 203, 237
investment, 226, 234
Isidore of Seville, 34, 138, 194
Islip, Simon, 126, 222

James II, 37
Jeake, Samuel, 60
Jews, 32, 37
jousting, 37

Kenfig, 46, 187, 188
Kent County Council, 61
Kent, Joan of, 85
Keyingham, 32
Killaraus, 183
King's Lynn, 57, 99, 151
Kings Langley, 100, 222
Kinneve, 137
Kirbymoorside, 126
Knights Templar, 26, 31
Krems, 197, 234

L'Aquilla, 231
labour, 100, 101, 120, 128, 212, 219–221, 227, 228
Lacey, Edmund, 167
Lacock Abbey, 125
Lamarsh, 127
landholding, 22, 23, 31, 119, 126, 210, 222–224, 226, 228, 232
landslides, 8, 92–95, 128, 130, 228, 235, 240
– 1893, 9
Lark (river), 117
laws, 232
lay belief, 34, 139
lead, 114, 206
Leask, 131
legumes, 105
Leiden, 202
Leirfall, 130
Leiston, 27, 132
LiDAR, 95
life expectancy, 229
lightning, 35, 45, 46, 125, 138, 147
lime, 103, 105
liminal areas, 168, 176, 179, 180
limpets, 29
Lincoln, Hugh of, 158

Lisbon, 9
Liskeard, 85
literacy, 184
Little Ice Age, 34, 210, 212, 239, 243
Little Livermere, 101
livestock, 46, 103, 104, 219, 222, 234, 237
living standards, 155, 210, 228, 239
Livonia, 33
loans, 30, 31
Lombok Island, 103
London, 82, 100, 103, 126, 129, 151, 179, 227
– Austin Friars, 112, 234
– Chapel of St Martin-le-Grand, 186
– St Paul's, 47, 142
– Tower of, 86
Long Sutton, 116
Long Wittenham, 91
Longstanton, 83, 142
Lostwithiel, 85
Lydd, 187

Mablesthrope, St Peter, 55
machair, 211
Magi, the, 146, 148–151, 240
magic, 33, 138
maintenance, 119–121, 219, 223
Malma, 165
Malmesbury Abbey, 165
mania, 230
Mare, Thomas de la, 114
markets, 100, 101, 128, 212, 219, 222–224, 227, 228, 243
Marshland, 118, 120–122, 124, 125, 219, 226, 228, 229, 237
Martham, 108
Mary, the Virgin, 93, 94, 142, 160, 165, 228
masons, 91, 192–195
mass graves, 102–104, 237
mass movements, *see* landslides
Maynburgh, Henry, 126, 227
measurement, 207
Medieval Climate Anomaly, 210, 240, 243
Melk, Abbey of, 188
memorials, *see also* commemoration, 183, 185, 199
memories, 10, 98, 182, 189, 201, 205–207, 209, 234
mental health, 229–233
Meols, 131
Merthyr Mawr, 46
metal detecting, 141, 158
Midley, 57
migration, 129–133, 200, 203, 212, 213, 225, 231
Milan, 149
miracles, 32, 157, 203
mitigation, 9
moated sites, 239
Molycourt, 128
Moon, the, 52
mordbrenner, 43
mortality, 102
Moses, 38
music, 190
Muslims, 32
myths, 183, 196

natural disasters, *see* disasters
natural hazards, *see also* disasters, droughts, earthquakes, floods, storms, 4, 33, 51
Nene (river), 118
Nevern, 168, 176
Neville's Cross, Battle of, 185
New Romney, 57, 58, 60, 61, 63–65, 67, 99, 132–135, 166, 197, 199, 214, 218, 237
– Southlands School, 64–66
– St Nicholas, 176, 197, 205
Newton, 119, 186, 187
Nicosia, 198
nightmares, 231
Noah, 38
Norfolk Archaeological Unit, 108
Norse religion, *see also* paganism, 137
North Atlantic Oscillation, 215, 217
Northeye, 56
Norton Priory, 147
Norwich Cathedral, 31, 88, 99, 128, 158
Norwich, William of, 158
Novgorod, 29
nutrition, *see also* famines, 182

Old Romney, 58
Ølst, 137
Oppenheim, 194
Öræfajökull, 87, 215
oral traditions, 183, 184, 187, 188, 195–197, 200, 201, 205, 206
orchards, 28
Ordinance of Labourers, 228

Otterburn, 185, 190
Ouse (river), 118, 125
oxen, 219
Oxford, 126, 222
Oxnead, John of, 49, 51, 54, 108, 197

paganism, 33, 137, 138
paintings, 151
palaeoenvironmental evidence, 210
Palladius, 138
Palmarsh, 51
panzootics, *see also* plague, 102–106, 219
Papacy, 128, 160–162, 186, 198, 199, 243
Paris, Matthew, 38, 58, 63
parks, 84
Parliament, 126
Passau, 139
pastoralism, *see also* livestock, 46, 104–106, 212
Paul II, 161
peasants, 184
Peasenhall, 28
peat, 219
Penmaen, 46
Penn, 114
Pennard, 31, 46
pentangles, 175, 176, 179
Perth, 139
Peterborough, 100
petitions, 124–126, 226
Pevensey, 135
Piccolomini, Enea Silvio, 38
Piers Plowman, 36, 191
pilgrim badges, 150–153, 155, 240
pilgrimage, 20, 31, 150–153, 155, 157, 158, 203
pipes, 118
placenames, 205
plague, *see also* panzootics, Yersinia pestis, 86, 101, 102, 180, 204, 211, 212, 219, 220, 222, 228, 230, 240
Plymouth, 86
Podimore, 106
poetry, *see also* verse, 41, 190
Pokedike, 118, 120
pollen, 212
Pompeii, 9
Pons Hugh IV de Ampurias, 37
population, *see* demographic change
Portable Antiquities Scheme, 140, 144, 146, 151, 152, 155, 157–159, 161–163, 169, 171–173, 175, 180, 181, 243, 244
Portchester Castle, 88
post-traumatic stress disorder, 229–233
Potter Heigham, 108
pottery, 133, 221, 238
poverty, 30, 210, 218, 232
prayer books, 153
prayers, 20, 142, 160, 200, 203–205, 229, 235, 244
predictions, 21, 34, 38, 39
prehistoric artefacts, 139
prices, 100, 101, 222, 223, 227, 228
priests, 127, 139, 200
processions, 20, 139, 143, 144, 198, 199, 233
profiteering, *see also* disaster capitalism, 222
property, *see* landholding
punctuated equilibrium, 238
purgatory, 30, 128, 161

rain, 7–9, 13, 15, 16, 22, 93, 95, 116, 143, 201, 211
Ravenser Odd, 87
Reformation, the, 139, 158, 175, 180
relics, 139, 141–145, 152, 167, 240
reliquaries, 142, 144, 145
rents, *see also* tenants, 125
repairs, 85, 86, 88, 91, 99–101, 107, 109, 112–114, 121, 122, 126, 205, 207, 221, 222, 226, 227, 234, 237
Repps with Bastwick, 108
resilience, 28, 98, 104, 182, 183, 197, 206, 207, 231, 232, 237, 245–247
return periods, 17, 182
reuse, 100, 162, 202, 205, 206
Rhee Wall, 58, 60
risk, 6, 17, 21–23, 25, 26, 31, 41, 116, 232, 247
river morphology, 109, 112
Robines, Nicholas, 85
Rochester Castle, 88
rodents, 211
Roes, Alexander of, 49
rogation, 198, 199
Rome, 160, 161, 197, 198
Romney Marsh, 56, 57, 120, 135, 180, 218, 221, 224, 226, 229, 232
roofing, 100, 114, 222, 227, 228
Rother (river), 58
royal commissions, 119, 124, 226, 228

royal involvement, 108, 119, 120, 124, 130, 131, 226, 228, 243
Rush, 204
Rye, 58, 60, 135

sacrifice, 167
sailors, 203
Saint-Sever, 125
Saint-Émilion, 197
saints, 198, 202–204, 208
salinity, 105
Salisbury Cathedral, 114, 128, 234
salt-working, 56
Salutati, Coluccio, 97
Samalas, 103, 215
San Romano, 223
sand, 45, 46, 131, 169, 211
Santiago de Compostela, 157
savings, 23
scapegoats, 37, 44
sea surface temperatures, 215
Seaford, 135
security, 239
settlements, 22, 213, 226
Shapwick, 26, 28, 103
shellfish, 29
Shetland, 210
ships, 50, 58, 64, 86
– construction, 212
– graffiti, 179, 180, 203
– wrecks, 187
Shipston-on-Stour, 31
Shoreham, 135
shrines, 157, 167, 202
Siena, 167
silk roads, 211
siltation, 58, 222
sins, 37, 161
skeletal remains, 102, 183
Skitter Beck, 123
Skriðuklaustur, 30
Snargate, 50, 60, 226
soils, 105, 212, 219
solar irradiance, 214, 217
South Denes, 56
South Kvinneby, 138
South Uist, 211
Spexhall, 28
spires, 206
Spitalfields Market, 103
Spurn Point, 87
St Adelm, 165
St Albans, 35, 100
– Abbey of, 82, 88, 113, 114, 167
St Andrews Cathedral, 47
St Barbara, 146–151, 165, 240
St Benet at Holme, Abbey of, 55, 108
St Briavel's Castle, 86
St Christopher, 145–147, 149, 151, 240
St Columba, 143
St Cuthbert, 159
St Elizabeth, 195, 203
St George, 179
St Modan, 143
St Nicholas, 203
St Pancras, 142
St Paul, 160, 161
St Peter, 160
St Sebastian, 139
St Thomas Becket, 155
St Thomas Cantilupe, 170
Stackpole, 46
Stanydelf, Thomas de, 222
stars, 177
Statute of Labourers, 100, 228
Stoke, Melis, 50, 51
Stone, 129
Stonehenge, 183
storage, 22, 25–27, 31
Storegga Slide, 9
storm surges, 6, 54
storms, 9, 35, 91, 170, 203, 206, 210–212, 222
– 1250s, 130
– 1287/88, 48–58, 60, 61, 63–65, 67, 105, 130–134, 136, 176, 179, 180, 187, 205, 214, 217, 218, 221, 225, 226, 234, 236, 237, 240
– 1210, 112
– 1287, 197
– 1362, 37, 39, 79, 81–91, 100, 101, 112–114, 125–129, 142, 186, 191–194, 199, 205, 214, 215, 217, 219, 221, 222, 227, 228, 234, 238, 240
– 1662, 80
– 1703, 9, 80, 86
– 1953, 54, 121
– 1987, 80, 84
– 2013, 83
Stour (river), 116

Stowmarket, 27, 91, 205, 212
stress, 30, 229
Strumpshaw, 163
subsidence, 113
Sudbury, 109, 112
sudden death, 145–150, 159, 160
sunspots, 214, 217
superstitions, 137–139, 155
support networks, *see also* communities, 232
Swavesey, 117

taboos, 138
Tarrit, the, 86
taxation, 29, 124, 125, 129, 218, 227, 228
teleconnections, 240
tempestarios, 33
tenants, 210, 219, 227, 228
Terrington, 125
Terrington St Clement, 122
Terrington St John, 116
Thatcher, John, 101
Thaxted, 83
Thor, 137
Thornton Abbey, 123
Thorsberg, 137
tides, 51, 52
Tiler, Thomas, 101
timber, 125–127
time, 220
tolls, 108, 109, 226
trade, 58, 179, 222
trees, 84, 125, 126
Trent (river), 112
Trimley Marshes, 121
Troston St Mary, 36, 175
Trubia, (river), 223
tsunami, 9
Tydd St Giles, 116
Tyler, John, 85
Tywi (river), 41

Udine, 98
Ullensaker, 205

Vale Royal Abbey, 203
Verdalen, 9
verse, *see also* poetry, 190, 191
vessels, 166
victims, 83, 101, 199, 200, 231, 232
Vila Franca do Campo, 20, 222
Villanueva, 223
Vincenza, 194
Vitruvius, 175
volcanic eruptions, 103, 214, 215, 217
– 1257, 103, 215
votive objects, 167, 203, 204
vulnerability, 21, 102, 210, 220

wages, 100, 101, 220–223, 227, 228
Walden St Paul's, 112
Walland Marsh, 57, 61, 123
Wallingford Castle, 84, 85
Walpole Cross Keys, 122
Walpole St Peter, 187, 205
Walsham-le-Willows, 218
Walton, 118
wars, 124, 135, 218, 222, 227
waste, 106
water management, *see also* floods, 118, 132
watersheds, 237, 238
Watlington, 84
Waxham, 108
weather, 6, 7, 9
– forecasting, 21, 233
Wells Cathedral, 113
West Walton, 116
Westminster, Abbey of, 186
wetlands, 104–106, 120–122, 219, 234
Wieldrecht, 203
Wiggenhall, 120, 124
Winchelsea, 56, 130, 135, 205, 225, 226
– St Thomas, 179
wind speeds, 80, 82
wind-blown sand, 131, 211
Windsor Castle, 86
wine, 234
Wisbech, 57, 99, 106
Wisbech High Fen, 106
witches, 44, 138
Wolf Minimum, 214
Woodeaton, 145, 147
wool, 222
Worcester Cathedral Priory, 85
Wygenhale, Symon, 101
Würzburg, 188, 195, 201

Yakeyama, 215
Yersinia pestis, 211
York, 144
– All Saints', 39, 147

Zürich, 201

www.ingramcontent.com/pod-product-compliance
Ingram Content Group UK Ltd.
Pitfield, Milton Keynes, MK11 3LW, UK
UKHW051022091026
15624UKWH00012B/629

9 783110 719574